BIOLOGIC MARKERS IN URINARY TOXICOLOGY

Subcommittee on Biologic Markers
in Urinary Toxicology
Committee on Biologic Markers

Board on Environmental Studies and Toxicology

Commission on Life Sciences

National Research Council

NATIONAL ACADEMY PRESS
WASHINGTON, D.C. 1995

NATIONAL ACADEMY PRESS 2101 Constitution Ave., N.W. Washington, D.C. 20418

NOTICE: The project that is the subject of this report was approved by the Governing Board of the National Research Council, whose members are drawn from the councils of the National Academy of Sciences, the National Academy of Engineering, and the Institute of Medicine. The members of the committee responsible for the report were chosen for their special competences and with regard for appropriate balance.

This report has been reviewed by a group other than the authors according to procedures approved by a report review committee consisting of members of the National Academy of Sciences, the National Academy of Engineering, and the Institute of Medicine.

The National Academy of Sciences is a private, nonprofit, self-perpetuating society of distinguished scholars engaged in scientific and engineering research, dedicated to the furtherance of science and technology and to their use for the general welfare. Upon the authority of the charter granted to it by the Congress in 1863, the Academy has a mandate that requires it to advise the federal government on scientific and technical matters. Dr. Bruce Alberts is president of the National Academy of Sciences.

The National Academy of Engineering was established in 1964, under the charter of the National Academy of Sciences, as a parallel organization of outstanding engineers. It is autonomous in its administration and in the selection of its members, sharing with the National Academy of Sciences the responsibility for advising the federal government. The National Academy of Engineering also sponsors engineering programs aimed at meeting national needs, encourages education and research, and recognizes the superior achievements of engineers. Dr. Harold Liebowitz is president of the National Academy of Engineering.

The Institute of Medicine was established in 1970 by the National Academy of Sciences to secure the services of eminent members of appropriate professions in the examination of policy matters pertaining to the health of the public. The Institute acts under the responsibility given to the National Academy of Sciences by its congressional charter to be an adviser to the federal government and, upon its own initiative, to identify issues of medical care, research, and education. Dr. Kenneth I. Shine is president of the Institute of Medicine.

The National Research Council was organized by the National Academy of Sciences in 1916 to associate the broad community of science and technology with the Academy's purposes of furthering knowledge and advising the federal government. Functioning in accordance with general policies determined by the Academy, the Council has become the principal operating agency of both the National Academy of Sciences and the National Academy of Engineering in providing services to the government, the public, and the scientific and engineering communities. The Council is administered jointly by both Academies and the Institute of Medicine. Dr. Bruce Alberts and Dr. Harold Liebowitz are chairman and vice chairman, respectively, of the National Research Council.

The project was supported by the Agency for Toxic Substances and Disease Registry under grant nos. U50/ATU30009-05 and U50/ATU300009-04 and the National Institute for Environmental Health Sciences under contract nos. NO1-ES-35366 and 273-MH-107389.

Library of Congress Catalog Card No. 94-69722
International Standard Book No 0-309-05228-9

Additional copies of this report are available from:

National Academy Press
2101 Constitution Ave., NW
Box 285
Washington, DC 20055
800-624-6242
202-334-3313 (in the Washington Metropolitan Area)
B-509

Copyright 1995 by the National Academy of Sciences. All rights reserved.
Printed in the United States of America

SUBCOMMITTEE ON BIOLOGIC MARKERS IN URINARY TOXICOLOGY

WILLIAM F. FINN *(Chair)*, School of Medicine, University of North Carolina, Chapel Hill, N.C.
GEORGE P. HEMSTREET III *(Vice-Chair)*, University of Oklahoma Health Sciences Center, Oklahoma City, Okla.
BRUCE C. ALLEN, ICF Kaiser, Morrisville, N.C.
ROBERT L. ANDERSON, Menlo Park, Calif.
WILLIAM O. BERNDT, University of Nebraska Medical Center, Omaha, Neb.
FRANKLIN H. EPSTEIN, Beth Israel Hospital, Boston, Mass.
ERNEST C. FOULKES, University of Cincinnati, College of Medicine, Cincinnati, Ohio
BRUCE FOWLER, University of Maryland, Baltimore, Md.
CARL W. GOTTSCHALK, School of Medicine, University of North Carolina, Chapel Hill, N.C.
ERNEST E. MCCONNELL, Raleigh, N.C.
BERNICE NOBLE, School of Medicine and Biomedical Sciences, State University of New York at Buffalo, Buffalo, N.Y.
JEAN C. PARKER, U.S. Environmental Protection Agency, Washington, D.C.
GEORGE A. PORTER, Oregon Health Sciences Center, Portland, Ore.

Technical Advisers

LAWRENCE LASH, Wayne State University, Detroit, Mich.
ROBERT SAFIRSTEIN, University of Texas Medical Branch - Galveston, Galveston, Tex.

Staff

CAROL A. MACZKA, Program Director
J. DAVID SANDLER, Project Director
ROBERT BELILES, Project Director *(until 1992)*
NORMAN GROSSBLATT, Editor
LINDA V. LEONARD, Senior Project Assistant

Sponsors

Agency for Toxic Substances and Disease Registry
National Institute of Environmental Health Sciences

COMMITTEE ON BIOLOGIC MARKERS

BERNARD GOLDSTEIN *(Chair)*, UMDNJ-Robert Wood Johnson Medical School, Piscataway, N.J.
JAMES GIBSON, Dow-Elanco, Indianapolis, Ind.
ROGENE F. HENDERSON, Lovelace Biomedical and Environmental Research Institute, Albuquerque, N.M.
JOHN E. HOBBIE, Marine Biological Laboratory, Woods Hole, Mass.
PHILIP J. LANDRIGAN, Mount Sinai Medical Center, New York, N.Y.
DONALD R. MATTISON, University of Arkansas for Medical Sciences and National Center for Toxicological Research, Little Rock, Ark.
FREDERICA PERERA, Columbia University, New York, N.Y.
EMIL A. PFITZER, Hoffmann-La Roche, Inc., Nutley, N.J.
ELLEN K. SILBERGELD, Environmental Defense Fund, Washington, D.C.

BOARD ON ENVIRONMENTAL STUDIES AND TOXICOLOGY

PAUL G. RISSER *(Chair)*, Miami University, Oxford, Ohio
MICHAEL J. BEAN, Environmental Defense Fund, Washington, D.C.
EULA BINGHAM, University of Cincinnati, Cincinnati, Ohio
EDWIN H. CLARK II, Clean Sites, Inc., Alexandria, Va.
ALLAN H. CONNEY, Rutgers University, Piscataway, N.J.
ELLIS COWLING, North Carolina State University, Raleigh, N.C.
JOHN L. EMMERSON, Eli Lilly & Company, Greenfield, Ind.
ROBERT C. FORNEY, Unionville, Penn.
ROBERT A. FROSCH, Harvard University, Cambridge, Mass.
KAI LEE, Williams College, Williamstown, Mass.
JANE LUBCHENCO, Oregon State University, Corvallis, Ore.
GORDON ORIANS, University of Washington, Seattle, Wash.
FRANK L. PARKER, Vanderbilt University, Nashville, Tenn.
GEOFFREY PLACE, Hilton Head, S.C.
DAVID P. RALL, Washington, D.C.
LESLIE A. REAL, Indiana University, Bloomington, Ind.
KRISTIN SHRADER-FRECHETTE, University of South Florida, Tampa, Fla.
BURTON H. SINGER, Princeton University, Princeton, N.J.
MARGARET STRAND, Bayh, Connaughton and Malone, Washington, D.C.
GERALD VAN BELLE, University of Washington, Seattle, Wash.
BAILUS WALKER, JR., Howard University, Washington, D.C.

Staff

JAMES J. REISA, Director
DAVID J. POLICANSKY, Associate Director and Program Director for Natural Resources and Applied Ecology
CAROL A. MACZKA, Program Director for Toxicology and Risk Assessment
LEE R. PAULSON, Program Director for Information Systems and Statistics
RAYMOND A. WASSEL, Program Director for Environmental Sciences and Engineering

COMMISSION ON LIFE SCIENCES

THOMAS D. POLLARD *(Chair)*, Johns Hopkins Medical School, Baltimore, Md.
BRUCE N. AMES, University of California, Berkeley, Calif.
JOHN C. BAILAR III, McGill University, Montreal, Quebec
MICHAEL BISHOP, Hooper Research Foundation, University of California Medical Center, San Francisco, Calif.
JOHN E. BURRIS, Marine Biological Laboratory, Woods Hole, Mass.
MICHAEL T. CLEGG, University of California, Riverside, Calif.
GLENN A. CROSBY, Washington State University, Pullman, Wash.
MARIAN E. KOSHLAND, University of California, Berkeley, Calif.
RICHARD E. LENSKI, Michigan State University, East Lansing, Mich.
EMIL A. PFITZER, Hoffmann-La Roche Inc., Nutley, N.J.
MALCOLM C. PIKE, University of Southern California School of Medicine, Los Angeles, Calif.
HENRY C. PITOT III, University of Wisconsin, Madison, Wisc.
JONATHAN M. SAMET, The Johns Hopkins University School of Medicine, Baltimore, Md.
HAROLD M. SCHMECK, JR., Armonk, N.Y.
CARLA J. SHATZ, University of California, Berkeley, Calif.
SUSAN S. TAYLOR, University of California at San Diego, La Jolla, Calif.
P. ROY VAGELOS, Merck & Company, Whitehouse Station, N.J.
JOHN L. VANDEBERG, Southwestern Foundation for Biomedical Research, San Antonio, Tex.

PAUL GILMAN, Executive Director

OTHER RECENT REPORTS OF THE BOARD ON ENVIRONMENTAL STUDIES AND TOXICOLOGY

Science and the Endangered Species Act (1995)
Wetlands: Characteristics and Boundaries (1995)
Biologic Markers in Urinary Toxicology (1995)
Review of EPA's Environmental Monitoring and Assessment Program (three reports, 1994-1995)
Science and Judgment in Risk Assessment (1994)
Ranking Hazardous Sites for Remedial Action (1994)
Review of EPA's Environmental Monitoring and Assessment Program: Forests and Estuaries (1994)
Review of EPA's Environmental Monitoring and Assessment Program: Surface Waters (1994)
Pesticides in the Diets of Infants and Children (1993)
Issues in Risk Assessment (1993)
Setting Priorities for Land Conservation (1993)
Protecting Visibility in National Parks and Wilderness Areas (1993)
Biologic Markers in Immunotoxicology (1992)
Dolphins and the Tuna Industry (1992)
Environmental Neurotoxicology (1992)
Hazardous Materials on the Public Lands (1992)
Science and the National Parks (1992)
Animals as Sentinels of Environmental Health Hazards (1991)
Assessment of the U.S. Outer Continental Shelf Environmental Studies Program, Volumes I-IV (1991-1993)
Human Exposure Assessment for Airborne Pollutants (1991)
Monitoring Human Tissues for Toxic Substances (1991)
Rethinking the Ozone Problem in Urban and Regional Air Pollution (1991)
Decline of the Sea Turtles (1990)
Tracking Toxic Substances at Industrial Facilities (1990)
Biologic Markers in Pulmonary Toxicology (1989)
Biologic Markers in Reproductive Toxicology (1989)

Copies of these reports may be ordered from the National Academy Press: (800) 624-6242; (202) 334-3313

PREFACE

There is an increasing need for accurate information on the health effects of pollutants. In keeping with that need, the Agency for Toxic Substances and Disease Registry of the U.S. Public Health Service and the National Institute of Environmental Health Sciences asked the National Research Council to examine the potential for use of biologic markers in environmental health research. The term biologic markers has been used by the National Research Council's Committee on Biological Markers to refer to indicators of toxicologic events in biologic systems or samples.

The Committee on Biological Markers was organized to consider which uses of biologic markers offered the greatest potential for major contributions. Four biologic systems were chosen: the reproductive system, the respiratory system, the immune system, and the urinary system. This report is the product of the Subcommittee on Biologic Markers in Urinary Toxicology, which comprised scientists with backgrounds in and knowledge of nephrology, urology, pathology, renal toxicology and metabolism, pharmacokinetics, immunology, risk assessment, pharmacology, renal physiology, and other disciplines. The subcommittee reviewed research on known biologic markers and identified and evaluated promising new technologies to find markers, important research opportunities in the field, and subjects in which interdisciplinary research is needed.

During the subcommittee's deliberations, several scientists were asked to provide information. The subcommittee especially wishes to recognize the contributions of Lawrence Lash, Wayne State University, and Robert Safirstein, University of Texas-Galveston.

This report could not have been produced without the efforts of the National Research Council staff, especially J. David Sandler, project director, and Linda V. Leonard, senior project assistant; Norman Grossblatt, editor, Carol Maczka, program director; Gail Charnley, acting program director (through September 1994); Robert P. Beliles and Joyce Walz (the project director and assistant during the early months of the project); and James J. Reisa, director of BEST.

William F. Finn, M.D., Chairman
Subcommittee on Biologic Markers
in Urinary Toxicology

CONTENTS

EXECUTIVE SUMMARY   1

1 INTRODUCTION   15

Biologic Markers, 16
Validity of Biologic Markers, 18
Ethical Issues, 21
Structure of the Report, 22

2 TOXIC EXPOSURE OF THE URINARY TRACT   23

The Urinary Tract, 25
Mechanisms of Renal Toxicity, 27
Host Factors in Renal Toxicity, 30
Clinical Effects of Chemical Exposure on the Kidney, 35
Cancer of the Bladder, Kidney, and Prostate, 40
Animal Models of Interstitial Cystitis, 51
Summary, 52

3 BIOLOGIC MARKERS OF SUSCEPTIBILITY AND EXPOSURE   53

Populations at Risk, 53
Markers of Susceptibility, 58
Markers of Exposure, 76
Summary, 78

4 BIOLOGIC MARKERS OF EFFECT	81

    External Visualization, 82
    Urinary Clearance Measurements, 82
    Urinalysis, 85
    Markers of Cytotoxicity and Cellular Response, 95
    Markers of Neoplasia, 103
    Markers of Interstitial Cystitis, 114
    Summary, 116

5 BIOLOGIC MARKERS IN EXTRAPOLATION	153

    Basis of Extrapolation, 156
    Techniques Used in Risk Assessment, 166
    Improved Risk-Assessment Extrapolation, 180
    Summary, 187

6 NEW TECHNOLOGIES	191

    Markers of Cell Injury, Regeneration, and Hypertrophy, 191
    Cancer of the Bladder, 199
    Differential-Display Polymerase Chain Reaction, 202
    Cancer of the Prostate, 203
    Cancer of the Kidney, 207

7 CONCLUSIONS AND RECOMMENDATIONS	217

    Toxic Exposure of the Urinary Tract, 217
    Biologic Markers of Exposure and Susceptibility, 220
    Biologic Markers of Effect, 221
    Use of Biologic Makers in Extrapolation, 223
    New Technologies, 224

REFERENCES	227

EXECUTIVE SUMMARY

Diseases of the kidney, bladder, and prostate exact an enormous human and economic toll on the population of the United States. According to the 1993 Annual Data Report from the U.S. Renal Data System, in the United States, some 165,000 people with irreversible renal failure received renal-replacement therapy for end-stage renal disease (ESRD) in 1990 with the aid of chronic dialysis therapy, and 9,800 renal-transplant procedures were performed in the same year. The federal government spent $5.22 billion in 1990 to provide maintenance dialysis, kidney transplantation, and all other health services to ESRD patients. With regard to cancer, the figures are equally striking. In 1989, 47,000 new bladder-cancer cases and 10,000 bladder-cancer deaths were reported. About 18,000-20,000 new cases of cancer of the kidney are diagnosed each year in the United States, and they result in 8,000 fatalities a year. Prostatic cancer, the most frequent cancer in men in the United States, resulted in about 36,000 deaths in 1992 and carries with it an annual cost for diagnosis and care of more than $1 billion.

With the sponsorship of the Agency for Toxic Substances and Disease Registry (ATSDR) and the National Institute of Environmental Health Sciences (NIEHS), the National Research Council formed the Subcommittee on Biologic Markers in Urinary Toxicology of the Committee on Biological Markers to undertake a study of biologic markers in the urinary tract. Previous subcommittees of the Committee on Biological Markers have published separate volumes on reproductive and developmental toxicology, pulmonary toxicology, and immunotoxicology. The Subcommittee on Biologic Markers in Urinary Toxicology, which prepared this report, comprised scientists with diverse backgrounds in and knowledge of nephrology, urology, pathology, renal toxicology and metabolism, pharmacokinetics, immunology, risk assessment, pharmacology, renal physiology, and other disciplines.

In response to the charge to the subcommittee, this report discusses currently known genitourinary tract biologic markers, emphasizes the need to identify and evaluate promising technologies to find new markers, and identifies important research opportunities. The report discusses the structure and function of the urinary tract, toxic effects associated with the urinary tract, and their risk factors. Relationships between exposure, susceptibility, and the associated markers are described, and the usefulness of markers in monitoring urinary diseases is discussed. The report presents a rationale, based on epidemiologic studies, for the use of biologic markers in the protection of human health and the extrapolation of data from animals to humans. Currently available biologic markers in the genitourinary tract are discussed throughout the report and summarized in Chapter 4, although exhaustive descriptions of these are readily available elsewhere.

Several characteristics of the normal genitourinary tract increase the risk of damage by toxic chemicals. For example, the total amount of noxious substances delivered to the kidneys can be high, owing to the large amount of blood flowing to them. Furthermore, the capacity of the kidneys to concentrate substances by processes of filtration, reabsorption, and secretion can increase the toxicity of agents that would otherwise not lead to tissue damage. This is particularly important in the bladder, which is routinely exposed to concentrated toxicants. Also important are the mechanisms of biotransformation by which the kidney and bladder epithelium can metabolize xenobiotics and produce substances that might be more toxic than the parent substance.

Biologic markers can be useful in confirming toxic exposures (i.e., biologic markers of exposure), estimating their results (i.e., biologic markers of effect), and identifying persons most likely to be adversely affected if exposures continue (i.e., biologic markers of susceptibility).

Markers of exposure. A biologic marker of exposure is a xenobiotic chemical or its metabolite or a product of interaction between the chemical and some target cell. Markers of exposure most commonly used are the concentrations of such materials in urine, blood, or other body tissue, including hair and nails. Markers of exposure alone give no indication whether an exposure has produced a biologically significant result. The same dose in persons who are susceptible and resistant to a given xenobiotic can have different results. Urine is one source of markers of exposure. High exposures to a toxic substance can result in increased concentrations of the substance or its metabolites in urine. When sufficient pharmacokinetic information is available, urinary markers of exposure can be used to estimate the total exposure of a person to a substance.

Markers of effect. A marker of effect is a measurable cellular, physiologic, or biochemical alteration within an organism caused by interaction with a toxicant. Markers of adverse effect can be biochemical or cellular signals of tissue

Executive Summary

dysfunction, increased enzyme activity, the appearance of excessive waste products in the urine or other sampling media, and physiologic signs of abnormal function, such as increased blood pressure or blood in the urine. The effects themselves might not be directly adverse, but rather might indicate a potential for health impairment (e.g., DNA adducts). Biologic markers of effect include changes in the components of urine itself (such as an increase in urinary protein excretion or cellularity) and changes in the volume or composition of other body fluids caused by kidney dysfunction. With impaired renal function, excretion products such as creatinine and urea can accumulate in the blood. Increased blood concentrations of these products can be markers of renal damage, although they are not particularly useful as markers except in cases of severe damage. Other markers of renal toxicity can be present in urine because of altered renal function or damage to the kidney. Many of these can be detected in examination of the urine, including color, volume, pH, specific gravity, albumin, electrolytes, enzymes, and elements in the urinary sediment. Markers that could show up there include white and red blood cells, kidney epithelial cells, casts, and crystals.

Markers of susceptibility. A marker of susceptibility is an indicator of an inherited or acquired limitation of an organism's ability to resist the adverse effects of exposure to a xenobiotic substance. A biologic marker of susceptibility can be a genetic characteristic or a pre-existing disease that results in an increase in the amount of a toxicant absorbed, an alteration in its metabolism, or an increase in the target-tissue response. It can be difficult to distinguish between susceptible and nonsusceptible persons, either because of multiple interactive influences and genetic polymorphisms or because of difficulties in measuring markers of susceptibility. Potential sources of normal variability in markers of susceptibility can be the result of concurrent disease or genetic, environmental, or biorhythmic influences other than the specified toxicologic events that are the object of study. Such influences can cause differences between populations, differences between individuals, and differences within individuals over time. Decreased detoxification and excretory function are often important factors in susceptibility. Thus, elderly persons with declining organ function and young persons with immature and developing organs are likely to be more vulnerable to toxic substances than healthy adults.

EFFECTS OF TOXIC EXPOSURE OF THE URINARY TRACT

Advances in understanding and using biologic markers should assist in identifying xenobiotics that are toxic to the urinary tract. The functional role of the urinary tract, including clearance of toxic substances from the blood, predisposes it to xenobiotic exposure and toxicity. Historically, identification of the type and amount of xenobiotic exposure has been difficult, frequently because of the interval between exposure and the onset of disease. Blacks and other minority groups, for reasons that are not entirely apparent, are at higher risk of disease.

In diseases such as bladder cancer, xenobiotics associated with particular occupations are strongly implicated, and their mutagenic effects may be important. However, in kidney cancer and other renal diseases, a number of host factors, as well as the typically low levels of exposure to multiple xenobiotics and such confounding variables as smoking and genetic susceptibility, often mask the epidemiologic importance of individual xenobiotics. A powerful approach toward unraveling the complexities of xenobiotic exposure is to integrate biologic markers of susceptibility with biologic markers of effect.

Some xenobiotics are known to cause acute renal failure. Heavy metals and organic solvents stand out in this regard. There are several well-established associations between xenobiotic exposure and the development of chronic renal failure, as exemplified by exposures to lead and cadmium. The association of bladder cancer and occupational exposure to aniline dyes serves as a paradigm for the potential adverse health effects of xenobiotics.

Environmental agents have also been implicated in the development of neoplasms in the kidney. Some of these can be facilitated by acquired or inherited genetic defects. The association of xenobiotic exposure and cancer of the prostate and interstitial cystitis is less certain but merits attention.

BIOLOGIC MARKERS OF SUSCEPTIBILITY AND EXPOSURE

For diseases of the urinary tract, the most efficient program for determining the importance of occupational and environmental toxicants and carcinogens requires the identification of susceptible populations and the correlation of disease processes with the magnitudes and durations of exposure to the agents. Various factors modify human susceptibility to the effects of occupational and environmental nephrotoxicants and carcinogens.

Although much of the information on nephrotoxic chronic renal failure is circumstantial and comes from epidemiologic surveys that started with ESRD patients, for some agents the evidence is substantial. The most obvious group at risk consists of persons exposed to known or suspected nephrotoxicants in the workplace. Also at risk are people who live in regions of documented contamination. The possible link between a family history of renal disease and the development of renal failure might be an inherited susceptibility or a common geographic exposure. Altered nutrition and some coexisting diseases, including addictive behavior, are additional characteristics that indicate increased risk associated with nephrotoxicants.

Gender, race, and socioeconomic status provide tantalizing clues for understanding risk, but much more information needs to be collected than is currently available. Targeting populations at risk for future evaluation and follow-up is the most efficient strategy for the identification of patients early in the course of their toxic injury, this strategy might make it possible to introduce protective measures to reduce the progression of renal disease and to decrease the rate of entry of patients into ESRD programs.

Susceptibility factors for cancer can be either hereditary or acquired. For example, various hereditary conditions are associated with the development of progressive renal disease. Moreover, specific genetic conditions that predispose people to develop disease, such as the absence of a tumor-suppressor gene, have been identified. People who inherit one defective copy of a tumor-suppressor gene are at much greater risk for cancer than those people who have two intact copies. Likewise, individual variations in the metabolic pathways play a large role in susceptibility to both urogenital cancer and nephrotoxicity.

The identification of specific markers of susceptibility and exposure is a daunting task. The susceptible individual might be characterized by specific genotypic or phenotypic markers. Identifying these markers is likely to require a more complete understanding of the biochemical and physiologic properties of the kidney and lower urinary tract, as well as better insight into factors that control the linkages between cell growth, differentiation, proliferation, and malignant transformation. Consequently, these issues are addressed in detail in this report. The goal of identifying these markers is not to separate one population from another, but rather to identify exposures that can be tolerated by all.

BIOLOGIC MARKERS OF EFFECT

The ideal cytologic marker of nephrotoxicity would represent a nonspecific, if not universal, cellular response to injury. Such a marker should be produced in the kidney and secreted into urine in a readily detectable form. Substantial changes in urinary concentration of the marker should correlate well with pathophysiologic or histopathologic manifestations of kidney injury. The marker should be expressed soon after injury is sustained. Persistence of marker expression would increase its clinical value. The marker should be relatively easy to measure. It should be stable in storage, resistant to degradation, and unambiguously identifiable. In the search for suitable markers, more emphasis should be placed on the response to injury of tubules and interstitium, because these compartments appear to be the major sites of susceptibility to toxic injury. Finally, extensive use must be made of appropriate animal models to evaluate potential markers, because complex responses of intact organisms (vis-a-vis isolated systems) can influence response to injury, inflammation, and repair in vivo.

Several markers that fulfill many of the above criteria have already been identified. Urinalysis and clearance measurements will continue to provide important functional markers of renal injury. Of particular use in that regard would be a nonisotopic technique for analyzing iodothalamate or chromium ethylenediaminetetraacetic acid (Cr-EDTA), both markers of glomerular filtration rate.

There is no "ideal" marker, i.e., a single marker able to provide all the information necessary to identify people at risk. The main priority for research should be the identification of strong markers that define preclinical, potentially dangerous disease. Addressing disease at this point minimizes costs, morbidity, and mortality and is now possible because of the availability of biologic markers.

BIOLOGIC MARKERS IN EXTRAPOLATION

Extrapolation from animal models is a common and necessary component of risk assessment for humans. To improve the validity of such extrapolations, a better understanding of the relationship between these markers and disease is needed. In most cases, the scientific basis for assuming that animals are good surrogates for humans, and therefore a suitable basis for extrapolation to humans, is overwhelming. It is reasonable to use animal models for extrapolation to humans unless specific information on specific chemicals indicates otherwise. Identification of chemical hazards should include assimilation and evaluation of all relevant information, including appraisal of physical and chemical properties and structure-activity relationships, which often can provide important indications of potential toxicity. Difficulties in diagnosing renal injury and predicting its health consequences are considerable, primarily because the kidneys can undergo substantial chemically induced injury without any clinical manifestation, and subtle injury can be negligible because of the considerable functional reserve of the kidneys. Standard diagnostic criteria are needed that are sensitive enough to serve as markers of renal damage in the presence of renal functional reserve.

Only whole-animal studies or observations in humans can provide information on the operation of multiple cells, tissues, and organs under the influence of complicated feedback mechanisms. Animals are necessary in the study of chemically induced toxicity, because studies that involve modulation of cellular responses and tissue-sampling cannot be appropriately performed in humans.

The first stage of any investigation of nephrotoxicity of a xenobiotic should be in vivo studies. In the absence of any knowledge about potential toxicity and target-organ specificity, the first step should be to determine whether toxicity occurs and the tissue distribution of the toxic response. More detailed studies, both in vivo and in a variety of in vitro models, can then be pursued to elucidate modes of chemical action, specific mechanisms of toxicity, and potential protective or preventive strategies.

A variety of experimental model systems are available for study of renal metabolism, function, and nephrotoxicity. Such systems range from whole-animal studies to those in the isolated perfused kidney, kidney slices, isolated nephron segments, isolated tubule fragments, and isolated renal cells. Each model has advantages and limitations that must be taken into account when developing conclusions and extrapolating animal data to human risk assessment. In vitro

Executive Summary

models of nonrenal urinary tract epithelia have also been developed and applied primarily to examination of carcinogenesis. Development of markers of exposure and susceptibility has not been addressed directly with such nonrenal models and should be pursued for better extrapolation of data for risk assessment.

The importance of enzymatic activation of toxic chemicals is central to an understanding of chemically induced renal injury. Species and strain differences in amounts and tissue distribution of various enzymes can be critical in determining the ultimate toxic response. Consequently, patterns observed and conclusions reached in one species might not apply to another species. The subcommittee recommends that species and strain differences in disposition and metabolism be evaluated for each chemical or class of chemicals. For assessing risk, any experimental model should account for individual variability in response and in health and environmental status. Differences in those factors will alter susceptibility to potentially toxic chemicals.

NEW TECHNOLOGIES

Two branches of study are central to the development of new markers: research into the mechanisms of cell growth, regeneration, and proliferation; and further study of the metabolic capacities of the kidney. Two categories of dispute are acknowledged: the problems that can emerge from the too-rapid and widespread use of a single marker for a specific disease, as typified by the introduction of the test for prostate-specific antigen (PSA) for the detection of prostatic cancer; and the problems associated with extrapolation from animal studies to human conditions, as illustrated by the importance of urinary excretion of $alpha_{2u}$-globulin.

Cell repair can occur in response to cell injury. Thus, markers of cell growth, regeneration, and proliferation can indicate injury and can be particularly useful when an injury is difficult to detect. In this circumstance, markers of repair might be the only indication that injury has occurred. This class of markers is not fully developed and holds promise as a new generation of markers.

Changes in metabolic pathways can also occur in response to cell injury. Thus, biochemical markers associated with these pathways can be of value in detecting nephrotoxicity and carcinogenesis. Like markers of cell growth, regeneration, and proliferation, this class of markers is not fully developed and holds substantial promise.

Improved understanding of the mechanisms of cell growth and metabolism will enable further definition of the steps in the initiation and progression of various urinary tract cancers. It is anticipated that parallels will emerge that will yield insight into the progression of parenchymal renal disease.

The technology that is likely to yield new markers is complex. Equally complex is the identification of susceptible populations with the appropriate clinical assessment of exposure and effect. The use of biologic markers is essential in the examination of xenobiotic-induced diseases and other diseases of the human kidney, bladder, and prostate. Comprehending the sequences of events is an iterative process that involves a complex data set derived from scientific advances in molecular biology, epidemiology, pathology, biochemistry, and clinical medicine. Assembly of those data into an organized framework will be a major step toward improving risk assessment and should be a long-term objective.

In an increasingly complex technical and industrial society, exposure to xenobiotics is unavoidable. Health hazards are undeniably associated with such exposure, and they should be monitored to prevent or modify disease. Some forms of parenchymal renal disease and cancer of the kidney and bladder are among the conditions associated with xenobiotic exposure. The development of reliable markers of susceptibility, exposure, and effect is among the first steps to be taken toward prevention of diseases of the kidney, bladder, and prostate. Indeed, the advent of new technologies in molecular biology and sophisticated understanding of metabolic pathways holds promise that markers can be developed and prevention of the diseases achieved. Classical approaches to the study of nephrotoxicants and carcinogens should not be disregarded, however; for example, the utility of animal studies in the study of xenobiotics has been emphasized in this report.

THE SUBCOMMITTEE'S RECOMMENDATIONS

The subcommittee's recommendations follow the contents of this report: *General* (toxic exposure to the urinary tract overall—kidney, bladder, and prostate); *Biologic Markers of Exposure and Susceptibility*; *Biologic Markers of Effect*; *Biologic Markers in Extrapolation*; and *New Technologies*.

General

The Kidney

• For patients entering programs for treatment of end-stage renal disease (ESRD), details of occupational history or other factors that would show the impact of patients' environments on their condition should be obtained. As a first step, available information in relevant databases should be examined. Studies should be undertaken to determine whether the higher incidences of ESRD among minority groups and the economically disadvantaged are related to occupational or environmental exposure to nephrotoxicants. Epidemiologic studies need to focus on the various populations at risk; this focus should include not only the identification of the populations but their continued monitoring.

• Studies should be performed to determine whether an association between anatomic or physiologic differences of the kidney at birth and a later response to environmental or occupational nephrotoxicants leads to susceptibility to disease or to progression once disease occurs.

• Data should be collected on the incidence of renal abnormalities among recreational-drug users to determine the influence of those substances on the rate

Executive Summary

of progression of renal disease due to other causes.

- Basic studies are needed to determine the effects of occupational and environmental toxicants on specific segments of the kidney. These effects should be correlated with biochemical and anatomic changes.

The Bladder

- Human bladder cancer induced by xenobiotic exposure in worker cohorts should be investigated to develop strategies of individual risk assessment, to formulate programs for prevention, and to evaluate new forms of therapy. Strategies of individual risk assessment need to be developed as the cornerstone of prevention. Once cohorts of at-risk persons are identified, they should be enrolled in long-term monitoring studies to assess the efficacy of prevention and treatment strategies.
- Further research on the direct effect of xenobiotics on the bladder and the interactions of xenobiotics with the protective mechanisms of the bladder is very likely to uncover additional evidence that the bladder is a target organ. Markers associated with susceptibility should be identified to define the higher relative risk of disease in an exposed subset of the population.

The Prostate

- Options for improving the efficacy of screening procedures should be studied. Tests with lower false-positive rates should be developed, as should tests able to detect premalignant changes and to separate quiescent from biologically active disease.

Biologic Markers of Exposure and Susceptibility

- Populations at risk of identifiable renal insults and carcinogenesis should be defined. Markers of human exposure and susceptibility should be sensitive (i.e., detectable before injury occurs), noninvasive, and chemically stable.
- Genetic and nongenetic factors that modify susceptibility to occupational and environmental genitourinary toxicants and carcinogens should be considered in the evaluation of individual susceptibility. These include sex, race, nutrition, socioeconomic factors, age, coexisting chronic disease, and drug abuse.
- Markers of exposure and susceptibility should be identified to determine the relationship between coincident exposure to nephrotoxicants and development or progression of chronic renal disease. Particular attention should be paid to the role of widespread and sometimes excessive use of analgesics, including nonsteroidal anti-inflammatory drugs, in diseases of the urinary tract. Clinicians should be aware of the danger associated with abuse of such agents and should query renal-disease patients about their use. In particular, patients with established renal disease should be wary of exposure to these agents and other potential nephrotoxicants.

Biologic Markers of Effect

- A battery of relatively simple and noninvasive tests should be used as a first step in screening populations at risk. On the basis of available information and technology, adequate initial screening results should be obtained by testing for proteinuria with dipsticks and then measuring urinary concentrating ability and serum creatinine and, for more sensitive measurements of tubular integrity, monitoring for an increase in urinary enzyme or low-molecular-weight protein excretion. Application of those tests to a population exposed, for instance, to diagnostic procedures or treatment with nephrotoxicants might identify early renal damage with adequate sensitivity.
- Several of the newer testing procedures hold promise of future usefulness and should be further investigated. Among them are tests of urinary excretion of various growth factors, such as epidermal growth factor (EGF), and other tubular enzymes, such as intestinal alkaline phosphatase (IAP); both reflect some specificity of localization along the nephron and of cellular origin.
- Molecular techniques have identified a variety of potentially useful markers of renal-cell injury. These include the products of expression of some early genes and changes in the expression of renal cytokines, growth factors, and growth-factor receptors. It is highly likely that studies of these and similar molecular events will yield better markers of effect, and continued research in this area should be encouraged.
- Understanding fundamental cellular and molecular mechanisms of growth control in human tissues undergoing carcinogenesis (e.g., high-risk occupationally exposed populations or patients with premalignant processes) should be emphasized, because it is highly likely that more specific markers of effect will be identified and allow early intervention.
- Markers that define preclinical disease should be identified. In addition, markers that detect xenobiotic-induced mutational events should be identified.

Use of Biologic Markers in Extrapolation

- Models for the identification and validation of markers should continue to be developed. The models must have sufficient sensitivity to distinguish between normal and abnormal function and must correlate well with known human toxicities. The models also must distinguish between functional alterations and pathologic changes. To obtain those characteristics, it will be necessary to develop and apply new technologies. Issues related to cost effectiveness should be considered.
- Whole-animal studies should be used to establish target-organ specificity and to assess renal function in relation to survival. Species, sex, and strain differences must be taken into account in selecting animal models for particular uses.
- In vitro methods should be used for mechanistic studies; the choice of

models should depend on compatibility and validation with whole-animal studies.

• Metabolic studies should be conducted to ascertain whether xenobiotics (or other agents) are biotransformed to reactive and toxic species and to identify sites of transformation, including renal tissue and other tissue in the urinary tract.

New Technologies

• Research should continue toward better understanding of the mechanisms of cell injury, because they can underlie the development of new markers. Emphasis should also be placed on understanding the mechanisms of cell growth, regeneration, and proliferation. Insight into the factors that control the cell cycle, regulate various growth factors, influence gene expression, and modulate nucleic acid synthesis might be critical in the development of new classes of markers.

• Research should continue toward better understanding of the metabolic pathways of the kidney in relation to the effects of xenobiotics and susceptibility to them.

• Attention should be directed toward a deeper understanding of the mechanisms by which proto-oncogenes, tumor-suppressor genes, and epigenetic factors regulate the cell cycle and how damage to these mechanisms is related to disease. Attention should also be directed toward the elucidation of metabolic pathways, particularly as they are related to the production of toxic metabolites. Additional markers should be identified that help to identify populations at risk and to study the mechanisms by which environmental and occupational toxicants promote cancer.

• Research on the relation of growth factors to the prostate requires rigorous experimental approaches and designs and must consider multiple variables. Studies of biochemical changes in the areas next to a prostatic tumor might be more informative than analysis of the cancer itself.

• The general relationship between nephrotoxicity and renal carcinogenesis should be explored.

• To achieve the desired goal of identifying more useful markers, cooperation between laboratory scientists, epidemiologists, and clinical researchers should be encouraged. Assays, particularly those involving enzymes or molecular probes, must be replicable in different laboratories.

BIOLOGIC MARKERS IN URINARY TOXICOLOGY

1
INTRODUCTION

At the request of several federal agencies, the Board on Environmental Studies and Toxicology in the National Research Council's Commission on Life Sciences convened the Committee on Biological Markers to examine the use of biologic markers in environmental health research. Biologic markers indicate events in biologic systems and are variations in the number, structure, or function of cellular or biochemical components of tissues and organs. Biologic markers afford a means to identify early stages of disease and to understand the basic mechanisms of the biologic responses to substances found in the environment (Committee on Biological Markers of the National Research Council, 1987). Four specific biologic systems were chosen for study: the reproductive system (NRC, 1989a), the respiratory system (NRC, 1989b), the immune system (NRC, 1992), and the urinary system. This is the report of the Subcommittee on Biologic Markers in Urinary Toxicology, which considered the kidney, bladder, and prostate to be within the scope of its charge.

The kidney's main function is excretion of waste products in the urine, but it has other functions. It plays an important role in homeostasis and is the predominant organ in the control of extracellular fluid volume and electrolyte composition. The kidney also produces hormones, including erythropoietin, renin, 1,25-dihydroxyvitamin D3, and several vasoactive prostanoids and kinins. The very processes by which the kidney maintains internal homeostasis make it highly susceptible to toxic damage; that is, because of the mechanisms by which large amounts of blood are filtered and waste products are concentrated within the kidney, its exposure to some toxic material can exceed that of other organs. In addition, the kidney is metabolically active and has the capacity to metabolize some materials to toxicants. The bladder is exposed to the waste products collected by the kidney, which at times reach high concentrations in the urine. Depending on the frequency of voiding, the bladder can have a longer exposure to toxic waste products than the kidney itself. Emptying of the bladder can be influenced by various diseases of the prostate. These organs can also be exposed to bacterial and viral infectious processes that modify the normal milieu and act as confounding variables. All the characteristics just enumerated help to explain why the urinary tract—in both laboratory animals and human beings—is a site of xenobiotic-induced cancer.

BIOLOGIC MARKERS

Humans are often and unavoidably exposed to hazardous environmental chemicals. Biologic markers can be useful in confirming such exposures, estimating their magnitude, and identifying persons adversely affected or most likely to be adversely affected if exposures continue. Biologic markers are indicators of events or conditions in biologic systems or samples. It is hoped that they will clarify the relationship between environmental exposures and human diseases and help to prevent such exposures and diseases. Interest in the use of biologic markers to study the health effects of exposure to environmental toxicants is growing among researchers in clinical medicine, epidemiology, toxicology, and related biomedical fields. Specifically, clinicians can use markers for early detection of disease; epidemiologists can use them as indicators of exposure to determine internal dose or health effects; and toxicologists can use them to estimate dose-response relationships and to facilitate assessment of risk associated with exposures. Biologic markers also can be helpful in clarifying the underlying mechanisms of specific diseases.

New developments in molecular genetic and biochemical approaches to medicine have yielded sensitive markers for assessing toxicant exposures (NRC, 1991). They have also increased our knowledge of disease, improved our ability to predict the outcome of disease, and helped to direct courses of treatment. Many diseases are defined not only by clinical signs and symptoms but also by the assessment of biologic markers at the subcellular and molecular levels. Diseases of the liver and kidney, for example, are often detected by measurement of enzymes in blood, proteins, or cells in urine; diabetes can be suspected if glucose is found in urine; and inborn errors of metabolism, such as phenylketonuria, are found early by biochemical analysis, rather than later as a result of clinical dysfunction. Concentrations of material in the urine are often used to determine the degree of exposure in an occupational setting, in monitoring therapeutic regimens, and in forensic medicine.

Biologic markers are indicators of exposure, effect, or susceptibility. That classification is a useful theoretical scheme by which to characterize biologic markers of any organ system; however, it must be qualified somewhat for practical application. A strict classification with respect to any of the three categories (exposure, effect, or susceptibility) is difficult, for the three categories often are related and can be thought of as elements in a continuum

Introduction

from environmental exposure to clinical disease.

The identification of a chemical or its metabolite in a biologic specimen is a marker of exposure. Cellular or molecular changes within an organism are biologic markers of effect. Biologic markers often are demonstrated by clinical laboratory measurements or clinical tests used in the differential diagnosis of various diseases. They can serve as surrogates for other methods of detection in determining the molecular and cellular events in the development of health problems. If such markers could be detected before exposed persons became obviously ill, a disease process might be reversed in those affected or prevented in others. Finally, some cellular or molecular measurements can identify groups of people or individuals who are more vulnerable than others to the effects of toxic exposure; the results of these measurements are markers of susceptibility.

Markers of Exposure

A biologic marker of exposure is a xenobiotic chemical or its metabolite or a product of an interaction between the chemical and some target cell. Markers of exposure most commonly are the concentrations of materials in urine, blood, or other body tissue, including hair and nails.

Several problems limit the usefulness of specific markers of exposure. For example, many environmental xenobiotics have similar metabolites, and this makes identification of the parent chemical difficult. Also, large doses of a substance, by inducing anatomic or physiologic change, can alter the retention or excretion of the marker and in this way decrease its utility as a marker of exposure. The body burden of a substance reflected by biologic markers might result from more than one route of exposure. It can be difficult to identify persons exposed to hazardous concentrations of substances commonly found in the body in lower concentrations. Such is the case in excessive exposure to essential minerals, such as copper and cobalt. Markers of exposure decay after exposure ceases; measurement soon after exposure is required for materials that are metabolized rapidly, such as trichloroethylene, but not after materials with long half-lives, such as cadmium. Markers of exposure alone give no indication whether an exposure has produced a biologically significant result. The same dose in persons who are susceptible and resistant to a given xenobiotic can have different results. Urine is one source of markers of exposure. High exposures to a material can result in increased concentrations of it or its metabolites in urine. When sufficient pharmacokinetic information is available, urinary markers of exposure can be used to estimate the total exposure of a person to a substance.

Markers of Effect

A marker of effect is a measurable cellular, physiologic, or biochemical alteration within an organism. Markers of adverse effect can be biochemical or cellular signals of tissue dysfunction, increased enzyme activity, the appearance of excessive waste products in the urine or other sampling compartments, and physiologic signs of abnormal function, such as increased blood pressure or blood in the urine. The effects themselves might not be directly adverse, but rather might indicate a potential for health impairment (e.g., DNA adducts). Biologic markers of effect include changes in the components of urine itself (such as an increase in urinary protein excretion or cellularity) and changes in the volume or composition of other body fluids caused by kidney dysfunction.

With impaired renal function, excretion products, such as creatinine and urea, can accumulate in the blood. Increased blood concentrations of these products can be markers of renal damage, although they are not particularly useful except in cases of severe damage as discussed in Chapters 2 and 4. Other markers of renal toxicity can be present in urine because of altered renal function or damage to the kidney. Many of these can be detected in examination of the urine, including color, volume, pH, specific gravity, albumin, electrolytes, enzymes, and elements in the urinary sediment. Markers that could show up include white and red blood cells, kidney epithelial cells, casts, and crystals.

Markers of Susceptibility

A marker of susceptibility is an indicator of an inherent or acquired limitation of an organism's ability to resist the effects of exposure to a specific xenobiotic substance. A biologic marker of susceptibility can be an intrinsic genetic or other characteristic or a pre-existing disease that results in an increase in the amount of a material absorbed, an alteration in its metabolism, or an increase in the target-tissue response. It can be difficult to distinguish between susceptible and nonsusceptible persons, either because of multiple interactive influences and genetic polymorphisms or because of difficulties in measuring markers of susceptibility.

Potential sources of normal variability in markers of susceptibility can be the result of concurrent disease or genetic, environmental, or biorhythmic influences other than the specified events that are the object of study. Such influences can cause differences between populations, differences between individuals, and differences within individuals over time. Decreased detoxification or excretory function are commonly involved in susceptibility. Thus, elderly persons with declining organ function and young persons with immature and developing organs are likely to be more vulnerable to toxic substances than healthy adults (Figure 1-1).

VALIDITY OF BIOLOGIC MARKERS

In the course of developing this document, the subcommittee noted that var

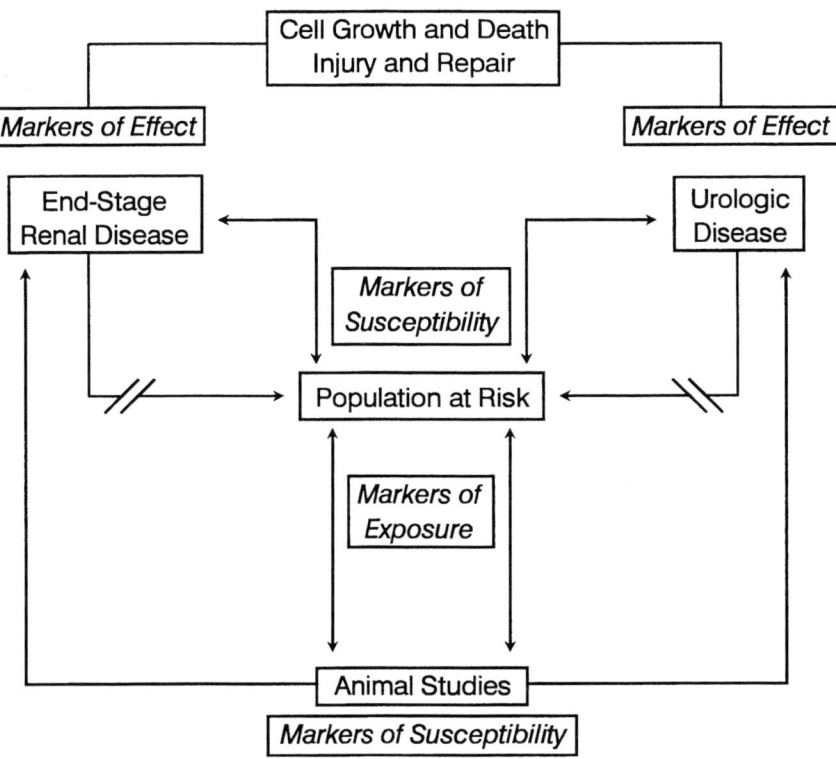

FIGURE 1-1. Flow diagram depicting the relationship between markers of exposure, effect and susceptibility for patients with either ESRD or UD. The central focus is on identification of the population at risk, that is, those most likely to be exposed to xenobiotics and susceptible to their adverse effects. The most devastating clinical outcomes are the development of chronic renal disease or urogenital cancer. Identification of the population at risk can be facilitated by retrospective analysis of those with ESRD or UD and by prospective use of animal studies. In both cases, the utility of markers of exposure and susceptibility can be established. Markers of susceptibility can vary between humans and animals, whereas markers of exposure must be shared if they are to be used to extrapolate between populations at risk and laboratory investigations. The aim is to provide insight into disease mechanisms and allow testing for prevention or modification. Before changes in organ function, adverse response to xenobiotics might be detected by understanding the mechanisms of cell growth, injury, repair, and death. It is anticipated that this will lead to identification of highly sensitive and specific markers of effect.

ious disciplines use *validity*, *sensitivity*, and *specificity* in different ways. The oversight committee discussed the validation of biologic markers for use in assessments of human health, particularly in epidemiologic studies. To validate the use of a biologic measurement as a marker, it is necessary to understand its relationship to the event or condition of interest, such as the potential for health impairment or susceptibility to disease. Sensitivity and specificity are critical in the process of validation (MacMahon and Pugh, 1970). In epidemiology, sensitivity is the ability of a test to identify correctly persons who have the disease, condition, or exposure of interest; specificity is the ability to identify correctly persons who do not have the disease, condition, or exposure of interest. The sensitivity and specificity of markers of exposure must be validated.

The use of the terms *sensitivity* and *specificity* in epidemiology must be distinguished from their use with respect to laboratory methods. *Laboratory sensitivity* refers to the lowest concentration of a substance that can be measured reliably, and *laboratory specificity* refers to the ability of an analytic technique to measure only the substance in question.

Predictive value is a measure of the potential usefulness of a test in identifying an exposed member of a population (positive predictive value) or identifying an unexposed member of a population (negative predictive value). For rare conditions (those with low prevalence in the population), even very good tests can have low positive predictive value. When there is high prevalence in a population, even substantially poorer tests can have high positive predictive value.

Sensitivity and specificity are determined by both intrinsic and extrinsic factors. Intrinsic factors include differences in the distribution of the marker between the unexposed population and the exposed population. An ideal biologic marker is absent in all unexposed persons and easily detectable in all exposed persons. No such marker is known to exist for any organ system, and, because of the variability of responses within and between individuals, the distributions of most markers are likely to show considerable overlap in unexposed and exposed individuals.

Extrinsic factors also influence specificity and sensitivity. For example, people who have not been exposed to a toxicant could have other conditions that cause the appearance of the biologic marker that is seen in people who have been exposed. In addition, errors in measurement also influence specificity and sensitivity. Analytic imprecision blurs the difference in distribution of a marker between exposed and unexposed populations, and analytic inaccuracy can lead to misclassification of individuals.

The criteria used to define an event or condition (especially a health effect) can affect the fraction of the population that is found to have it. If, for example, the event is a disease, the finding of a "case" is often based on a subjectively determined collection of signs and symptoms. Adding or excluding a particular symptom can result in finding a larger or smaller number of cases. This will influence estimates of the sensitivity and specificity of the method used to find cases.

If being above (or below) some index value or threshold on a test is what determines whether a finding is "positive," the sensitivity of the test can sometimes be increased by adjusting the threshold. However, because sensitivity and specificity are linked for a given test, increasing one will inevitably result in decreasing the other. There are circumstances, however, in which adding another test (or tests) can be of value. For example, a test with poor specificity (that is, one that incorrectly identifies persons as having the condition of interest) can sometimes be followed by a test with greater specificity that will eliminate many (if not all) false-positive results.

In developing a test, sensitivity and specificity are usually estimated as a result of a trial on a known or (often) an easily acquired population, such as medical students or nurses. The sensitivity and specificity often shift with the population tested, so persons using the test need to be aware that the first published sensitivities and specificities are almost certain to change when different populations are used.

Specificity of markers is important in the consideration of validity. For example, urinary concentrations of trichloroethanol and trichloroacetic acid, which are metabolites of trichloroethylene, are used as indicators of occupational exposure to trichloroethylene. But, they are not specific for trichloroethylene, inasmuch as metabolism of tetrachloroethylene and chloral hydrate also produces these metabolites; in addition, their half-lives are relatively short, so the ability to detect them decays over time.

The variability between individuals in the factors that affect sensitivity greatly influences the ability of markers to detect exposure. Some markers that are validated generally have been established only for obvious clinical events. A major purpose of markers in environmental health research is to identify exposed persons, so that risk can be predicted and disease prevented. Validation involves both forward and backward processes of association—from the marker back to exposure and from the marker forward to effect.

ETHICAL ISSUES

The availability of highly sensitive assays that can identify effects resulting from the interplay between small exposures and genetic or acquired susceptibility raises several ethical questions. The National Research Council's Committee on Biological Markers (NRC, 1987), Schulte (1987), and Weiss (1989) have discussed these issues. Primary among them is the use made of information derived from biologic markers. Consider the hypothetical case of a biologic marker known to reflect susceptibility to the effects of a chemical. Should a worker in a workplace that contains the chemical who tests positive for that marker or has an increased measurement of it in his or her blood be removed from the workplace? If so, should the worker be offered an equivalent job in the same industry? Or should the workplace be cleaned up to protect even the most sensitive worker?

The relationship of group and individual risk to biologic markers is crucial. It is important, therefore, to inform research-study participants in advance of the degree to which the results will be interpretable at the group level as opposed to the individual level. Participants in such studies should be given test results that are presented and discussed in the context of variability within and between people in the normal (nonexposed) population and the study group. Participants in epidemiologic studies might resist or refuse invasive techniques for obtaining markers or resist providing markers obtained by techniques perceived to be not fully safe. Although the use of urine samples for marker detection might be universally accepted, attention should not be directed away from blood, other body fluids, or tissue as reliable sources of markers; accuracy and not convenience should be the driving force. Whatever the choice, it must be economically feasible.

It is anticipated that some characteristics or markers of susceptibility will define a population at high risk for developing renal or urologic disease in response to environmental or occupational nephrotoxicants. Identification of the susceptible populations and appropriate markers may be greatly facilitated by comparative studies in animals with the understanding that in both cases it is necessary to consider precise definitions of markers of exposure, because susceptibility and exposure can be closely associated. In many cases before renal or urologic disease is clinically apparent, various markers of effect are present. These markers—which reflect cell injury, repair, growth, differentiation, or death—facilitate early intervention with the hope that disease progression can be slowed, halted, or reversed.

STRUCTURE OF THE REPORT

Chapter 2 provides information on the structure and function of the urinary tract, the diseases associated with the urinary tract, and their risk factors. In Chapter 3, relationships among exposure, susceptibility, and the associated markers are developed. The usefulness of markers in monitoring urinary diseases is discussed in Chapter 4. Chapter 5 develops a rationale based on epidemiologic studies for the use of biologic markers in protection of human health and extrapolation of data from animals to humans. New technologies and related issues are reviewed in light of research needs in Chapter 6. Chapter 7 summarizes the subcommittee's findings and recommendations.

2
TOXIC EXPOSURE OF THE URINARY TRACT

Many environmental and industrial pollutants that appear to be well tolerated at low doses can produce renal damage at high doses; and some nephrotoxic agents that are poorly tolerated at any dose are used injudiciously or otherwise find their way into the environment. In addition, a growing list of therapeutic and diagnostic agents are capable of causing renal injury, and the problem of renal disease due to recreational drug use is growing.

The extent to which those and other agents result in clinical renal insufficiency and the nature of the population at risk are incompletely defined. However, many forms of acute and chronic renal failure occur for unknown reasons, and the incidence of end-stage renal disease (ESRD) has marked racial and regional differences, so nephrotoxicants might well present a serious health hazard. It has been estimated that one in five patients hospitalized with acute renal failure has been exposed to one or more nephrotoxic agents (Rasmussen and Ibels, 1982).

The identification of agents with nephrotoxic potential is hampered by several obstacles. Foremost is a constantly changing environment of chemical hazards. In addition, variations in diagnostic criteria make it difficult to identify conditions due to prolonged exposure to nephrotoxicants.

The lack of availability of simple and reliable tests for early renal injury and the long period between exposure to environmental nephrotoxicants and the onset of definable disease seriously limit the ability to define a cause-effect relationship. Standard indexes of renal function are rather insensitive markers of injury. Each human kidney contains some 800,000-1,000,000 nephrons, and a nephron can have an average filtration rate of 50 μl/min. The total glomerular filtration rate of both kidneys (1,600,000-2,000,000 nephrons) exceeds 100 ml/min. After acute or chronic exposure to a nephrotoxicant, hypertrophy of less severely damaged nephrons tends to counterbalance the atrophy of the most severely damaged nephrons. It is conceivable that one-third of the nephrons of the two kidneys could be lost without a noticeable reduction in the whole glomerular filtration rate if the remaining 1,330,000 nephrons (assuming a normal total of 2,000,000) hypertrophied so that their filtration rate increased by 50%. Such fine adjustment rarely occurs; the presence of a substantial amount of structural and functional damage might be impossible to determine precisely with standard tests of renal function.

Finally, there are almost no epidemiologic data on the number of people who develop ESRD as a result of acute or chronic exposure to environmental nephrotoxicants. All the estimates depend on inferences drawn from inconclusive sources, such as surveys of patients entering dialysis and transplantation programs. The results of several of the surveys indicate that substantial gaps exist in the ability to identify the primary abnormality leading to ESRD (Burton and Hirschman, 1979; Easterling, 1977; Evans et al., 1981; NIH, 1990; Rostand et al., 1982). For example, data compiled by the U.S. Renal Data System (USRDS) for the years 1987-1990 indicate that disease of unknown cause made up 6.6% of the cases of ESRD (NIH, 1993). Patients with interstitial nephritis not due to analgesics abuse were 3.4% of the total. Because those groups of patients include some who have been exposed to xenobiotics, such as heavy metals, it is possible that for 10.0% of the patients with ESRD, environmental and occupational nephrotoxicants might be of primary importance in the etiology of the disease. Even in other persons with ESRD, exposure to environmental pollutants might have been a factor in the onset or progression of the disease. Information on occupational history or other factors that would implicate a patient's environment in his or her potentially catastrophic illness is rarely available.

On the basis of the available social and demographic data, several notable groupings might be relevant to the incidence of renal diseases. ESRD is found in a disproportionately high percentage of minority, ethnic, and racial groups in the United States. Native Americans, blacks, and Hispanics, especially Mexican Americans, have overall ESRD rates about 3-4 times greater than the rate in whites (USRDS, 1991). Although the reasons for the increased susceptibility of minority groups to developing ESRD are unknown (Rostand, 1992), several possibilities have been suggested (Feldman et al., 1992). They can be separated into two broad categories: differences in the access to preventive health care and renal-replacement therapy, and physiologic heterogeneity among racial groups that might increase renal sensitivity to toxic exposures.

A study of 9,390 black and white New York state residents who began treatment between 1982 and 1988 sought to determine whether the incidence of ESRD due to the three most frequent causes (diabetic glomerulosclerosis, hypertensive nephrosclerosis, and glomerulonephritis) was related to socioeconomic status (Byrne et al., 1994). A clear effect of socioeconomic status on the incidence of ESRD due to diabetes or hypertension was demonstrated in whites, but, perhaps because of overriding factors, no such effect was seen in blacks. It was suggested that vigorous pursuit of other epidemiologic factors in the development of the progression of renal disease in blacks and of the possible relevance of the different structural, physiologic, and vascular renal responses between blacks and whites is indicated.

The hypothesis that treated ESRD is associated with socioeconomic status—independently of the known associations with race, age, and sex—and the hypothesis that the higher incidence of treated ESRD among blacks could be explained by differences in socioeconomic status have been examined in a study that linked the information from the USRDS and the Bureau of Health Professions Area Resource File (Young et al., 1994). An inverse association between the incidence of treated ESRD and socioeconomic status, as estimated by average income of county of residence, was found after adjustment for race, sex, and age. Differences in socioeconomic status appeared to explain some of the difference between blacks and whites in the incidence of treated ESRD. That many patients who develop ESRD do so for unknown reasons, are proportionately more likely to be members of minority groups, and come from economically marginal backgrounds is consistent with the possibility that environmental factors influence the development of such disease.

In the United States in 1990, 165,000 people with irreversible renal failure received renal therapy for ESRD with the aid of chronic dialysis, and 9,800 renal transplantations were performed (NIH, 1993). In that year, total medical payments to provide maintenance dialysis, kidney transplantation, and all related health services to ESRD patients were in excess of $6.39 billion. In view of the limited rehabilitation achieved by dialysis, the complications associated with transplantation, and the tremendous costs involved in each, substantial efforts are required to identify the specific causes of renal disease and the factors that determine progressive and irreversible decline in renal function.

THE URINARY TRACT

To understand the inherent limitations in the detection of early renal injury, we consider here several elements of normal renal function. In the sections that follow, we treat the mechanisms that lead to renal toxicity once exposure to a nephrotoxic substance occurs, the host factors that modify the response and the nature and extent of the physiologic adjustments to injury; it is presumably the responses to injury that give rise to the various markers described in later chapters.

Renal Blood Flow

The kidneys are highly vascular organs with a blood flow of about 1,000 and 1,200 ml/min in women and men, respectively, of average height and weight. That flow is about one-fifth of the resting cardiac output. Within the kidney, 85-90% of blood flows through the cortex and only 10-15% through the medulla.

Glomerular Filtration

The initial step in the formation of urine is the production of an ultrafiltrate of plasma (filtration under pressure results in the retention of colloids but permits the passage of crystalloids). The unique interposition of the glomerular capillaries between the afferent and efferent arterioles is fundamental to the formation of an ultrafiltrate. Each minute, the kidneys produce 100-140 ml of glomerular filtrate with an osmolality of 280-290 mOsmol/L. In 24 hours, this amounts to 150-200 L of filtrate and over 40,000 mOsmol of solute. The amount and characteristics of the filtrate are influenced by the area available for filtration and the electric charges on the glomerular capillaries.

Tubular Reabsorption and Secretion

Once the glomerular filtrate is formed, it passes through a complex series of tubular structures where it is modified in such a fashion that waste products are excreted in the urine, critical body constituents are conserved, and the body's fluid volume is regulated. More than 99% of the filtered solute and water is reabsorbed. The principal oxygen-consuming work performed by the kidney is electrolyte reabsorption.

Urine volume depends on the dietary intake of water, endogenous water production, insensible water losses, and the ability to concentrate or dilute the urine. The final osmolality of the urine can be as high as 1,400 mOsmol/L or as low as 40 mOsmol/L. The mechanisms by which the kidneys adjust the final composition of the urine are varied. Some substances, which are protein-bound, escape filtration only to be added to the urine by the process of tubular secretion. Other substances that are freely filtered—such as amino acids and glucose—are in normal circumstances completely reabsorbed by the tubules. These processes generally require the expenditure of energy and are particularly vulnerable to the effects of toxicants.

Hormonal Action

Renal function is modified by several extrarenal and intrarenal hormones. The major extrarenal hormones—aldosterone, vasopressin, and parathyroid hormone—modulate the excretion of sodium, water, and phosphorus, respectively. Intrarenal hormones—such as renin, prostaglandins, and kallikreins—affect renal blood flow, glomerular filtration, and tubular function. The kidneys also produce erythropoietin, which stimulates red-cell production; synthesize vitamin D from its precursor; and participate in the metabolism of several hormones, such as insulin. A complex group of peptide mediators, cytokines, influence cell growth and function;

these are produced locally or systemically and have the potential to influence the response to injury.

MECHANISMS OF RENAL TOXICITY

Susceptibility to Injury

The susceptibility of the kidney to toxic damage is related to various aspects of renal function. First, because blood flow to the kidney per gram of tissue is greater than that to most other organs, the total amount of toxicant delivered can be disproportionately high. Second, the processes of glomerular filtration, tubular reabsorption, and secretion tend to concentrate a toxicant that reaches the kidney. Third, the high metabolic rate of tubular epithelial cells leaves the kidney vulnerable to the actions of metabolic inhibitors. Fourth, the kidney can metabolically alter various endogenous and exogenous chemicals; this generally produces compounds with reduced biologic activity, but occasionally compounds with increased biologic activity are formed. Fifth, the mechanism of countercurrent exchange, which allows the kidney to form a concentrated urine, can prolong the residence time of a toxicant in the kidney.

Direct Toxic Effects

The nephrotoxicity of environmental pollutants is determined by their particular chemical properties, the duration and extent of exposure, and the nature of the host response. Manifestations of toxicity are related to the site of action in the kidney, the degree of damage produced, and the ability of the kidney to compensate for the loss of function or to repair injury.

Renal Vascular Injury

Involvement of the renal vasculature leads to changes in renal vascular resistance with a redistribution of blood flow in the kidney, a decrease in total blood flow, or both. The kidney can also lose its ability to autoregulate its blood flow. To the extent that renal plasma flow determines the rate of glomerular filtration, a decrease in the clearance of a number of substances can be expected to accompany changes in renal vascular resistance. Those effects might be mediated by anatomic changes in the renal vasculature, by changes in the sensitivity to systemic or local vasoactive substances, or by changes in muscular reflexes within the vascular walls themselves.

Glomerular Capillary Injury

The primary effect of a toxicant might be to change the ultrafiltration coefficient of the glomerular capillary membranes. That coefficient is a product of the glomerular capillary surface area and hydraulic conductivity. A decrease in either results in a proportional decrease in the filtration rate. Changes in pore size or configuration and neutralization of the fixed negative charges can have generalized or selective effects on the ability of various substances to pass the glomerular barrier.

Renal Tubular-Cell Injury

The most firmly established effects of nephrotoxicants are on renal tubular epithelial cells, in particular those of the proximal tubule. Indeed, toxicity might be limited to a specific cell type or to a particular organelle in the cell. For example, some compounds have their major effect on tubular epithelial cell membranes, and others selectively alter the function of lysosomes, mitochondria, nuclei, or the endoplasmic reticulum (Fowler, 1982). Disruption of some organelles—particularly those which provide energy for cellular respiration—can lead to cell death.

The medullary countercurrent multiplication system provides an efficient mechanism for eliminating waste products and minimizing body-water losses. The system is such that drugs and their metabolites can accumulate in the medullary interstitium. Their chemical properties determine whether the accumulated substances initiate an inflammatory response.

The proximal tubule's organic acid and base transport systems provide an important route of elimination of molecular species that, as a result of their charge or size, do not undergo glomerular filtration but still require urinary elimination. Chemicals that interact with the organic ion-transport system can accumulate in cells or achieve high concentrations in the urine.

Other tubular mechanisms that can be impaired include those involved in electrolyte excretion and water metabolism. In addition, the mechanism of pinocytosis—whereby high molecular-weight molecules, if filtered, are recaptured from the proximal tubule fluid—can be disrupted. Finally, mechanisms of tubular epithelial cell regeneration can be compromised by nephrotoxicants.

Indirect Toxic Effects

Immunopathologic Mechanisms

Evidence has accumulated that the toxicity of environmental agents can in part be mediated by immunologic mechanisms that result in glomerular or tubulointerstitial disease (Wilson, 1982). There are four major categories of immunologic mechanisms. In Type I, or immediate hypersensitivity, damage results from the binding of antigen to IgE antibodies fixed to mast cells and basophils. In Type II, or cytotoxic reaction, damage results from the reaction of antibodies with cell-bound antigens and leads to activation of the complement cascade and cell death. Type III, or immune-complex reaction, stems from the formation of immune complexes in situ or in the circulation and leads to tissue damage. Type IV, or delayed hypersensitivity, is mediated primarily by T lymphocytes.

Data are insufficient to implicate a Type I response in mediating the effects of nephrotoxicants. But, various agents, including xenobiotics, might alter glomerular or tubular basement-membrane structures so that autoantigens are produced, autoantibodies or sensitized lymphocytes are formed and come to rest in the glomerulus through the process of filtration arrest, and a Type II response occurs. The Type III response is likely to be more important and might involve either the deposition of immune complexes formed in the intravascular compartment producing a serum-sickness-like reaction, or an antibody-antigen reaction in the extravascular compartment that results in inflammation produced by antigen-reactive cells, rather than antibodies—the so-called Arthus reaction. In either Type III case, the toxicant can act either as a full antigen or as a hapten. The haptens are small antigenic determinants that are covalently coupled to larger carrier molecules. Alternatively, various antibody-antigen complexes can be formed in situ. Once bound to tissue, the complexes fix complement, activating the complement cascade and triggering an in situ inflammatory response. In this situation, material previously trapped or planted in renal structures serves as the antigen; this material can be cationic proteins that are sequestered in the glomerulus or in vascular cells, where they become *planted* antigens. Later, a circulating antibody can attach to such antigens and result in in situ immune-complex formation. In addition, when structural damage is produced as a result of direct toxic effects, local autoantigens can be produced. Environmental agents can also have a primary effect on other antibody-antigen interactions, favoring the formation of complexes of such a size or composition that nephritogenic immune reactions occur.

Apart from those considerations is the possibility that the Type IV mechanism, once thought to be of little importance in the development of renal injury, plays a role in some forms of interstitial disease. In the Type IV, macrophages, either leukocytes or monocytes, invade glomeruli and initiate a local inflammatory response mediated by cytokines, thromboxanes, and leukotriene prostaglandins.

Nephrotoxicants might also produce novel antigens that are capable of stimulating an autoimmune response in keeping with any or all of the four mechanisms.

Xenobiotic Metabolism

Under most circumstances, foreign substances (xenobiotics), once absorbed, are distributed to various tissues where they undergo biotransformation with the production of innocuous metabolites, which are then eliminated. The enzymes responsible for biotransformation include various mixed-function oxidases in microsomes. Nonmicrosomal biotransformation can also occur. At times, these reactions result in the augmentation of toxicity. Although activity of these enzymes in the kidney as a whole is at a low level, certain specific cell types show considerable activity in this regard. The site of transport of many xenobiotics is the proximal tubule. During transport, they must cross two cell membranes: the contraluminal and the luminal. Most xenobiotics are transported by the transport system for hydrophobic anions, some by the system for hydrophobic cations, and some by both (Ullrich and Rumrich, 1993). Several environmental pollutants—most notably the polybrominated biphenyls (PBBs), the polychlorinated biphenyls (PCBs), some halogenated dibenzodioxins, and hexachlorobenzene—have the capacity to alter activity of mixed-function oxidases (Kluwe and Hook, 1980). A result of this activity is that renal injury occurs after exposure to metabolites that by themselves would be otherwise well tolerated.

Necrosis and Apoptosis

There are two discrete types of cell death. Most commonly described is necrosis, an inflammatory process in response to cell injury from a variety of causes, such as exposure to nephrotoxicants or ischemia. The other type of cell death, apoptosis or programed cell death, is a finely controlled, active process that affects scattered individual cells, rather than tracts of contiguous cells, as occurs with necrosis. Various external and internal stimuli regulate apoptosis. Low levels of stimuli, such as ionizing radiation and toxins, can initiate apoptosis (Duvall and Wyllie, 1986). As described in the final section of this chapter, oncogenes are genetic loci ordinarily carried on tumor viruses that are responsible for neoplastic transformation. Under some circumstances, the proto-oncogene *c-myc* can drive cells into apoptosis (Evan et al., 1992). The *bcl-2*-proto-oncogene protects against apoptosis (Hockenbery et al., 1992). The tumor-suppressor gene *p53* promotes differentiation, maturation, and apoptosis (Clarke et al., 1993).

Although the precise control mechanisms are unknown, cell injury can induce synthesis of genes that stimulate the apoptosis process. Apoptosis has been demonstrated in oxidant injury to renal tubular epithelial cells, renal arterial stenosis, ischemia-reperfusion injury, hydronephrosis, polycystic kidney disease, glomerulonephritis, lead nitrate injury, radiation nephropathy, and analgesic nephropathy. Further research on the mechanism of apoptosis might identify means of reversing this process and preventing atrophy.

HOST FACTORS IN RENAL TOXICITY

Various host factors can alter renal toxicity by influencing the metabolism of xenobiotics, by minimizing the concentrations of toxic substances in the kidney, or by decreasing susceptibility to cell injury by other mechanisms. The various host factors are important in determining populations at risk, as discussed in detail in Chapter 3. Physiologic variation associated with age alters the response to various toxicants, as do some forms of underlying renal insufficiency. Other environmental concerns, such as nutritional status, can also influence toxicity.

Changes with Renal Growth and Aging

The susceptibility of the kidneys to toxic damage increases with age. This increased susceptibility can be related to a number of anatomic and functional changes that occur over time (Darmady et al., 1973; Epstein, 1979; Friedman et al., 1972; Spitzer, 1982). Those changes can be separated into phases of growth, maturation, and aging. At birth, the human kidney contains a full complement of nephrons, those of the outer cortex being relatively small and incompletely differentiated. During the first year, the glomeruli mature and grow as renal blood flow and cardiac output increase and renal vascular resistance decreases. The length and volume of the tubular structures rapidly increase, especially the convolutions of the proximal tubules. The increase in tubular mass is reflected in a pronounced increase in kidney weight. After the first year, growth slows. By the age of 4 years, the nephrons in the outer cortex have longer proximal convoluted tubules than those of the inner cortex and the ability to concentrate urine, to excrete excess acid, to maintain sodium balance, and to transport organic ions increases substantially.

By age 18-20 years, maturity is reached, and a plateau in structural and functional development is obtained that persists until around the age of 40. Thereafter, regressive changes take place. Prominent among those are changes within the renal vasculature and glomeruli; in particular, glomeruli appear to shrink with proportional decreases in the length and volume of the proximal convoluted tubules. Renal blood flow decreases at about 10% per decade and faster after the age of 60 years. The glomerular filtration rate also decreases with age, as does the ability to conserve sodium and the ability to produce maximally concentrated urine. With regard to the decrease in the glomerular filtration rate, it should be noted that muscle mass and creatinine production also decrease, so the serum creatinine concentration remains constant; this is why the serum creatinine concentration cannot be used reliably as a measure of glomerular function in the elderly.

The kidney's susceptibility to injury increases with age. This is most apparent with renal ischemic injury and has been demonstrated in both clinical settings (Balslov and Jorgenson, 1963; Groeneveld et al., 1991; Kiley et al., 1960; McMurray et al., 1978; Swan and Merrill, 1953) and laboratory settings (Kunes et al., 1978). Similarly, kidneys of young animals are relatively resistant to ischemic insults (Kunes et al., 1978). Changes in subcellular structures, such as mitochondria and lysosomes, and variation in xenobiotic biotransformation might be important determinants in age-related responses to environmental nephrotoxicants. Age-related alterations in the immune system might also play a role.

Changes with Diet

Malnutrition by itself does not seem to lead to parenchymal renal disease, but dietary inadequacy might result in developmental abnormalities in the very young and physiologic defects in adults. For instance, if caloric deficiency in the mother occurs early in the growth phase in the fetus, when cell multiplication is rapid, the kidneys might not achieve their proper weight or number of nephrons. Under normal circumstances humans have a full complement of nephrons at birth, and malnutrition after birth would not necessarily be expected to have an adverse effect on kidney size. However, the kidneys of some infants that died from protein-calorie malnutrition reportedly showed signs of chronic contraction and scarring.

In adults, physiologic defects predominate; these can be acute or chronic and are generally reversible. For example, fasting is associated with a natriuresis that can be abolished by ingestion of carbohydrate. The natriuresis has been related to alterations in glucagon concentration (Spark, 1975) and the need to excrete anions produced as a consequence of continuing metabolism (Sigler, 1975). With prolonged malnutrition, although renal blood flow and glomerular filtration rate are thought to be normal, other physiologic measures might be adversely affected. In particular, abnormal responses to salt and water loads are reflected in a propensity for edema.

When normal kidneys are stimulated to undergo compensatory hypertrophy, dietary protein restriction retards the response. In chronic renal disease in both experimental animals (Alfrey and Tomford, 1982) and humans (Walzer, 1982), reducing phosphorus intake slows the progressive decline in function, presumably by minimizing the adverse effects of hypertrophy of residual nephrons. Nutritional status clearly can be an important determinant of the ultimate effects of exposure to environmental nephrotoxicants.

Influence of Pre-existing Renal Function

Nephron Number at Birth

Injury to a population of nephrons can be underestimated because of the so-called reserve capacity of the kidney (i.e., the compensatory increase in function of normal or less severely injured nephrons). It might be expected that an inherited reduction in the number of nephrons in a kidney could explain in part the highly variable rates of expression and progression of human renal disease. Persons with a greater number of nephrons at birth might be able to sustain renal function after initial injury better than those with fewer nephrons at birth. Indeed, it has been postulated that those born with nephron numbers at the low end of the distribution curve can demonstrate accelerated declines in renal function after initial renal injury (Brenner and Anderson, 1989). Females have smaller kidneys and 10% fewer glomeruli than males (McLachlan et al., 1977), and age-related loss of renal function is faster in North American blacks than in whites (Boyle, 1970).

Acute Renal Disease

A combination of insults, which by themselves are mild and individually well tolerated, can result in unexpectedly severe acute renal failure. For example, mild tubular injury produced by gentamicin (Zager and Sharma, 1983) or amino acid infusion (Zager and Venkatachalam, 1983) sometimes potentiates the effect of ischemia. Even relatively mild renal ischemic injury increases the sensitivity of the kidney to damage by a number of nephrotoxic agents, including aminoglycosides (Zager, 1988) and radiocontrast agents (Humes et al., 1987). Similarly, studies in rats subjected to renal ischemia (Ding et al. 1991) and studies in human renal-transplant patients have shown that the later administration of cyclosporine has a deleterious effect on renal function.

Chronic Renal Disease

Patients with diabetes, severe atherosclerosis, or any type of pre-existing renal disease and patients who are hypovolemic but otherwise healthy are all at risk for the development of renal injury from nephrotoxicants. Perhaps the most compelling data on the influence of exposure to potential nephrotoxicants and the rate of progression of renal disease come from studies of the relationship between hydrocarbon exposure, glomerulonephritis, and other forms of nonneoplastic renal disease.

Physiologic Responses to Renal Injury

Physiologic Changes with Compensatory Hypertrophy

To some extent, the changes in individual nephron structure and function observed during the phases of growth and aging (Tucker and Blantz, 1977) can occur in a kidney undergoing compensatory hypertrophy (Deen et al., 1974; Finn, 1982; Hayslett, 1979), recovering from an acute insult, or adapting to chronic disease. The consequences of exposure to environmental nephrotoxicants can be expected to differ from the consequences in normal kidneys. For example, in chronic disease, some of the reserve capacity of the kidney has been used; that is, adaptive changes have taken place in individual nephrons in response to abnormalities in others.

It is helpful to consider the response that occurs in the nephrons of a normal kidney after the surgical or traumatic loss of its mate. The anatomic response is marked by a combination of hypertrophy and hyperplasia without the formation of additional nephrons. There is a homogeneous response of individual nephrons that is most marked by growth in the proximal convoluted tubule and an increase in glomerular size. The corresponding functional responses are marked by an increase in renal blood flow with parallel decreases in preglomerular and postglomerular vascular resistances. The glomerular filtration rate increases but tends to lag behind the increase in renal blood flow. Consequently, the filtration fraction falls. The increase in renal plasma flow is a major factor in the increase of the glomerular filtration rate. Increases in the glomerular capillary hydrostatic pressure and in the surface area or hydraulic permeability of the glomerular capillary membranes have also been found. Before any of the changes in renal blood flow or glomerular filtration, tubular function changes so that salt and water excretion quickly increase to equal that from both kidneys while two were still functioning—an appropriate response aimed at maintaining homeostasis. As noted earlier, many of these changes resemble those observed during the normal growth phase and are more prominent in young and mature kidneys than in aged kidneys.

In contrast with the homogeneous response of individual nephrons in a normal kidney undergoing compensatory hypertrophy, the individual nephrons in a diseased kidney demonstrate a considerable degree of heterogeneity. Indeed, within such a kidney, regressive features (marked by necrosis and atrophy) coexist with progressive features (marked by hypertrophy and hyperplasia). The anatomic changes are accompanied by marked functional changes, the most prominent of which is the great variation in the filtration rates of individual nephrons (Allison et al., 1973). In many ways, hypertrophy of some nephrons tends to counterbalance atrophy in others (Finn, 1983). Nephrotoxic damage to hypertrophied nephrons of a diseased kidney causes a greater decline in the whole kidney glomerular filtration rate than does damage to an equal number of nephrons in a normal kidney.

Factors Associated with Progressive Loss of Renal Function

A concern over the early detection of renal injury is appropriate and is heightened by the propensity of many forms of renal disease, once established, to progress relentlessly. It has been suggested that some proportion of nephrons are irreversibly injured after an initial injury. Adaptive changes in the remaining nephrons allow the whole kidney filtration rate to remain at near normal, so an acute injury might go undetected.

Those adaptive changes can lead to premature demise of nephrons that have undergone the greatest degree of hypertrophy. Because a prominent finding in markedly hypertrophic kidneys is an increase in the number of sclerotic glomeruli, it has been proposed that the hyperperfusion or increase in the glomerular blood flow that is associated with compensatory hypertrophy is deleterious to the glomerular capillaries (Brenner et al., 1982; Schimamuria and Morris, 1975). It is also possible that an increase in the glomerular capillary hydrostatic pressure or a primary alteration in the glomerular capillary membranes themselves can contribute to progressive glomerular damage and tubular atrophy. Whatever the cause, the loss of those nephrons contributes to the progressive decrease in whole kidney filtration rate and stimulates other, less-involved nephrons to undergo a similar process, thus repeating a destructive cycle. Eventually, the compensatory processes in some nephrons are unable to keep up with the progressive loss of other nephrons, and major alterations in renal function occur.

CLINICAL EFFECTS OF CHEMICAL EXPOSURE ON THE KIDNEY

Acute Renal Failure

Acute renal failure is marked by a progressive rise in the serum concentration of creatinine and other nitrogenous compounds. It has several causes, many of which are due to ischemic or nephrotoxic agents. Acute renal failure is often but not always accompanied by a reduction in urinary output to less than 500 cm^3/day. Even if urinary flow is unaltered, inevitable compositional changes in the urine reflect parenchymal injury and tend to separate postischemic and nephrotoxic renal failure from reduction in renal function that occurs merely as a result of alterations in systemic hemodynamics.

When acute renal failure occurs in association with drugs and toxicants, the overall mortality rate is about 37%. In the majority of those who survive, life-sustaining renal function can be expected to return, but recovery is usually not complete. In a small percentage, recovery of renal function does not occur. In the remainder, various degrees of structural and functional impairment persist indefinitely.

Tubulointerstitial Nephritis

Acute tubulointerstitial nephritis is marked by interstitial edema and infiltration of the interstitium with inflammatory cells, some of which appear later in the urine. Both structural evidence and functional evidence of tubular epithelial cell injury is present. Chronic tubulointerstitial nephritis is distinguished by the presence of interstitial fibrosis and tubular atrophy. The manifestations of tubulointerstitial nephritis depend on the extent of injury, the tubular segments most severely involved, and the degree of compensation achieved by the less severely involved nephrons.

The proximal tubule is responsible for the reabsorption of some 60-70% of filtered sodium and water and nearly all the filtered glucose, amino acids, and low-molecular-weight proteins. The predominant site of phosphate reabsorption and bicarbonate reabsorption and regeneration is also the proximal tubule. Individual cell types in the proximal tubule are identified with one or more of these processes. Damage to these cells can be expected to result in the appearance in urine of the substances ordinarily reabsorbed or metabolized. Damage to more distal structures—including the loop of Henle, the distal convoluted tubule, and the collecting duct—is accompanied by abnormalities in the ability to concentrate and dilute urine. The former effect is more pronounced and can result in polyuria. Acidification of urine also occurs at distal sites, where damage can lead to metabolic acidosis.

There is continuing interest in the urinary presence of various cellular enzymes and low-molecular-weight proteins as markers of recent renal damage. One such enzyme is *N*-acetyl-beta-glucosaminidase (NAG), which is released from injured renal tubular cells; the activity of this enzyme is often the only clinical-chemistry value that is increased in the urine in workers exposed to inorganic mercury, a classical renal poison. The appearance in the urine of low-molecular-weight proteins, such as beta$_2$-microglobulin, is also a marker of renal injury. Under normal conditions, low molecular weight proteins are filtered through the glomerulus and undergo complete reabsorption by the action of the renal tubules; in tubular disorders, as can be found in people exposed to cadmium, reabsorption can be incomplete and high urinary concentrations can result.

Exposure to environmental cadmium can cause chronic interstitial nephritis. Proteinuria and other signs of renal dysfunction have been found in people who live in cadmium-polluted areas or near smelters. Qualitative analysis of the urinary protein most commonly reveals a small albumin fraction and large alpha$_2$, beta, and gamma protein fractions. Cadmium-produced injury to proximal tubular epithelial cells also results in the presence of aminoaciduria, enzymuria, and glycosuria. A decrease in the tubular reabsorption of phosphorus, an increase in the fractional excretion of uric acid, and an increase in cadmium excretion can be present. Those abnormalities were highlighted by the results of recent CADMIBEL studies in Belgium (Buchet et al., 1990). The studies compared the relative sensitivities of various urinary biologic markers, which included retinol-binding protein, NAG, beta$_2$-microglobulin, amino acids, and calcium as biologic markers of cadmium-induced nephropathy among environmentally exposed human populations. All were found to be significantly increased in association with increased 24-hour urinary cadmium excretion. There were, however, differences in the relative sensitivities. For example, increased excretion of retinol-binding protein, beta$_2$-microglobulin, and amino acids occurred at lower urinary cadmium excretion levels than the other markers.

Two types of lead nephropathy can

be found in association with lead poisoning. The first is an acute form marked by generalized defects of proximal tubular function with aminoaciduria, glycosuria, and phosphaturia. These abnormalities most often occur in children after several months of heavy lead ingestion. The defects are generally rapidly reversible. Some patients will develop the chronic form. This condition is an indolent disease that is difficult to separate from other forms of chronic, slowly progressive renal insufficiency. Its incidence is difficult to determine. Evidence of prior excessive lead absorption might be found by administering EDTA and then measuring urinary lead excretion.

Chronic Glomerulonephritis

Chronic glomerulonephritis disease is an insidious process generally accompanied by albuminuria and microscopic hematuria. Its onset is often impossible to date, and its diagnosis is usually delayed until a secondary complication occurs, such as hypertension, anemia, or metabolic bone disease. Lacking such a complication, the diagnosis might not be suspected until an abnormal urinary sediment is seen during routine examination. In patients with chronic renal failure, the progressive decline in the glomerular filtration rate is too slow and the deviation from the steady-state condition too small to result in day-to-day changes in the serum creatinine or blood urea nitrogen concentration. A decrease in kidney size confirms the chronic and irreversible nature of this condition. Examination of tissue obtained by percutaneous renal biopsy can confirm that the primary process involved glomerular structures. Such a distinction is often clouded by nonspecific changes that occur in all forms of chronic renal disease.

Rapidly Progressive Glomerulonephritis

Occasionally, a more aggressive form of glomerular disease occurs in which renal function is lost over a period of weeks or months. This so-called rapidly progressive glomerulonephritis can be identified by the presence of glomerular epithelial crescents in renal-biopsy specimens. The kidney is sometimes the only organ affected; at other times the renal abnormalities are part of a systemic disease that can result from severe vasculitis.

The combination of prolonged, excessive exposure to hydrocarbons with unidentified host factors can predispose to glomerular injury or aggravate injury due to other causes (Yaqoob and Bell, 1994). An apparent association of Goodpasture's syndrome—a form of rapidly progressive glomerulonephritis—with exposure to petroleum products has been reported (Beirne and Brennan, 1972; Bombassei and Kaplan, 1992). It has also been claimed that previous exposure to hydrocarbon solvents is a common feature in some groups of patients with crescentic glomerulonephritis or proliferative glomerulonephritis (Zimmerman et al., 1975). Additional presumptive evidence that hydrocarbons can produce glomerular damage has come from the observation that remissions and exacerbations of the nephrotic syndrome follow removal from and re-exposure to solvents (Cagnoli et al., 1980). Other studies have noted a historical relationship between exposure to organic solvents and a wide spectrum of renal disorders, including tubular necrosis, interstitial disease, glomerulonephritis, and neoplasia (Nelson et al., 1990).

In a case-control study, hydrocarbon exposure was significantly higher in those with primary glomerulonephritis than in a group of normal subjects and an internal control group (Yaqoob et al., 1992a). Those with glomerulonephritis had a significantly greater exposure to petroleum products, greasing and degreasing agents, and paints and glue with resulting estimated risks of developing glomerulonephritis 15.5, 5.3, and 2.0 times greater than normal, respectively. In another study of patients with diabetes mellitus, hydrocarbon exposure was found to be significantly greater in those with incipient (microalbuminuria) and overt (macroalbuminuria) diabetic nephropathy than in those with no clinical evidence of nephropathy, with odds ratios of 4.0 and 5.8, respectively (Yaqoob et al., 1992b).

The role of hydrocarbon exposure on the progression of renal failure in patients with primary glomerulonephritis has also been studied (Bell et al., 1985; Ravnskov, 1986). Patients with primary glomerulonephritis and progressive renal failure have heavier hydrocarbon exposure and worse renal impairment at presentation than those with stable or improving function (Yaqoob et al., 1993). Moreover, patients with declining renal function were more likely to have continued occupational hydrocarbon exposure after the diagnosis of glomerulonephritis.

Several cross-sectional studies comparing measures of renal dysfunction in hydrocarbon-exposed and -nonexposed workers have suggested an association between hydrocarbon exposure and renal injury (Askergren et al., 1981; Askergren, 1984; Franchini et al., 1983; Hotz et al., 1991; Lauwerys et al., 1985; Mutti et al., 1992; Solet and Robins, 1991; Viau et al., 1987). For example, exposed subjects have been found to have a slight but significant increase in the abnormalities found at urinalysis (Askergren, 1984).

Increased proteinuria and tubular enzymuria (lysozyme and beta-glucuronidase) in the absence of albuminuria, indicative of tubular rather than glomerular dysfunction, has been seen in a large group of subjects exposed to aliphatic and acyclic hydrocarbons (Franchini et al., 1983). A separate study of 20,000 workers showed that the prevalence of proteinuria was higher in those with hydrocarbon exposure than in nonexposed subjects. Others have found a higher mean albuminuria and urinary excretion of renal antigen and a higher prevalence of antilaminin antibodies in a group of 53 male refinery workers (Viau et al., 1987).

Nephrotic Syndrome

The nephrotic syndrome is marked by the presence of heavy proteinuria (generally in excess of 3 g per 24 hours), hypoalbuminemia, and edema. In addition, abnormalities in lipid metabolism with hypercholesterolemia, hypertriglyceridemia, and lipiduria are common. At the onset, the glomerular filtration rate might not be reduced and is occasionally increased. The nephrotic syndrome can occur in conjunction with a variety of systemic diseases. Or it can be a manifestation of a primary glomerular injury without a definable etiology and thus be termed idiopathic; in this situation, the condition is classified according to the appearance of glomeruli on light and electron microscopic sections and on the basis of various immunofluorescent patterns. Chronic occupational exposure to gold, bismuth, and mercury salts produces pathologic lesions similar to those found in some forms of the idiopathic nephrotic syndrome. One of the most common types, membranous glomerulonephropathy, is associated with an increased frequency of the histocompatibility leukocyte antigen (HLA) DR3 (Klouda et al., 1979). It has been suggested that exposure to environmental or occupational agents, such as formaldehyde, can act as a "triggering" agent in genetically susceptible persons (Breysse et al., 1994).

Hypertension

The association between kidney function and hypertension is based on both clinical and experimental observations reported over the last 150 years. Renal disease in the form of diabetic nephropathy, glomerulonephritis, interstitial nephritis, obstructive uropathy, polycystic kidney disease, pyelonephritis, and vasculitis is the most common cause of secondary hypertension in humans. Conversely, primary hypertension ranks second only to diabetes as an etiology for patients entering treatment for ESRD in the United States (NIH, 1993). It was recently proposed that ischemic nephropathy accounts for the rising incidence of hypertension-induced ESRD. Thus, the age-old question of cause versus effect appropriately characterizes the interaction between the kidney and hypertension.

It is germane to this report that some drugs or toxicants, often through their action on the kidney, are recognized causes of hypertension. Examples include amphetamines, estrogens and oral contraceptives, steroids, sympathomimetic drugs, tricyclic antidepressants, cisplatin, cyclosporine, licorice, lead, and ethanol.

Clinically, hypertension is usually detected as an incidental finding during a health evaluation, inasmuch as mild to moderate blood-pressure increases are usually asymptomatic. However, in a small percentage of patients, hypertension has a sudden severe onset accompanied by headache, nausea, vomiting, and mental confusion, which demand immediate emergency treatment; in these cases, underlying renal disease is often present.

It is estimated that one-fourth of

adults in the United States have hypertension and suggested that treatment decisions should be based on the associated evidence of target-organ damage that is traceable to increased blood pressure. Today, a multitude of antihypertensive drugs are available, so virtually any hypertensive patient can be successfully treated without intolerable side effects of administered drugs. Long term studies are being conducted to evaluate the effectiveness of antihypertensive treatment in preserving renal function and preventing renal damage. For patients whose hypertension is the result of chemical or drug nephrotoxicity, recognition and withdrawal of the offending agent are the appropriate clinical strategies. In addition, coexisting hypertension often emerges as a statistically significant risk factor in studies of nephrotoxicity.

CANCER OF THE BLADDER, KIDNEY, AND PROSTATE

Tumor-Suppressor Genes, Oncogenes, and Growth Factors

According to Barrett and Huff (1991), the multistep process of carcinogenesis can be operationally divided into initiation, promotion, and progression. Initiation consists of the first heritable alterations that predispose a cell to neoplastic transformation; promotion is the clonal expansion of initiated cells; and progression is the acquisition of other changes that are required for a cell to become fully malignant. It is now believed that more than two changes are required for neoplastic conversion of a cell; additional clonal evolutions most likely occur in the later stages and make the distinction between promotion and progression difficult. However, it is important not to confuse the two processes. Promotion involves the multiplication of the initiated cell, whereas progression involves the acquisition of additional, heritable changes in the initiated cell. Promotion can lead to progression, although with a low frequency. Progression can occur as a direct result of a chemical on an initiated cell. Increasing the target size of the population of initiated cells by promotional mechanisms will increase the probability of secondary spontaneous or chemically induced changes and therefore progression. The rate-limiting step in malignant development is the acquisition of additional genetic changes in an initiated cell. Therefore, a weak mutagenic effect can be at least as important as a potent tumor-promoting effect for cancer development.

A better way to define the multistep process of carcinogenesis is to identify the genetic alterations in tumor cells and attempt to determine how chemical carcinogens affect the neoplastic process that leads to these changes. There is now convincing evidence of the importance of two classes of genes in the carcinogenic process: proto-oncogenes and tumor-suppressor genes. Proto-oncogenes are a family of cellular genes with at least 40 members, which appear to be involved in normal cellular growth and development; activation or inappropriate expression of these genes results in proliferative signals involved in neoplastic growth. Tumor-suppressor genes are less well defined, but they might also function in the control of normal cellular division and possibly differentiation. For a tumor cell to emerge, suppressor genes must be inactivated or lost. The number of tumor-suppressor genes is unknown.

The control of cellular growth resides in a complex, interacting system of positive (oncogenes) and negative (tumor-suppressor genes) controls. Each cell in a tissue responds to unique rules in the form of specific genes that control its growth. The subversion of these systems is responsible for the development of cancer. Indeed, specific tumor-suppressor genes and oncogenes peculiar to colon cancer have been identified.

Growth factors are peptides that regulate cellular growth and usually function in paracrine or endocrine modes rather than in an autocrine mode. The complex communication among various cells becomes subverted during tumorigenesis, because cells often move from paracrine to autocrine control (Nathan and Sporn, 1991). In some tissues, such as the prostate, growth is also under strong hormonal influences. For example, prostatic differentiation is under endocrine control during embryologic development. That is followed by a second sequence of changes associated with puberty. During adulthood, both hormonally dependent and hormonally independent cells can be found. Finally, the prostate undergoes hypertrophy as a result of dysregulated growth control among its various constituent cells. Each factor that controls growth can function as an important biologic marker. Also to be considered are cell surface receptors for the growth factors and the many proteins that are triggered by the growth peptides.

Cancer of the Bladder

In 1989, there were 47,000 new cases of bladder cancer and 10,000 deaths due to bladder cancer (Smart, 1990). In that year, bladder cancer accounted for 5% of all new cancer cases and 2.2% of all cancer deaths. The incidence of bladder cancer was 29.1 per 100,000 males and 7.7 per 100,000 females. Male bladder-cancer mortality was 5.9 per 100,000.

Bladder cancer presents an interesting paradigm of the mechanism by which environmental or occupational toxicants can be initiators or promoters of cancer development or factors in its progression. Indeed, bladder cancer is highly correlated with occupational exposure to xenobiotics. For example, Silverman (Silverman et al., 1989a,b, 1992) reported that 20-25% of cases of bladder cancer in white males, 27% in nonwhite males, and 11% in white females were associated with occupational exposure to toxicants. The relationship is not new: In 1895, Rehn first reported that bladder cancers could be caused by specific carcinogens in a group of workers exposed to aniline dyes. Various occupations have since been associated with bladder cancer. Table 2-1 contains national estimates, according to occupation, of the number of workers exposed to agents known to cause bladder cancer in animals. Over 200 chemicals have been suggested as associated with bladder cancer, and a few are documented human carcinogens (Anonymous, 1990). A list of compounds classified by IARC as human bladder carcinogens is contained in Table 2-2.

The bladder is a specialized neuromuscular organ whose muscular layer, or muscularis, is protected by a unique impermeable epithelium, the urothelium. This layer is three to seven cells thick and is capable of distention. The luminal layer consists of terminally differentiated cells, the so-called umbrella cells,

TABLE 2-1 Exposure to Animal Bladder Tumorigens in Selected Occupations During 1980s

Occupation	No. of Workers
Painters	65,469
Machinists	503,074
Truck drivers	471,742
Nurses (RNs)	927,434
Garage workers	156,657
Electricians	251,440
Janitors, cleaners	905,291
Assemblers	1,508,060
Miscellaneous machine operators	64,768
Machine operators, NEC[a]	517,471
Welders	389,822
Sewing-machine operators	817,956
All workers	19,571,942

[a]Not elsewhere classified.
Source: Adapted from Ruder et al., 1990.

TABLE 2-2 Compounds Classified as Human Bladder Carcinogens[a]

Benzidine
4-Biphenylamine
2-Naphthylamine
   2-Naphthylamine, N,N-bis(2-choloroethyl)-
   2-Naphthylamine, 1-methoxy-
   2-Naphthylamine, 1-methoxy-, HCl
   2-Naphthylamine, 3-methyl-
2H-1,3,2,-Oxazaphosphorine, 2-(bis(2-chloroethyl)amino)tetrahydro-, 2-oxide
Purine, 6-((1-methyl-4-nitroimidazol-5-yl)thio)-

[a]Compounds classified by International Agency for Research on Cancer as Group 1 carcinogens, having sufficient evidence of human carcinogenicity.
Source: Adapted from Ruder et al., 1990.

which have a lifetime measured in months. This cell layer seems to be responsible for the bladder's impermeability to urinary solutes, perhaps because of prominent desmosomes, tight junctions between cells, specialized ion pumps, and a thick layer of highly charged glycosaminoglycan (Parsons et al., 1990). The bladder also contains components of the cytochrome P-450 system (Vanderslice et al., 1985), so it might have some role in xenobiotic detoxification or bioactivation, in addition to being subjected to the effects of carcinogens excreted in the urine (Kadlubar et al., 1992).

The urothelium is normally quiescent, but it is capable of re-entering the cell cycle in response to injury. Epidermal growth factor (EGF), which is present at high concentrations in the urine, has been suggested as a major factor in urothelial growth and differentiation. Only the basal cell layer normally expresses EGF receptor (Messing et al., 1987). It is possible that injury to these cells and exposure to urine trigger rapid growth.

Bladder cancer, like many other cancers, occurs as a result of the interaction of genetic predisposition, occupational exposure, and a variety of cofactors. One of the prime nonoccupational causes of bladder cancer is cigarette-smoking, with an increasing incidence of disease in women that correlates with increased tobacco use (Cole and Hoover, 1971). Whether low-level arsenic exposure is synergistic with known carcinogens, such as cigarette-smoking, is unknown. These examples illustrate the complex interaction between genetic factors and multiple occupational or environmental toxicants.

Cancer of the Kidney

About 18,000-20,000 new cases of renal cancer are diagnosed each year in the United States. This malignancy, which makes up 2-3% of all cancers, ranks eleventh in cancer incidence and results in 8,000 deaths a year in the United States. White males have the highest mortality, 4.8 per 100,000; black females have the lowest mortality, 2.0 per 100,000.

About 85% of the renal cancers diagnosed are renal-cell cancers, whose incidence is about twice as high in men as in women. In 1986, the incidence of renal-cell carcinoma per 100,000 was 11.3 in white males, 11.9 in black males, and 5.6 in white females and black females. Kidney cancer is increasing in the United States; Huff reported in 1991 that percentage age-adjusted increases (per 100,000) in mortality (and in incidence) were 9.1% (21.7%) in whites, 38.2% (19.8%) in blacks, 7.8% (21.3%) in white males, 44.0% (26.9%) in black males, 13.5% (23.0%) in white females, and 34.7% (11.9%) in black females (Huff and Haseman, 1991).

Environmental agents have been implicated in the development of neoplasms of the renal parenchyma. Renal adenomas and adenocarcinomas account for about 85% of renal neoplasms. Commonly referred to as hypernephromas, these tumors arise from cells of the proximal convoluted tubule. They account for 2.1% and 1.6% of all cancer deaths in males and females, respectively. Squamous-cell carcinomas are much less common and account for 5-6% of renal neoplasms. Neither nephroblastomas nor renal sarcomas have been associated with renal carcinogens. Transitional-cell carcinomas of the renal pelvis and ureter can be induced by the same carcinogens that produce bladder tumors. Workers in the aniline-dye, rubber, textile, and plastic industries have a higher incidence of these tumors, which overall account for 7-8% of renal neoplasms. A list of occupations found to have excess risks for kidney cancer is contained in Table 2-3.

Renal cancers in humans have been associated with exposures to tobacco smoke, some environmental and occupational factors (e.g., coke-oven emissions

TABLE 2-3 Occupations Found to Have Excess Risks for Kidney Cancer[a]

Occupation and Industry	Risk (statistical significance)[b,c]
Steel (coke plant)	5.0 ($p < 0.01$)
Chemical mfg.	50% incidence
Leather workers	4.8 (95% CI = 1.8-121)
Cadmium	2.5 ($p < 0.05$)
Wholesale and retail	3.3 ($p < 0.05$)
Entertainment	2.4 (NS)
Transportation	2.2 (NS)
Asbestos	2.3 (NS)
Pharmaceutical	PMR 2.85 ($p < 0.05$)
Dry cleaning	PMR 200
Dry cleaning	PMR 257 ($p < 0.05$)
Dry cleaning	3.8 (95% CI = 1.9-7.6)
Dry cleaning	SMR 200 (95% CI = 55-517)
Dockyard workers	1.9 ($p < 0.001$)
Metallurgy workers	3.78 ($p < 0.05$)
Ironsmiths	1.83 (NS)
Coppersmiths	1.63 (NS)
Fitters	2.38 (NS)
Autogenous welders	5.06 ($p < 0.05$)
Smiths/shipwrights	3.7 (NS)
Electric welders	2.5 (NS)
Joiners	1.2 (NS)
Carpenters	1.02 (NS)
Caulkers	1.85 (NS)
Masons	6.06 (NS)
Oil refinery	SMR 152 (NS)
Petroleum refinery	PMR 2.14 ($p < 0.05$)
Petroleum refinery	SMR 155 (71-294); SMR 205 (94-389)
Petroleum	SMR 1.23 (95% CI = 1.77-1.86)
Petroleum	1.7 (95% CI = 1.0-2.9)
Lumberjacks	1.93 (90% CI = 1.03-3.61)
Plastics	SMR 203 (NS)
Aluminum prebake	151. (NS)
Embalmers	PMR 247 ($p < 0.05$); PCMR 142
Chlorine production	3.8 (90% CI = 1.32-10.93)
Smelter	SMR 204 (95% CI = 75-444)
Artists	PMR 280 ($p < 0.001$)

[a]Each entry represents a single study.
[b]Unless otherwise specified, relative risks.
[c]CI = confidence interval; NS = not statistically significant; PMR = proportional mortality ratio; SMR = standardized mortality ratio; PCMR = proportional cancer mortality ratio.
Source: Adapted from Schulte et al., 1987.

and possibly rubber-industry byproducts), and therapeutic agents, particularly analgesic mixtures containing phenacetin (Amico et al., 1991). A large epidemiologic study confirmed the association of lead and renal cancer in humans (Steenland et al., 1992). It is also suspected that arsenic is related to cancers of the kidney (Scandinavian Committee on Enzymes, 1985). However, most causes of kidney cancers are unknown, and these malignancies remain an important human health problem. The true incidence and mortality in the human population may be considerably higher than those reported. As evidence of that, more than one-third of the reports of the relatively few routine autopsies in the United States disclose undiagnosed cancers (Azzopardi and Evans, 1971). According to Holm-Nielsen and Olsen (Bauer, 1988), renal adenomas (minute cortical foci of proliferating tubular or papillary epithelium) often are present in 15-22% of all adult kidneys. Whether these small tumors (typically 2-3 mm, up to 6 mm) should be regarded as carcinomas or as benign precursors of renal-cell carcinomas remains controversial (Bauer, 1988).

Data from a number of animal and some human studies (IARC, 1987) generally support a relationship between lead-induced chronic renal disease and renal adenocarcinoma (Steenland et al., 1992), but an increased incidence of renal cancer has also been reported among lead-exposed workers without statistically increased rates of chronic renal disease (Steenland et al., 1992). The role of renal lead-binding proteins that both animals and humans mediate individual susceptibility to renal cancer from lead has been hypothesized (Fowler et al., 1994) as a mechanism for explaining observed variability among lead-exposed persons.

Recent evidence suggests that the original Knudson hypothesis—that the chromosomal regions often lost or mutated in tumors harbor tumor-suppressor genes—is correct. Several forms of renal cancer should be considered, including those with Wilms's tumor, the Von Hippel-Lindau disease, and the Li-Fraumeni syndrome, all of which represent the inheritance of a deleted or otherwise inactive suppressor gene (Latif et al., 1993). Studies of Wilms's tumor were the first to suggest that chromosomal defects in cancer cells harbored tumor-suppressor genes (Knudson and Strong, 1972). The Wilms's tumor-locus gene is a tumor-suppressor gene on chromosome 11p13. Germline mutations in *WT-1* are associated with both the heritable and sporadic forms of Wilms's tumor and urogenital abnormalities. People heterozygous for mutations of the *WT-1* gene are predisposed to Wilms's tumor (Haber and Housman, 1992). The tumor cells have lost heterozygosity and contain mutants at both alleles; this is consistent with the Knudson model of genetic predisposition to cancer. The *WT-1* gene encodes a nuclear protein that possesses a so-called zinc finger domain governing DNA-binding specificity.

Von Hippel-Lindau disease is a relatively rare, dominantly inherited tumor disorder characterized by retinal angiomatosis and cerebellar hemangioblastoma (Latif et al., 1993). Renal-cell carcinoma is a frequent cause of death in this disease. It is inherited as an autosomal dominant trait. The gene has been mapped to the short arm of chromosome 3.

The Li-Fraumeni syndrome is a familial tumor syndrome associated with malignant tumors in various organs. Inherited mutations of the tumor-suppressor *p53* gene have been described. Like several other previously described tumor-suppressor gene products, *p53* is thought to regulate transcription of genes critical to the control of cell growth and differentiation and in this way to be involved in the regulation of the cell cycle. The *p53* gene is a promising candidate marker for cancer susceptibility (Harris, 1993). It is on chromosome *17p* and is the most frequently mutated gene of human tumors in the United States. Evidence that the *p53* gene is an anti-oncogene came from the observations that the wild-type form of the protein inhibits oncogene-mediated transformation of cells and that the growth of human cancers with endogenous *p53* mutations was inhibited. In a detailed analysis of a group of families with multiple cancers first described by Li and Fraumeni, it was shown that the *p53* gene was inherited in a mutated form and that cancer resulted only when additional mutations accumulated; this is consistent with the clinical appearance of cancer at ages 10-40 (Harris and Hollstein, 1993). That different carcinogens cause different characteristic mutations in the *p53* gene suggests that the location and characteristics of these mutations can reveal clues about etiology and molecular pathogenesis of cancer. Thus, screening for germline mutations of the *p53* gene could be used to identify a population at risk for cancer after exposure to occupational or environmental toxicants.

The retinoblastoma gene product regulates the cell cycle by maintaining cells in a non-proliferating state. It does so by binding to transcription factors presumably related to the cell cycle, thereby inactivating them (Marx, 1991). Binding is regulated by phosphorylation and dephosphorylation events—events also regulated during the cell cycle. The retinoblastoma gene binds to the adenoviral E1A oncogenic protein, thus inactivating it and allowing the cells to enter the cell cycle continuously. Both alleles at the retinoblastoma-gene locus on chromosome 13 are defective in retinoblastoma and other cancers, including prostatic and bladder cancer (DeCaprio et al., 1989).

The importance of gasoline exposure and various aliphatic hydrocarbons in the induction of renal-cell carcinoma is controversial (Kadamani et al., 1989). The linkage to gasoline and hydrocarbon exposure stems from earlier studies in male rats that had an increased incidence of renal-cell carcinoma; however, the extension of the information obtained in the rats to humans has not proved informative (see Chapter 5). For example, the deposition of alpha$_{2u}$-globulin, which has been observed in rats, has not been observed in humans (see Chapter 6). A review of the pathologic changes that were present in cases of renal-cell carcinoma in which hydrocarbon exposure was thought to play a role (Pitha et al., 1987) did not detect important changes in the normal kidney cells adjacent to the carcinomatous cells. Those results do not substantiate the importance of hydrocarbon exposure in either the development of subclinical nephrotoxicity or the pathogenesis of renal-cell carcinoma.

Unsubstantiated evidence that occupational exposure is a factor in renal carcinoma indicates that development of biologic markers for individual risk assessment might assist in identification. The increased incidence in males might be attributed to the hormonal milieu or to smoking, which has been more frequent in men in the past. Recent studies indicate only a weak association with exposure to dry-cleaning agents, but there is a further need for individual risk assessment. The potential causative agents for renal-cell carcinoma have been extensively reviewed by Schulte and Kaye (1988).

Keys to the understanding of renal cancer might come from such seemingly diverse disciplines as toxicology, molecular biology, and cancer biology, which share an interest in understanding the regulation of the cell cycle. For example, the interaction of inherited defects in the *p53* tumor-suppressor gene and environmental toxicants has proved to be an important lesson in the induction of cancer. Similarly, inheritance of the disabled *WT-1* tumor suppressor gene is a well-known cause of a form of renal-cell cancer.

Cancer of the Prostate

Prostatic cancer is the most frequent cancer in men in the United States, with an annual incidence of 87.8 per 100,000. Its mortality is much lower, 23.8 per 100,000, but it is still the second most common cause of cancer death in men and the most common cause of cancer death in older men. The annual cost of diagnosis and care is more than $1 billion. In 1992, there were about 132,000 diagnosed cases of prostatic cancer and 36,000 related deaths. On the basis of autopsy studies, it is believed that more than 40% of men over 50 have undiagnosed prostatic cancer. It is projected that 10-12% of men alive today will have clinically manifested prostatic cancer and that 2% will die of it.

Blacks in the United States have a higher incidence of prostatic cancer than do whites in the United States or blacks in some other countries. For example, several studies show an odds ratio of 1.8-2.0 for prostatic cancer for blacks compared with whites in the United States. The incidence of prostatic cancer is markedly higher among blacks in the United States (100 per 100,000) than among blacks in Nigeria (10 per 100,000). Not only is the incidence of the disease higher, but prostatic cancer mortality is higher among blacks than among whites. That has been attributed to later diagnosis and increased risk among blacks, which might be related to genetic factors that influence the response to occupational or environmental exposures. The increased incidence of prostatic cancer associated with the migration of ethnic groups and the modulation of disease by diet support a xenobiotic etiology for prostatic cancer. The importance of environmental factors has been brought into focus by differences in the incidence of disease in various countries, and several hypotheses have been offered to explain the etiology. For instance, the low incidence of prostatic cancer and high incidence of breast cancer in males in Egypt have been attributed to increased estrogen levels in patients with bilharziasis infection of the liver (Bouffloux, 1980; El-Aaser et al., 1985; Sherif et al., 1980). A further reflection of environmental or occupational exposure comes from a 28-year Japanese study of latent prostatic cancer found in surgical specimens. During the study, an increase in overall incidence of prostatic carcinoma was observed; the number of high-grade (Gleason 3-5) cases increased, and the incidence of low-grade (Gleason 1-2) cases remained the same. These observations indicate that the changes are related to two diseases that develop along separate tracks.

In Hawaii, with its many ethnic groups, Kolonel (Kolonel et al., 1983) found a high correlation between dietary fat (both total fat and animal saturated and unsaturated fats) and incidence of prostatic cancer. The importance of saturated fats was substantiated by Hankin (Hankin et al., 1992), and of both saturated and unsaturated fats by Hursting (Hursting et al., 1990). Although cigarette-smoking as an etiologic factor for prostatic cancer has been considered, recent evidence suggests that the risk associated with cigarette-smoking, if any, is small (Hsing et al., 1990).

Increasingly, evidence indicates that prostatic cancer is a multistep process in which a series of events are required for a normal malignant cell to give rise to a fully malignant cancer cell. A large number of clinically undetectable cancer cells are present in a prostatic cancer. Although defined histologically as cancer, they might not have undergone all the steps to malignancy (Carter at al., 1990). That is an important consideration because widespread screening for prostate-specific antigen (PSA)—see Chapter 6—has led to an increase in detection of the occult form of prostatic cancer. Consequently, many men are being offered the option of surgical intervention or radiation therapy without proof from clinical trials that screening with PSA will enhance patient survival. Screening is detecting many cases of prostatic cancer, some of which might be biologically inactive.

Alterations in biologic markers that are associated with prostatic cancer reflect the subversion of growth control. The complex organization of the prostate is an example of the multiplicity of cellular interactions critical to homeostasis. The gland contains neuroendocrine cells, basement-membrane cells, associated stromal cells, and, of great importance for the continued reproductive success of the species, functional epithelial cells. Any of the several cellular types can abort normal regulatory control and participate in a new program of tumorigenesis. In the past, the influences of various hormones on the control of growth have been the primary focus of research and therapy. With the recognition that the tumor cells that progress are hormone-independent, increased attention has been given to other mechanisms of tumorigenesis.

The scientific challenge is to determine which of the cells are primary targets of xenobiotics; to understand the genetic factors that allow the transition from controlled paracrine growth to independent autocrine growth; and to define the early histopathologic malignant alterations and correlate them with early biochemical markers. Each of the histopathologic manifestations of disordered cell growth—proliferation, invasion, and metastasis—may be accompanied by subtle, but detectable and quantitative alterations. A key to the recognition of early biochemical profiles might come from study of cells surrounding those which have undergone malignant transformation, given the assumption that a gradient of changes can be detected.

In the adult male, homeostasis in the prostate is under the influence of androgens and is mediated by a series of diffusible peptides, or growth factors (EGF, transforming growth factor- (TGF) beta, insulin-like growth factors (IGFs), and fibroblast growth factor). The interaction of these and other growth factors is complex. Two examples are illustrative. First, TGF inhibits prostatic epithelial cell growth in the presence of EGF, but this inhibitory effect is abrogated in the presence of fibroblast growth factor, FGF, (McKeehan and Adams, 1988). Second, low concentrations of TGF are involved in a negative-feedback mechanism for the control of growth of prostatic stroma and epithelial cells. Obliteration of androgens—either surgically by orchiectomy or medically by use of a dihydroxysterone antagonist—results in prostatic-cell involution; this is followed by an increase in the prostatic growth factors, receptors, and cofactors, such as EGF and IGFs (Fiorelli et al., 1991a,b). In animals, administration of exogenous androgen decreases the production of TGF messenger RNA and its receptor to normal levels. The epithelial cells in turn regulate stromal cells through secretion of fibroblast growth factor, which influences angiogenesis, chemokinesis, and extracellular matrix-protein production (Gospodarowicz, 1991; Rifkin and Moscatelli, 1989). Alterations in one or more of those factors probably influence the metastatic potential of prostatic cancer or facilitate or inhibit the growth of transformed cells. Furthermore, it is not surprising that the growth of this highly regulated cellular system can be influenced by aging, by alterations in hormonal control, or by exogenous factors, such as xenobiotics or diet. One of the features of prostatic cancer is that the cells eventually become resistant to androgen suppression, proliferate rapidly and metastasize.

The prostate constitutes a unique environment that imposes special constraints on the development of biologic markers. Two of the most important constraints are that most men will eventually develop some degree of benign prostatic hyperplasia (BPH) and that in later life there is a high incidence of microfocal cancer. Implications of those facts are that there is almost always some confounding pathology and the problem is not to detect cancer, but rather to detect cancer that will become aggressive.

As with other organs, clues to the selection of markers lie in the molecules involved in the regulation of cell growth and differentiation, because it is the subversion of these systems that leads to cancer. The prostate is subject to complex growth and differentiation control. There are interacting stromal and epithelial elements and both androgen-dependent and androgen-independent mechanisms. The androgen-independent mechanisms seem to be operative during embryonic development; at puberty, there is a switch to androgen-dependent mechanisms. Clinically dangerous cancers seem to involve a reawakening of the androgen-independent mechanisms.

The biologic controls seem to involve several important principles: There are both positive (or stimulatory) and negative or (inhibitory) signals; an example of the former is platelet-derived growth factor, and an example of the latter is TGF although, to complicate the picture, TGF can inhibit epithelial cells and stimulate fibroblasts. Receptor processing at the membrane is complex, involving a receptor, a transmembrane domain, and an intracellular domain; the latter produces changes in cellular second messengers and modulates proteins, such as the *ras p21* protein and other G proteins. Receptors are likely to be hormone-dependent. Extracellular-matrix proteins transmit growth or differentiation control signals to cells through linkages of integrins and other surface molecules. Autoregulation can liberate cells from exogenous controls. Growth peptides can be modified postranslationally; for example, TGF-alpha exists in the prepro form, which can be activated by appropriate proteases. The growth peptides can have actions on cells other than their main target cell.

It is difficult to know which factors are important to study clinically in these circumstances. For example, it can be difficult to determine whether the decrease in EGF or the increase in FGF is the important factor. Given the increasing number of growth factors identified, it is not feasible to examine all peptide growth factors in clinical studies. It is important to examine alterations in prostatic histopathology and cytology, as well as to correlate them with biochemical events, such as changes in the androgen receptor in the stroma, in androgen concentrations in plasma, in the amount being expressed by the testes or adrenal cortex, or even in the mechanisms that control the processing. One of the features of BPH is the increase in the stroma, compared with the endothelial or epithelial parts of the tissue, which results in such clinical manifestations as difficulty urinating or, in severe cases, hydronephrosis. Biochemical and structural analysis might provide clues as to the mechanisms in the increased amount of stromal formation in BPH.

Interstitial Cystitis

The bladder has several potential targets for toxic xenobiotics. First, the protective glycosaminoglycan layer is subject to inactivation by amine compounds, which are among the most potent bladder carcinogens. Inactivation of this layer by any one of several mechanisms increases the effect of later carcinogen exposure, whereas the administration of other glycosaminoglycans or heparin mitigates the carcinogenic effect of xenobiotics (Bodenstab et al., 1983). Second, primary and secondary alterations in the neuromuscular function of the bladder can produce changes in bladder physiology; any abnormality that causes obstruction of urine flow from the bladder alters bladder physiology and can lead to hypertrophy and altered neuromuscular function (Levin et al., 1990). Third, because there is a link between nerve function and inflammation, abnormalities in the nerves that innervate the bladder—similar to that described after exposure to organophosphate pesticides (Hohenfellner et al., 1992)—can produce an inflammatory response. Such a mechanism might also be involved in interstitial cystitis (described later). That is, considerable evidence suggests that at least some cases of interstitial cystitis result from urinary toxicants (Messing et al., 1992) that inactivate the glycosaminoglycan layer and expose the muscularis to various urinary solutes. It is not known whether this inactivation is a result of exposure to endogenous or exogenous agents.

Interstitial cystitis is a syndrome characterized by pain, urgency and frequency of urination, and cystoscopic abnormalities, all of which occur without known causes. Studies of the epidemiology of the disorder have not pointed to any one causal agent (Held et al., 1990; Koziol et al., 1993). One estimate, which might be low, placed the number of diagnosed cases in the United States at 43,500 in 1985. The reported incidence is much higher in women, but the reported incidence in men could reflect underdiagnosis. A subpopulation at special risk has not been identified, but Jewish women are overrepresented and black women are underrepresented. Although the etiology of interstitial cystitis is not known and there is no firm evidence that it results from xenobiotic exposure, the economic and human import of this condition merits attention and consideration of all possible causes.

ANIMAL MODELS OF INTERSTITIAL CYSTITIS

Several animal models of various aspects of interstitial cystitis have been reported recently. Ratliff and co-workers reported that some strains of mice develop an autoimmune cystitis that mimics the ulcerative form of interstitial cystitis, even to the point of showing bladder-permeability changes (Bullock et al., 1992). Stein and Parsons (1991) reported that the chronic instillation of protamine into rabbit bladders produced inflammation and a breakdown of the permeability defenses of the bladder that could persist after removal of the challenge and therefore provide an example of the "toxic urine" and "epithelial dysfunction" models. Buffington and co-workers reported that a spontaneous syndrome of cats manifested by frequent, apparently painful urination with sterile urine is apparently similar to interstitial cystitis in showing decreased urinary glycosaminoglycan excretion.

SUMMARY

Advances in understanding and using biologic markers should assist in identifying xenobiotics that are toxic to the urinary tract. The functional role of the urinary tract, including clearance of toxic substances from the blood, predisposes it to xenobiotic exposure and toxicity. Historically, identification of the type and amount of xenobiotic exposure has been difficult, often because of the interval between exposure and the onset of disease. Blacks and other minority groups, for reasons that are not entirely apparent, are at higher risk.

In diseases such as bladder cancer, xenobiotics associated with particular occupations are strongly implicated, and their mutagenic effects may be important. However, in kidney cancer and other renal diseases, a number of host factors, the typically low levels of exposure to multiple xenobiotics, and such confounding variables as smoking and genetic susceptibility often mask the epidemiologic significance of individual xenobiotics. A powerful approach toward unraveling the complexities of xenobiotic exposure is to integrate biologic markers of susceptibility with biologic markers of effect.

Some xenobiotics are known to cause acute renal failure. Heavy metals and organic solvents stand out in this regard. There are several well-established associations between xenobiotic exposure and the development of chronic renal failure, as exemplified by exposures to lead and cadmium. The association of bladder cancer and occupational exposure to aniline dyes serves as a paradigm for the potential adverse health effects of xenobiotics.

Environmental agents have also been implicated in the development of neoplasms in the kidney. Some of these can be facilitated by acquired or inherited genetic defects. The association of xenobiotic exposure and such conditions as prostate cancer and interstitial cystitis is less certain but merits attention.

3
BIOLOGIC MARKERS OF SUSCEPTIBILITY AND EXPOSURE

Biologic markers of susceptibility and exposure are intimately related in the evaluation of populations at risk for effects of xenobiotics. Although objective indicators of exposure, such as excreted metabolites or DNA adducts, are often convenient to identify a population at risk, the problem might be much more complex. For example, members of a population can vary widely in their susceptibility; a population defined as being at risk in the absence of knowledge of susceptibility might consist mostly of people who are not susceptible and therefore are not at risk or are susceptible to various degrees. This complication is particularly pertinent for diseases like cancer which have a long latency period and can involve a sequence of biologic changes. The situation can be less complex in the case of toxic responses that are related directly to a toxicant or its metabolites. Because of those fundamental principles, linking exposure to disease is difficult when large fractions of the population are not susceptible. For example, the tobacco industry maintains that smoking does not cause cancer, on the grounds that 95% of smokers never develop cancer. The power of biologic markers of exposure can be increased if they are linked to biologic markers of susceptibility and effect. Making that linkage provides a means to identify a real high-risk group among those exposed and can also provide an understanding of the mechanisms of disease.

POPULATIONS AT RISK

A major thrust of environmental-health research has been in the definition of acceptable magnitudes of exposure in the workplace and environment —usually without adequate data on effects on individuals or populations. Definition of risk with biologic markers can provide objective information concerning the effects of exposure on individuals and populations. This information can, in turn, be used to design cost-effective strategies of mitigation and is related directly to the ethical and practical issues discussed in Chapter 1.

The section following immediately discusses nephrotoxicity. Discussions of genitourinary cancer follow later in the chapter.

Hereditary Susceptibility to Nephrotoxicity

Hereditary renal conditions are a documented but infrequent cause of end-stage renal disease (ESRD). The most prevalent hereditary renal disease is cystic kidney disease, which accounts for 3.4% of the cases of ESRD. Other hereditary or congenital renal disease accounts for 0.9% of the cases of ESRD (NIH, 1993). An intriguing observation regarding the relationship between hereditary factors and ESRD comes from a case-control study of 325 men in which occupational exposure was sought as an etiologic explanation of their ESRD. Only patients whose diagnoses were compatible with toxicant-induced renal injury were included in the analysis; patients with other known causes of renal failure were excluded. That ESRD was most strongly associated with a family history of renal disease (odds ratio, 9.30:1) (Steenland et al., 1990), not with occupational exposure, suggested the presence of hereditary susceptibility.

Substantial evidence supports a sex-related predilection for susceptibility to various nephrotoxicants. For example, male rats are more sensitive than female rats to the nephrotoxic effects of carbon tetrachloride and aminoglycoside antibiotics (Bennett et al., 1991). In contrast, Moore et al. (1984) demonstrated a higher susceptibility of women than of men to the nephrotoxic effects of aminoglycoside antibiotics. In any case, there seems to be a sex-related effect in both rats and humans; whether these differences are genetic in origin remains to be determined.

Direct evidence of race as a risk factor in toxicant-induced renal injury is lacking, but blacks and some other minority groups are highly susceptible to other forms of renal disease, such as has been demonstrated for the renal disease due to hypertension and diabetes mellitus (see Chapter 2) (NIH, 1992).

Inherited renal disorders might influence susceptibility to toxic injury. The potential impact of genetic factors on the renal response to environmental agents has not been widely appreciated or reviewed. One important and complicating aspect is the highly variable penetrance or expression of most of the genetic abnormalities that involve the kidney. Many people who carry genes for renal abnormalities might be only mildly affected or remain completely asymptomatic for many decades. Although it might be relatively easy to identify the first person in a genetic line with overt clinical manifestations of genetic kidney disease, a much larger pool of asymptomatic people might also be at higher risk than normal for damage from exposure to biohazards.

A number of inherited disorders affect renal development or structure; these disorders have been extensively documented, and their clinical features are well described, as are the various modes of inheritance (Brenner and Rector, 1986; Fisher and Brenner, 1989). The best-studied among those diseases are autosomal dominant (adult) polycystic kidney disease, autosomal recessive (infantile) polycystic kidney disease, hereditary nephritis (Alport's syndrome), and hereditary osteoonychodysplasia (nail-patella syndrome).

Many inborn errors of metabolism can also have a major, if not primary, impact on the kidney. A variety of inherited disorders result in compromise of the secretory or reabsorptive functions of the renal tubule system (Brenner and Rector, 1986). Prominent among them are defects in phosphate transport, amino acid transport, and glucose-handling. The clinical characteristics of people affected by these genetic defects of metabolism are, for the most part, well reviewed in standard medical texts, and inheritance patterns have also been well studied. Affected persons with tubular impairment that does not reach clinical significance and those with late onset of disease might well be at increased risk for toxic injury. The issue deserves investigation.

Heredity also plays an important part in a wide variety of systemic diseases that can damage the kidney and thereby increase the risk of renal injury from biohazards. Among the most important are diabetes, hereditary amyloidosis, and alpha-antitrypsin deficiency (Brenner and Rector, 1986; Fisher and Brenner, 1989). Autoimmune diseases, many kinds of vasculitis, and systemic lupus erythematosus can also be considered in this susceptibility category. Again, attention should be paid to family members of persons with diagnosed, clinically significant disease to identify the possible increased risk to apparently unaffected carriers of the defects.

Finally, hereditary aspects of immune responsiveness appear to contribute to the susceptibility to a number of renal diseases (Ballardie, 1992; Oliveira, 1992). That finding is not surprising in light of the great importance of immune and inflammatory responses in mechanisms of glomerular and tubulointerstitial disease. Long-term effects of toxic injury might involve immunopathologic mechanisms (see Chapter 4), and genetic aspects of immune responsiveness could contribute substantially to susceptibility to kidney damage from exposure to biohazards.

Susceptibility to develop Goodpasture's syndrome, with anti-glomerular-basement-membrane antibodies, appears to be strongly associated with a very small number of Class II major histocompatibility antigens; other Class II histocompatibility antigens have been implicated in susceptibility to membranous nephropathy (Oliveira, 1992). The link with immune response genes is of special importance in susceptibility to toxic injury, inasmuch as organic solvents, heavy metals, and drugs have also been suggested to play a role in the pathogenesis of those immune disorders of glomeruli (see Chapter 4). Class II major histocompatibility antigens have also been evaluated in the heredity of

IgA nephropathies, membranoproliferative glomerulonephritis, minimal change disease and tubulointerstitial nephritis (Oliveira, 1992). Although some claim to have established significant associations, the results remain controversial, requiring study of larger populations and more investigation. Evidence has suggested a role of genes of Class I major histocompatibility antigens in susceptibility to injury by immunological mechanisms. At present, however, it seems that linkage-disequilibrium phenomena can explain the link to Class I antigens, given the strong association with Class II antigens of the histocompatibility complex. Genetic deficiencies of the complement system, many of them also mapping in the major histocompatibility complex, have been shown to be predisposing factors in lupus nephritis and IgA nephropathy. Studies with animal models have identified highly significant genetic components of susceptibility to experimental tubulointerstitial nephropathies, but little or no similar evidence is available on humans (Ballardie, 1992).

Nutrition

The glomerular hyperfiltration that regularly follows the ingestion of a protein-rich diet can induce glomerulosclerosis and chronic renal failure in animals deprived of their renal reserve. Furthermore, variation in the body's mineral content has been linked with chronic renal injury, as in the case of severe hypokalemia induced by eating disorders, and shown to augment toxicant-caused injury, as in the association of calcium depletion with lead nephropathy or of salt depletion with analgesic nephropathy.

Socioeconomic Factors

The relationship between income and the incidence of ESRD has previously been described (see Chapter 2). It is not clear whether income is a true independent variable or is closely associated with race or other factors.

Age

Age is a well-recognized factor in determining the severity of acute renal failure—particularly that acquired in hospitalized patients (Porter, 1989). In older patients, not only is there an increased susceptibility to injury, but once injury has occurred the rate of recovery is decreased. For example, weanling rats, as opposed to adult rats, are relatively resistant to the nephrotoxic effects of aminoglycoside antibiotics and have a greater capacity for tubular epithelial-cell repair (Fernandez-Repollet et al., 1992).

There is indirect support of the proposition that the elderly are at increased risk for the development of toxicant-induced chronic renal failure. In a study of patients who were 70 years old or older (Chester et al., 1979), 29% of the patients were classified as having chronic interstitial nephritis, a diagnosis quite compatible with toxicant-induced renal failure. The proportion was much higher than the 10.4%, observed when patients 50 and older were included (Marcias-Nunez and Cameron, 1987). Because toxicant-induced chronic renal failure is theorized to occur after years of low-level exposure, it stands to reason that the incidence of chronic renal failure would be clustered in elderly patients.

Coexisting Chronic Disease

Pre-existing renal insufficiency is well documented as a risk factor in acute nephrotoxic renal failure. For chronic renal failure, the information is circumstantial. Patients with sickle-cell disease who have a high incidence of renal papillary necrosis as a result of their underlying disease process also have a predisposition to analgesic use because of the pain associated with "sickle crisis." In this situation, it is difficult to determine whether the analgesic use increases the severity of the papillary necrosis. Another example of the relationship between pre-existing renal insufficiency and acute nephrotoxic renal failure is the increased risk of nephropathy associated with contrast medium in patients with diabetes mellitus or multiple myeloma. It has been suggested (Mudge, 1980) that in up to 25% of diabetic patients with contrast-medium-induced renal injury, the serum creatinine concentration does not return to baseline, and further deterioration of renal function occurs. The role of hypertension was alluded to in the discussion of race. Presence or absence of coexisting chronic disease in other organs can modify the effects of some urinary tract toxicants.

Addictive Behavior and Recreational Drug Use

Drug abuse is increasingly common among young people, and it is not surprising that it has been linked to renal injury. Heroin use is associated with a severe form of nephropathy and is a recognized cause of focal sclerosing glomerulonephritis with associated nephrotic syndrome. The resulting glomerular injury often progresses to ESRD and might account for up to 10% of the cases of ESRD in cities with large addicted populations (Cunningham et al., 1983). Renal ischemia can be an acute effect of cocaine inhalations, although cardiac ischemia and cerebral ischemia are more common (Pogue and Nurse, 1989; Singhal et al., 1990). Rhabdomyolysis and acute renal failure can accompany free-basing inhalation of cocaine (Horst et al., 1991). Various acid-base and electrolyte abnormalities can result from solvent abuse, as occurs with exposure to toluene from glue-sniffing (Carlisle et al., 1991; Gupta et al., 1991). When intravenous amphetamine (speed) was a popular street drug, a form of drug-induced polyarteritis nodosa with progressive renal failure and severe hypertension was a recognized outcome.

Occupational or Environmental Exposure

Drugs and environmental toxicants have in some instances induced acute renal failure but evidence of their causing the development or progression of chronic renal failure is circumstantial and thus less compelling. That is not surprising, given the insidious and progressive nature of chronic renal failure and the long latency between exposure and the onset of disease. Compounding this is the superimposition of other chronic conditions that are also associated with progressive renal failure and the lack of a uniform system of classifying renal disease. Finally, the presence of many potential nephrotoxicants in our environment suggests that the causes of many forms of renal failure are multifactorial (Sandler, 1987).

It has been estimated that nearly 4 million workers were exposed to known or suspected nephrotoxicants in the workplace in 1971-1972 (Landrigan et al., 1984). It is of interest to note that the specific nephrotoxicants that were cited in preparing that estimate are those believed to be capable of producing chronic renal failure and eventually ESRD. They include heavy metals (e.g., lead, mercury, uranium, and cadmium), solvents (especially light hydrocarbons), silica, beryllium, pesticides, and arsenic.

Solvents have been implicated as inducers of glomerulonephritis (Sandler and Smith, 1991), and the association between chronic interstitial nephritis and analgesic abuse is widely recognized (Gregg et al., 1989). The association between hypertensive renal disease (nephrosclerosis) and lead nephropathy continues to be explored (Staessen et al., 1990). In evaluating the occurrence of lead nephropathy in the general public, Staessen et al. (1992) concluded that although lead exposure could impair renal function, they were unable to demonstrate a cause-effect relationship. Examples of environmental contamination that have renal consequences are many. One that stands out is the poisoning by methyl mercury in industrial effluents that occurred in the Minamata Bay region of Japan and resulted in neurologic and renal impairment in several hundred adults who ate tainted fish (Iesato et al., 1977).

Table 3-1 provides a breakdown of some common chemical agents that cause nephrotoxicity.

MARKERS OF SUSCEPTIBILITY

One of the most important factors in the development of a xenobiotic-induced disease process is susceptibility. It would be of great advantage to be able to predict an individual's susceptibility to the adverse effects of a xenobiotic. Given the broad definition of biologic markers in general as indicators of variations in cellular or physiologic components or processes that alter structure or function, it is reasonable to

Markers of Susceptibility and Exposure

TABLE 3-1 Common Chemical Agents That Cause Nephrotoxicity

Industrial and Environmental Substances
- Glycols
- Heavy metals
- Organic solvents
- Insecticides, herbicides, fungicides

Drugs
- Prescription
    - Antibiotics
    - Antibacterial agents
    - Antiviral agents
    - Antifungal agents
    - Immunosuppressive agents
    - Antineoplastic agents
- Nonprescription, including nonsteroidal anti-inflammatory drugs
- Illicit (recreational), including heroin and cocaine

extend the definition to include genotypic (reflecting genetic constitution of the individual) or phenotypic (reflecting the entire physical, biochemical, and physiologic makeup of an individual as determined both genetically and environmentally) markers as indicators of susceptibility. Genetic changes can result from exogenous exposures to occupational or environmental toxicants. These changes or mutations in DNA are usually considered markers of effect but under some circumstances can serve as markers of susceptibility.

The objective of this section is to provide a framework for identifying markers of susceptibility and determining their relative value for individual risk assessment. Ideally, the relationship between the presence of the marker and the incidence of disease has high degrees of sensitivity and specificity (see Chapter 1). If that is not the case, many people with a given marker of susceptibility might be monitored unnecessarily.

It is reasonable to use the techniques of molecular biology to identify new or more precise markers of susceptibility. Care must be taken to distinguish between the effects of acute high-level exposure and chronic low-level exposure. For example, in two separate population studies of the relationship of exposure to aromatic amines and the development of bladder cancer, outcome could not be predicted on the basis of the industrial-hygiene guidelines for estimates of peak exposure, but outcome and duration of exposure were statistically correlated. Epidemiologic studies are useful for identifying xenobiotic substances with overt health effects and to set standards for exposure, but it might be difficult to determine the percentage of people who suffer adverse health effects of low-level exposure or of exposures to multiple agents in population studies. The interaction of multiple low-level toxicants might be difficult to elucidate even in a multivariate analysis.

Modifying Factors of Susceptibility

In assessing potential risk, it is important in any model to account for individual variability of drug-metabolizing enzymes and health or environmental status. Differences in those factors will alter susceptibility to potential chemical-induced injury. Three main conditions can alter susceptibility: underlying disease or altered physiologic function, nutritional status, and renal workload. Whether each of those conditions affects susceptibility depends on the chemical in question and on the route of exposure.

Pharmacokinetics can play an important role because metabolism in other organs might be required for the generation of the eventual nephrotoxic metabolite. For example, many nephrotoxicants are metabolized by hepatic enzymes (e.g., cytochrome P-450 or GSH *S*-transferase) to generate a substrate for a renal enzyme; this substrate is transported to the kidneys, where the eventual reactive and toxic metabolite is produced. Hence, any disease state that compromises liver function (e.g., cirrhosis) can alter delivery of protoxicant to the kidneys and thus modify susceptibility to nephrotoxicity. Xenobiotics that are ingested depend on proper intestinal function for delivery to the site of action; intestinal disease or another defect that diminishes absorption can modify susceptibility to injury. Similarly, xenobiotics that are inhaled depend on proper pulmonary function; any form of pulmonary disease that results in decreased absorption across the lung epithelium can modify susceptibility to injury. Other general disease states that can modify the response to a nephrotoxic chemical include diabetes and ischemia.

Reduction of functional nephron number, such as occurs after removal of a diseased kidney or in renal failure, can markedly alter drug metabolism, energy metabolism, and susceptibility to injury (see Fine, 1986, and Wolf and Neilson, 1991, for recent reviews on the physiologic and biochemical effects of reduced nephron number). Unilateral nephrectomy and the compensatory growth that follows increased renal GSH concentrations, particularly in the proximal straight tubule, where GSH concentrations are increased by more than 50% after compensatory growth, compared with normal conditions (Lash and Zalups, 1992; Zalups and Lash, 1990; Zalups and Veltman, 1988). The mechanism by which this increase occurs appears to be an increase in GSH synthesis in renal proximal tubules (Lash and Zalups, 1994).

Nutritional status can be an important determinant of responsiveness to toxicants. For example, essentially all the major drug-metabolism (both bioactivation and detoxification) pathways are directly or indirectly energy-dependent. Changes in nutritional status, such as starvation or vitamin deficiencies, can have profound effects on activities of numerous drug-metabolism pathways.

Finally, renal workload is particularly relevant for renal function and susceptibility to nephrotoxicants because of the high energy requirements for maintaining basal renal function. Changes in transport work (e.g., altered sodium or potassium ion loads) can directly lead to changes in the supply of energy available for drug metabolism and detoxification. Hence, such changes in renal function can produce changes in susceptibility to chemical injury.

A major question to be addressed is the scientific approach to be adopted for identifying markers of susceptibility. The methods involved can include in vitro studies, in vivo animal models, study of patients with clinical disease that might have been associated with a xenobiotic, and studies of populations at risk. Identifying a specific marker of susceptibility to a xenobiotic substance requires the identification of the active xenobiotic. Clues to the identification of an active xenobiotic might come from epidemiologic data or from the finding of the xenobiotic in a tissue.

Parenchymal Renal Disease

Although the liver has generally been the focus of most drug-metabolism studies and is one of the most important sites of metabolism, numerous studies over the last 2 decades have shown that the kidneys are capable of extensive oxidation, reduction, hydrolysis, and conjugation (Table 3-2) (Anders, 1980; Jones et al., 1980). Enzyme systems similar to those found in other tissues are involved in renal drug metabolism, including both Phase I enzymes (which catalyze oxidation, reduction, and hydrolysis) and Phase II enzymes (which generally catalyze conjugation). Moreover, the kidneys are critical sites of biotransformation of many classes of xenobiotics because some metabolic pathways that are present at low activities in other tissues are present at high activities in specific regions of the nephron. One of the best examples is the mercapturic acid pathway.

Pharmacokinetics and interorgan metabolism are important to consider in studying the role of metabolism in chemically induced nephrotoxicity (Cohen, 1986). Interorgan pathways can depend on the specific chemical involved and on the route of administration. The simplest scheme involves a chemical, such as cyanide or carbon monoxide, that does not require enzymatic activation to elicit nephrotoxicity and is delivered, via the circulation, directly to the kidneys; but there are few such nephrotoxicants—most toxic (or carcinogenic) chemicals require bioactivation to elicit their effects (Anders, 1985). In an alternative scheme, a chemical is delivered directly to the kidneys and metabolized by renal enzymes to a toxic form. Other, more complex schemes of interorgan processing of nephrotoxicants, involving both renal-hepatic and enterohepatic pathways, can also occur.

Species and strain differences in interorgan levels and distributions of enzyme activities are also important. An example of how species differences contribute to differences in interorgan patterns of metabolism is the processing of GSH and GSH *S*-conjugates, which undergo a series of metabolic transformations involving enzymes of the liver, biliary epithelium, small intestinal epithelium, and renal proximal tubular epithelium (Lash et al., 1988). In mam-

TABLE 3-2 Drug-Metabolism Enzymes in Kidney[a]

Phase I Enzymes	Phase II Enzymes	Ancillary Enzymes
Cytochrome P450	Esterase	GSH peroxidase
Microsomal FAD-containing mono-oxygenase	N-Acetyltransferase	GSSG reductase
Alcohol and aldehyde dehydrogenases	GSH S-transferase	Superoxide dismutase
Epoxide hydrolase	Thiol S-methyltransferase	Catalase
Prostaglandin synthase	UDP glucuronosyl transferase	DT-diaphorase
Monoamine oxidase	Sulfotransferase	NADPH-generating pathways

[a] Phase I enzymes catalyze oxidation, reduction, or hydrolysis. Phase II enzymes generally catalyze conjugation. Ancillary enzymes function in a secondary or supporting manner to facilitate drug metabolism.

mals, gamma-glutamyltransferase activity is at its highest in the kidneys; there is also a wide range of activities in the liver and biliary epithelium (Ballatori et al., 1988; Hinchman and Ballatori, 1990). The relative contributions of the various organs can differ substantially in different species. Both intrarenal and interorgan metabolic pathways are important in the determination of susceptibility and the development of markers of toxic exposure.

Each cell population possesses a distinct complement of drug-metabolism pathways, so the bioactivation mechanism for a specific chemical might differ among regions of the nephron. For example, one mechanism might occur in proximal tubular cells and another in the medullary thick ascending limb cells. Such biochemical heterogeneity probably contributes to the targeting of nephrotoxic chemicals to particular nephron cell types, and a given chemical might be a potent toxicant in one cell population and relatively inert in another (Lash, 1990).

Renal drug-metabolism pathways and their role in bioactivation have been the subject of several excellent reviews over the last decade (Anders, 1980, 1989; Anders et al., 1988; Commandeur and Verneulen, 1990; Dekant et al., 1988a; Gram et al., 1986; Jones et al., 1980; Kaloyanides, 1991; Lash et al., 1988; Rush et al., 1984; Walker and Duggin, 1988). The major renal bioactivation pathways are cytochrome P-450, prostaglandin synthase, GSH conjugation, and the cysteine conjugate beta-lyase.

Cytochrome P-450-Dependent Activation of Nephrotoxicants

The basic biochemistry of the renal cytochrome P-450 system is essentially the same as that of the more-studied hepatic system, but there are important differences involving substrate specificity patterns and inducibility (Jones et al., 1980). The various cytochrome P-450 isozymes and the associated electron-transport systems composed of both NADPH- and NADH-dependent cytochrome P-450 reductases are not uniformly distributed in the kidney but are found predominantly in the proximal tubular region (Guder and Ross, 1984; Jones et al., 1980). The toxicologic implication of this observation is that the cellular localization of the bioactivating enzyme system determines the site specificity of the toxicity. The basic principle is that reactive metabolites are generated by this enzyme system and are responsible for toxicity near the site at which they are produced. Examples of nephrotoxicants that are bioactivated by renal cytochrome P-450 are chloroform (Smith, 1986), acetaminophen and *p*-aminophenol (Newton et al., 1982), and cephaloridine (Tune, 1986).

Prostaglandin H Synthase-Dependent Activation of Nephrotoxicants

The ability of renal medullary tissue to oxidize drugs even though it lacks the cytochrome P-450 mono-oxygenase system (Guder and Ross, 1984) suggested that another system brings about these reactions (Spry et al., 1986). Prostaglandin H synthase (PHS) is a heme-containing protein found predominantly in the interstitial and collecting duct cells of the renal medulla and in smaller amounts in Henle's loop and medullary thick ascending limb, and it is associated with the endoplasmic reticulum and nuclear membranes. Two reactions are catalyzed: a fatty acid cyclo-oxygenase step and a prostaglandin hydroperoxidase step. The cyclo-oxygenase activity, which is specifically inhibited by aspirin and indomethacin, is responsible for the initial bis-dioxygenation of unsaturated fatty acid substrates, such as arachidonic acid. The product, a hydroperoxy cyclic endoperoxide, is reduced by the hydroperoxidase activity to the hydroperoxy form.

The heme moiety of the peroxidase loses two electrons or gains oxygen during the reduction of peroxides. Re-reduction of the peroxidase heme is accomplished by enzymatic removal of two electrons or donation of oxygen to a suitable electron donor, which acts as a reducing cosubstrate and is thereby oxidized (Spry et al., 1986). Although the specific endogenous cosubstrate has not been identified, tryptophan, ascorbate, and uric acid might function in vivo. Of toxicologic importance is the ability of some xenobiotics to function as reducing cosubstrates (Spry et al., 1986; Zenser and Davis, 1984). The cooxidations include dehydrogenations, demethylations, epoxidations, sulfoxidations, *N*-oxidations, *C*-oxidations, and dioxygenations (Anders, 1989). Once oxidized to reactive metabolites, the cosubstrates might covalently bind to critical cellular macromolecules and thereby produce cytotoxicity. Several examples of xenobiotics known to undergo renal PHS-dependent cooxidation are acetaminophen, benzidine, nitrofurans, and diethylstilbestrol. Benzidine and nitrofurans are both nephrotoxicants and bladder carcinogens and diethylstilbestrol is a nephrocarcinogen. This bioactivation pathway might thus be a risk factor for humans exposed to those or similar chemicals in the workplace or in the environment.

Glutathione-Dependent Activation of Nephrotoxicants

The traditional view of the mercapturic acid pathway is that GSH forms conjugates with reactive electrophiles, the conjugates are processed to highly polar mercapturates, and these are readily excreted in the urine (Chasseaud, 1979). This pathway can also lead to bioactivation, however, and the role of conjugation of xenobiotics with GSH as a mechanism of nephrotoxicity has been the subject of numerous recent reviews (e.g., Anders et al., 1988; Dekant et al., 1988a; Elfarra and Anders, 1984; Lash et al., 1988; Stevens and Jones, 1989).

Extrarenal Conjugation and Interorgan Metabolism

Xenobiotics that undergo bioactivation by this pathway include a variety of chemically unrelated compounds; many are halogenated alkanes or alkenes. The metabolism of GSH S-conjugates demonstrates several important principles of pharmacokinetics that are determinants of interorgan metabolism. Many of the byproducts in this pathway might be useful as markers of exposure. Depending on the route of exposure to the parent chemicals, different patterns can occur. Although these chemicals are potent and specific nephrotoxicants, the kidney might not be the first site of exposure. A substantial portion of most chemicals reaches the liver (the first-pass effect). The liver will usually be the predominant site of GSH S-conjugate formation because it contains particularly high amounts of GSH S-transferase activity (the cytosolic forms constitute up to 5% of cytosolic protein). But the kidney has the highest levels of metabolism, and the general pattern observed is that GSH S-conjugates are excreted from liver, either by transport across the canalicular membrane into bile or by transport across the sinusoidal membrane into plasma; the specific pattern of interorgan metabolism differs among species because levels of hepatic GSH and GSH S-conjugate metabolism differ (Hinchman and Ballatori, 1990). It is not known what determines the membrane across which S-conjugate efflux occurs. Many studies have found the biliary route to be predominant (Gietl and Anders, 1991; Koob and Dekant, 1990), possibly because of the hydrophobicity and size of most GSH S-conjugates; in some cases, possibly because the S-conjugates are more polar, excretion occurs into both bile and plasma (Grafström et al., 1979).

A unique feature of the GSH and GSH S-conjugate degradation pathway is that the two enzymes, gamma-glutamyltransferase and cysteinylglycine dipeptidase, that catalyze the reactions whereby cysteine and cysteine S-conjugates are formed, are membrane-bound with their active sites facing the tubular lumen. GSH S-conjugates that are transported into bile are delivered to the small-intestinal lumen intact or are acted on by gamma-glutamyltransferase and cysteinylglycine dipeptidase in the biliary epithelium to yield the corresponding cysteine S-conjugates (Ballatori et al., 1988; Lash et al., 1988; Stevens and Jones, 1989). Both the intestinal epithelium (Grafström et al., 1979) and the intestinal microflora (Larsen, 1985) can metabolize both the GSH and the cysteine S-conjugates to other sulfur-containing metabolites. Those metabolites are excreted in the feces or undergo enterohepatic circulation via the portal vein and are thereby returned to the liver for additional metabolism or for translocation to the kidneys. The N-acetylcysteine S-conjugates (i.e., mercapturates) and the cysteine S-conjugates (Stevens and Jones, 1989) are the predominant metabolites that are delivered to the kidneys, although GSH S-conjugates can also be delivered, depending on the route of parent-chemical administration and the metabolic or nutritional state of the organism (Lash et al., 1988).

Conjugates that undergo glomerular filtration are either excreted in the urine (generally the mercapturates) or reabsorbed into renal proximal tubular cells by active, sodium-dependent transport across the brush-border membrane (Schaeffer and Stevens, 1987). Because only 30% of plasma is filtered through the glomeruli during a single pass through the renal circulation, a substantial portion of chemicals cleared by the kidneys is taken up by processes localized to the basal-lateral membrane (Lash et al., 1988). Transport processes have been identified for several conjugates and their metabolites on the basal-lateral membrane in various in vitro systems, including renal cortical slices, membrane vesicles, isolated perfused kidney, isolated proximal tubules, and isolated proximal tubular cells (Boogaard et al., 1989; Inoue et al., 1981, 1984; Lash and Anders, 1989; Lash and Jones, 1983, 1984, 1985; Lock and Ishmael, 1985; Lock et al., 1986; Ullrich et al., 1989; Wolfgang et al., 1989; Zhang and Stevens, 1989). The plasma membrane transport systems play key roles in regulating the flux of metabolites and in determining the specific tissue patterns of accumulation. An important factor in the nephrotoxicity of S-conjugates and, indeed, in the proximal tubular specificity of that nephrotoxicity, is the presence of these transport systems for uptake into the cell (Lash et al., 1988; Monks and Lau, 1987). These transport systems can also deliver GSH to protect epithelial cells from oxidative injury (Hagen et al., 1988; Lash et al., 1986b) or to deliver prodrugs specifically to their sites of metabolism, where they can then exert their therapeutic effects (Hwang and Elfarra, 1989).

Formation of the cysteine S-conjugates constitutes a branch point in the metabolic pathway, in that N-acetylation to form the mercapturate is a detoxification reaction, whereas further metabolism by other renal enzymes, including the cysteine conjugate beta-lyase, is a bioactivation process that leads to the formation of reactive, sulfur-containing metabolites that produce nephrotoxicity (Table 3-3). Regulation of flow through the competing pathways is not completely understood, but the chemical properties of the specific conjugates and the kinetics of the interactions of these conjugates with the various enzymes involved are important in producing nephrotoxicity and are thus of fundamental concern in the determination of susceptibility and in developing markers of exposure to the parent compounds.

GSH Conjugation of Halogenated Alkanes and Alkenes by Cytosolic and Microsomal GSH S-Transferases

Enzymatic conjugation of nephrotoxic halogenated alkanes and alkenes with GSH is the initial step in the bioactivation process. Hepatic GSH S-transferases are found as a family of isoenzymes in cytosol and as a single, membrane-bound form in the endoplasmic reticulum that is distinct from any of the cytosolic forms. The liver is considered the primary site of GSH conjugation. Indeed, hepatic conjugation of GSH with several halogenated alkanes and alkenes—including tetrachloroethylene (Dekant et al., 1986b, 1987a; Green et al., 1990), trichloroethylene (Dekant et al., 1986c), tetrafluoroethylene (Odum and Green, 1984), hexachlorobutadiene (Dekant et al., 1988b; Gietl and Anders, 1991; Reichert et al., 1985; Wallin et al., 1988), hexafluoropropene (Koob and Dekant, 1990), and chlorotrifluoroethylene (Dohn and Anders, 1982; Dohn et al., 1985a)—has been demonstrated.

Because of the interorgan pathway of GSH and GSH S-conjugate metabolism, it is generally assumed that the hepatic GSH S-transferases are responsible for the initial conjugation reaction between parent compound and GSH and that later reactions in the mercapturate or beta-lyase pathways occur in extrahepatic tissues, including the biliary and small-intestinal epithelium and the kidney. There has been little focus on the potential role of renal GSH S-transferases in the initial conjugation reaction, even though the ultimate target site for toxic metabolites is the kidney and the kidney contains GSH S-transferase activity. Koob and Dekant (1990) found that S-conjugates of hexafluoropropene formed by the hepatic enzyme are eliminated in bile and are not translocated to the kidney and that only intrarenal conjugation of hexafluoropropene is associated with nephrotoxicity. That suggests the importance of some additional complicating factor in the pharmacokinetics of GSH S-conjugates.

Cysteine Conjugate Beta-Lyase-Dependent Bioactivation

In Vivo Toxicity of S-Conjugates

The toxicity of cysteine S-conjugates

TABLE 3-3 Renal Enzymes Acting on Cysteine S-Conjugates and Their Metabolites

Enzyme	Reaction Catalyzed
Cysteine S-conjugate N-acetyltransferase	RCyS → RNAcCyS
Deacetylase	RNAcCyS → RCyS
Cysteine conjugate beta-lyase	RCyS → RS⁻ or RCyS → RSPyr → RS⁻
L-2-Amino (L-2-hydroxy) acid oxidase	RCyS → RSPyr → RS⁻
Cysteine conjugate S-oxidase	RCyS → RCyS(O) → RSOH
3-Mercaptopyruvate S-conjugate reductase	RSPyr → RSLact
Thiol S-methyltransferase	RS⁻ → RSCH$_3$
Flavin-containing monooxygenase	RSCH$_3$ → RS(O)CH$_3$ → RS(O)$_2$CH$_3$

Abbreviations of metabolites: RCyS, cysteine S-conjugate; RNAcCyS, mercapturic acid; RS⁻, reactive thiol metabolite; RSPyr, 3-mercaptopyruvate S-conjugate; RCyS(O), sulfoxide of cysteine S-conjugate; RSOH, sulfenic acid of cysteine S-conjugate; RSLact, 3-mercaptolactate S-conjugate; RSCH$_3$, thiomethyl metabolite; RS(O)CH$_3$, sulfoxide of thiomethyl metabolite; RS(O)$_2$CH$_3$, sulfone of thiomethyl metabolite.

was first reported in 1957, when McKinney et al. identified the toxic factor in trichloroethylene-extracted soybean meal as S-(1,2-dichlorovinyl)-L-cysteine (DCVC). DCVC produces aplastic anemia and nephrotoxicity in cattle (McKinney et al., 1957, 1959; Schultz et al., 1959) and is a potent nephrotoxicant in all other mammalian species tested (mice, rats, guinea pigs, and rabbits). Minor hepatic and pancreatic damage was observed in some animals, but nephrotoxicity was the primary toxic response (Terracini and Parker, 1965). More recent studies on the nephrotoxicity of various GSH and cysteine S-conjugates in rats found evidence of proximal tubular damage (Dohn et al., 1985b; Elfarra and Anders, 1984; Elfarra et al., 1986a).

Many halogenated hydrocarbons that are bioactivated by this pathway are used in industrial processes and are environmental contaminants. A complete understanding of the bioactivation pathway is therefore essential for assessing human risk of injury and for developing useful markers of exposure.

Role of Cysteine Conjugate Beta-Lyase in Bioactivation of Cysteine S-Conjugates

Early studies with the model nephrotoxicant cysteine S-conjugate DCVC described its enzymatic breakdown by a C-S lyase in liver and kidney to pyruvate, ammonia, and an unidentified sulfur-containing fragment that formed covalent adducts with GSH and protein (Anderson and Schultze, 1965). The cysteine S-conjugates that are beta-lyase substrates are either haloalkyl or haloalkenyl S-conjugates. Although the reaction mechanism of the cysteine conjugate beta-lyase does appear not to differ in those two classes of substrates, the fates of the reactive metabolites formed are markedly different (Vamvakas et al., 1989a). Nonhalogenated, alkyl-substituted cysteine S-conjugates, such as S-methyl-L-cysteine, are not substrates for mammalian beta-lyases. Some of the biochemical and physiologic properties of the renal enzymes are summarized in Table 3-4.

Although the kidney is the target organ, cysteine conjugate beta-lyase activity is present in liver. The hepatic cytosolic cysteine conjugate beta-lyase activity depends on pyridoxal phosphate (PLP) and is a catalytic property of kynureninase (Stevens, 1985a). Bacteria in the intestinal microflora also contain cysteine conjugate beta-lyase activity (Larsen, 1985). That indicates that factors besides a bioactivating enzyme are necessary to determine the tissue and cell-type specificity of cysteine S-conjugate toxicity because nontarget tissues can also produce reactive metabolites from cysteine S-conjugates.

Renal cysteine conjugate beta-lyase activity is PLP-dependent and is found in the cytosolic and mitochondrial fractions of renal proximal tubule (Dekant et al., 1987b; Elfarra et al., 1986a,b, 1987; Lash et al., 1986c, 1990a; Stevens, 1985b; Stevens et al., 1986, 1988). The renal and hepatic forms are immunologically distinct (Stevens, 1985b). Immunocytochemical studies have localized the renal beta-lyase to the pars recta segment (i.e., S3 segment) of the proximal tubule, which is coincident with the nephron site specificity of cysteine S-conjugate nephrotoxicity (Jones et al., 1988; MacFarlane et al., 1989). No beta-lyase activity was detected in glomeruli or in distal tubular segments of the nephron, but this does not preclude cytotoxicity in these nephron segments by additional bioactivation mechanisms. In fact, immunocytochemical and other biochemical data are consistent with the presence of multiple cysteine conjugate beta-lyase activities in rat kidney (Jones et al., 1988; MacFarlane et al., 1989; Stevens, 1985b).

Parent chemicals for which an unequivocal role of beta-lyase-dependent bioactivation has been established include trichloroethylene, perchloroethylene, hexachlorobutadiene, and chlorotrifluoroethylene. An additional critical proof that the beta-lyase pathway functions in bioactivation and nephrotoxicity is to isolate and demonstrate formation of reactive metabolites; this has recently been achieved through chemical

TABLE 3-4 Biochemical and Physiologic Properties of Mammalian Renal Cysteine Conjugate Beta-Lyase Activities

Property	Description
Subcellular localization	Cytosolic and mitochondrial
Intramitochondrial localization	Matrix and outer membrane
	Intrarenal localization Predominantly in $S3$ segment of pars recta of proximal tubule
Cofactor	Pyridoxal phosphate
Substrate specificity	Good leaving group on beta carbon; haloalkyl or haloalkenyl cysteine S-conjugates are substrates; nonhalogenated, alkyl-substituted conjugates are *not* substrates
Cosubstrate specificity	2-Keto acids with relatively hydrophobic substituents on beta carbon (e.g., 2-keto-4-mercaptobutyrate, phenyl pyruvate, 2-keto octanoate)
Reaction mechanism	Beta-elimination or transamination followed by reverse Michael elimination
Inhibitor	Amino-oxyacetic acid
Other reactions catalyzed	Cytosolic and mitochondrial matrix forms identical with glutamine transaminase K

trapping experiments for chlorotrifluoroethylene (Dekant et al., 1987b), hexachlorobutadiene (Dekant et al., 1988c), and for trichloroethylene and perchloroethylene (Dekant et al., 1988d).

The metabolism of nephrotoxic and nephrocarcinogenic cysteine S-conjugates to thioacylating intermediates is consistent with DNA and protein alkylation and inhibition of thiol-dependent enzymatic activities as mechanisms of toxicity (Banki and Anders, 1989; Chen et al., 1990a; Lock and Schnellmann, 1990; Vamvakas et al., 1989b,c). Adducts of reactive S-conjugate metabolites with cellular protein, lipid, and DNA have been detected and characterized chemically (Hargus and

Anders, 1991; Harris et al., 1992; Hayden et al., 1992; Hill et al., 1978; Inskeep and Guengerich, 1984). Those adducts might be useful as markers of exposure because they are generally chemically stable. The chemical nature of the thioacylating intermediate is important in determining the nephrotoxic response in that S-conjugates of chloroalkenes are both acutely nephrotoxic and mutagenic, but S-conjugates of fluoroalkanes are acutely nephrotoxic and not mutagenic (Green and Odum, 1985; Vamvakas et al., 1989c).

Other Enzymes Involved in Cysteine S-Conjugate Metabolism

Although the cysteine conjugate beta-lyase is the best characterized of the enzymes known to be involved in cysteine S-conjugate bioactivation, recent work has shown that several other enzymes can metabolize cysteine S-conjugates, and some of them also produce nephrotoxicity (Anders et al., 1988; Dekant et al., 1988a). These alternative metabolic pathways include the L-2-amino (hydroxy) acid oxidase found in rat kidney cytosol and peroxisomes, pathway 3 (Lash et al., 1990b; Stevens et al., 1989; Tomisawa et al., 1986; Webster and Anders, 1989); the cysteine conjugate S-oxidase, pathway 6, which is a flavin-containing mixed-function oxygenase found in both renal and hepatic microsomes (Lash et al., 1994; Sausen and Elfarra, 1990); the thiol S-methyltransferase, pathway 8 (Jakoby and Stevens, 1984); and N-acetylation to form the mercapturic acid, pathways 1a and 1b. The latter two are detoxification pathways that generate stable, nontoxic, and readily excreted products. As the kinetics and regulation of those interacting pathways become better understood, they will increase the ability to define susceptibility. Moreover, many of the terminal metabolites or products of reactive metabolites and cellular macromolecules can be developed into early markers of exposure.

Susceptibility Factors in Cancers of the Genitourinary Tract

Susceptibility factors for cancer can be either hereditary or acquired. Hereditary factors can be divided into factors that result from inheritance of one inactive allele for a tumor-suppressor gene and factors that result from inheritance of a metabolic type that places the individual at higher risk by virtue of how carcinogens or procarcinogens are metabolized. Heritability of defective alleles follows the original hypothesis of Knudson (1971), who demonstrated that retinoblastoma followed such a heritable pattern. Recently, Wilms's tumor has been shown to follow this pattern of heritance with the *WT1* gene. One important finding has been that although such inherited cancers as retinoblastoma and Wilms's tumor are rare, the genes involved are often involved in sporadic cancers that develop later in life. In these cases, somatic damage to both alleles occurs. The human genome comprises 23 pairs of chromosomes, including one pair of sex chromosomes. Females have a pair of X chromosomes;

males have one X chromosome and one Y chromosome. The total display of these 46 chromosomes is called the human karyotype. Conventional karyotyping and interphase karyotyping with fluorescence in situ hybridization (FISH) are being used to identify common chromosomal abnormalities, which might identify chromosomes that contain tumor-suppressor genes or oncogenes. An inherited genotype can be a marker of susceptibility, but additional somatic alterations can provide markers of effect or disease that have utility as markers of prognosis.

Cancer of the Kidney

Familial and Sporadic Nonpapillary Renal-Cell Carcinoma

Karyotypic analysis of the individual chromosomes at mitosis, in cases of familial renal-cell carcinoma, has revealed translocations involving chromosomes 3 and 12 and chromosomes 3 and 6. Those translocations involve the exchange of segments of two chromosomes without the loss of any material. As described in Chapter 2, karyotypic analysis of patients with the von Hippel-Lindau disease has also identified abnormalities at chromosome 3 (Latif et al., 1993). The studies strongly implicate the short arm of chromosome 3 in the pathogenesis of renal-cell cancer; that is, the short arm might harbor a suppressor oncogene.

The nonfamilial form of adenocarcinoma of the kidney is also linked to the short arm of chromosome 3 but in a different fashion (Gnarra et al., 1994; Herman et al., 1994). In the sporadic form of nonpapillary renal-cell carcinoma, there are deletions in chromosome 3 (Gnarra et al., 1994; Herman et al., 1994). In addition to deletions in chromosome 3, 65% of the cases of renal-cell carcinoma involve another chromosome, most frequently chromosome 5.

Papillary Renal-Cell Carcinoma

Papillary renal-cell carcinoma is a second type of renal parenchymal tumor. In contrast with the nonpapillary type, papillary renal-cell carcinomas do not demonstrate abnormalities principally on chromosome 3. The tumors have some of the characteristic changes consistent with duplications in chromosomes 7 and 17 and a loss of the Y chromosome in men. Polysomy of chromosomes 7 and 17 has been reported in 70-75% of the cases studied (Kovacs et al., 1991). Other findings include trisomy of chromosomes 12, 16, and 20, and the *p53* tumor-suppressor gene is on chromosome 17.

Cancer of the Bladder

Karyotypic Studies

To date, no gene has been specifically linked to bladder cancer, but gene mutations have been identified on a number of chromosomes that are commonly involved. The most frequent anomalies in bladder cancer are associated with chromosomes 1, 7, 9, 5 and 11 and the Y chromosome. The most frequently observed anomaly is loss of all or part of chromosome 9, which is believed to be the most important cytogenetic event; it has been hypothesized that this chromosome contains the tumor-suppressor gene in bladder cancer, particularly cancer that develops along the low-grade papillary-tumor pathway. It has been proposed that markers of chromosomal loss, such as the deletion of chromosome 9, can occur not only in the tumor, but in dysplastic areas in the bladder.

Metabolic Pathways and Susceptibility to Bladder Cancer

The factors that activate or inactivate xenobiotic toxicants might be primary in bladder carcinogenesis. For example, acetylation inactivates aromatic amines, which are important carcinogens. Consequently, slow acetylators exposed to those compounds have a higher incidence of bladder cancer than do fast acetylators. The *N*-acetyltranferase (NAT) gene occurs as a single copy, and the rate of acetylation seems to depend on both transcriptional and translational events. Phenotypic analysis has revealed marked differences among ethnic groups. For example, there is a 1:1 distribution of slow to fast acetylators among North American Caucasians but a 1:9 distribution in the Japanese population.

The metabolic pathways for acetylation and the differences in mutagenicity of the products complicate the scheme of slow and fast acetylators. For example, acetylation of some compounds, such as arylamines, reduces their carcinogenicity because the *N*-acetylated arylamines are poor substrates for PHS—a key enzyme in the oxidation of 2-naphthylamine and the expression of its carcinogenicity. But benzidine, which requires oxidation for its activation as a carcinogen, is a poor substrate for PHS in its unacetylated form. Thus, to characterize the susceptibility of a population to bladder cancer as a consequence of its exposure to those compounds, one should design a strategy that includes characterization of the metabolic pathways and the particular compounds to which the population is exposed. The NAT gene, which controls acetylation, and the CYP2D6 gene, which is a member of the P-450 oxidative enzymes, have each been a focus of such studies. Epidemiologic studies (Lower et al., 1979) support the metabolic information noted above.

DNA Adducts

DNA adducts constitute a direct, measurable chemical change in DNA in susceptible cells and can result from carcinogen exposure (Hemminki, 1993). DNA adducts are a more specific indication of exposure than is the excretion of metabolites; because the presence of DNA adducts demonstrates that the target molecule has been affected by the exposure, they can also be considered as the first marker of effect. In the bladder, the urothelial cells normally sloughed into the urine are convenient for monitoring bladder carcinogen exposure (Talaska et al., 1993). Although the identification of adducts does demonstrate exposure, it does not demonstrate that particular critical genes have been damaged. Adducts are often found in superficial cells, which will eventually be sloughed, so only the presence of adducts in bladder stem cells is significant for long-term carcinogenesis. The adducts indicate sites where mutations might occur when the cell divides. If these mutations persist, they potentially can serve as markers of effect.

DNA adducts can be identified with very high sensitivity in the range of one adduct per 10^7 base pairs by ^{32}P-postlabeling or immunoassays, but the concentration of adducts does not necessarily correlate with disease, nor has a dose-response been documented. The lack of a dose-response relationship might reflect the efficiency of DNA repair processes or the absence of adduct formation in critical basal cells. Alternatively, in the bladder it might reflect the efficiency of the bladder's protective mechanisms that limit exposure of the critical basal layer (Bodenstab et al., 1983). The efficiency of these mechanisms suggests that carcinogens delivered through the blood can be more effective than those delivered through the urine unless there is some pre-existing damage to the protective layer or urinary retention. As with all markers of exposure, interpretation in terms of individual risk requires linkage to susceptibility and specific effects.

Acquired Conditions Predisposing to Bladder Cancer

Some congenital and acquired clinical conditions influence susceptibility to bladder cancer. The most common and important might be urinary stasis and infection, both of which can result from bladder-outlet obstruction secondary to benign prostatic hyperplasia or neurologic dysfunction. An increase in the incidence of transitional-cell carcinoma by a factor of 15 has been associated with diverticulum of the bladder, which can be congenital or acquired. Stone disease, which also can be congenital or acquired, can predispose the bladder and renal pelvis to the development of cancer, which often is manifested in squamous cell carcinoma (Catalona, 1992).

Special cases associated with the increased risk of cancer include inflammatory conditions, such as cystitis cystica and cystitis glandulares (Catalona, 1992). Patients with ureterosigmoidostomy (anastomosis of the ureter to the colon) can develop cancer at the site of anastomosis (Ambrose, 1983; Bergerheim et al., 1991; Filmer and Spencer, 1990). A common feature of the increased risk of these conditions might be the formation of nitrosourea compounds (which can be mutagenic) as a consequence of urinary stasis and infection. These special cases might provide clues to the identification of biologic markers of susceptibility.

The linkage between cellular differentiation, proliferation, and carcinogen exposure is complex. Cohen and Ellwein

(1991) point out that the target for carcinogens is the stem cells of any epithelial layer because differentiating cells are fated to die apoptotically. Alterations in the ratio of differentiating to stem cells will lead to clonal expansion of an initiated cell. The role of differentiation can be complex. With 2-acetaminofluorene (2-AAF) carcinogenesis, initiated bladder cells dedifferentiate sufficiently so that they no longer produce the enzyme that metabolizes 2-AAF to the proximate carcinogen and are therefore resistant to additional carcinogenic "hits." Carcinogenesis must then depend on endogenous processes. In the liver, that is not true, so the two organs show quite different dose-response curves. Loss of apoptosis in differentiating cells will result in retention of an altered genotype that can continue to progress to the malignant phenotype. In summary, the balance between differentiating cells fated for apoptotic death and initiated cells, the differential effects of xenobiotic exposure on foci of initiated cells and on normal cells, and any differential proliferative responses among stem and differentiating cells or among initiated cells can lead to carcinogenesis.

Cancer of the Prostate

Few data support a prominent role for occupational or environmental toxicants in the genesis of prostatic carcinoma, but the sharp differences in prevalence of prostatic cancers around the world and the convergence toward a prevalence in migrated ethnic groups similar to that in indigenous populations suggest prominent environmental factors. Whether these factors are protective or promotive is not known. Familial studies indicate the presence of predisposing genetic factors, and families can be useful for defining markers of susceptibility in the form of suppressor genes or other genetic markers. Karyotypic changes that occur in prostatic cancer suggest the presence of a tumor-suppressor gene, as discussed below. An additional important factor is the introduction of prostate-specific antigen (PSA) as a marker of prostatic cancer that serves as a useful model of the advantages and disadvantages of identifying an early marker to detect disease; this is discussed in Chapter 4.

Population Studies

A finding of increased risk of prostatic cancer in probands of family members with prostatic cancer indicates the importance of genetic factors in disease susceptibility. In a case-control study, Spitz et al. (1991) observed an increased risk of 2.41 if a first-degree relative (i.e., father or brother) had prostatic cancer. Carter et al. (1992) conducted a case-control study of 691 men with prostatic cancer and 640 spouse controls. Fifteen percent of the cases but only 8% of the controls ($p<0.001$) had a father or brother with prostatic cancer. A proband with two or three first-degree relatives with the disease experienced a risk of developing prostatic cancer increased by a factor of 5-11. In the same study, 29 "cancer families" were identified; they had an earlier age of onset of malignancies of multiple organ sites. Other cancers, including multiple myeloma and head and neck tumors, have been associated with prostatic cancer; this suggests the potency of the underlying carcinogenic process.

In another study, Umbas et al. (1992) interviewed 355 patients with histologically proven prostatic cancer and 339 controls. Of the cancer patients, 21 (4%) had a family member with prostatic cancer, but 9.5% of the controls. With a first-degree relative affected, the incidence is 2.8 times that of the general population; with both a first- and a second-degree relative, incidence is 6.1 times. In an earlier study, Woolf reported a statistically significantly increased incidence of prostatic cancer in a Mormon population compared with controls (Woolf, 1960). That was also true of the family members of those originally identified as having prostatic cancer. For deaths related to cancers other than prostatic cancer, there were no significant differences between the two groups.

In another study reported by Ghadirian et al. (1991), 140 prostatic cancer patients' families and families of 101 controls were interviewed. Of the prostatic-cancer patients, 15% gave history of the same cancer among first-degree relatives, compared with 2% of the controls (OR=8.7; CI=95%). The major limitation and the most difficult factor to control in these studies was the influence of the environment.

Karyotypic Analysis

Bergerheim et al. (1991) found deletions in chromosomes 8, 10, 16, and 18. The most frequent deletions (in 65% of cases) occurred on 8p, and the long arm of chromosome 16 had deletions 56% of the time. The results suggest that tumor-suppressor genes are localized on chromosomes 8, 10, and 16. Research to identify the genes associated with susceptibility clearly should have a high priority.

Histopathologic findings confirm the importance of genetic alterations in chromosome 16. Additional studies in animals and humans have focused on the *E-cadherin* gene, on chromosome 16, as an important site. *E-cadherin* is a member of the cell-adhesion gene family, which is important in formation of gap junctions between cells. Abnormalities in this gene could be responsible for the histologic appearance of prostatic cancer cells, as reflected by the Gleason scoring system, and could be important in determining the tumor aggressiveness. Taken together, the findings suggest the existence of a tumor-suppressor gene that could be altered by an environmental toxicant so as to permit the development of cancer, as proposed in the Knudson hypothesis (see Chapter 2).

MARKERS OF EXPOSURE

A critical issue in assessment of human exposure to environmental pollutants or other toxic chemicals is development of a marker of exposure that is noninvasive and sensitive. The goal is to enable early detection and therapeutic intervention and thereby prevent progression to chemical-induced injury. Measurements are typically made in urine to indicate exposure that involves renal metabolism or later renal toxicity or correlates with tumorigenesis in the genitourinary tract. Table 3-5 lists of environmental agents associated with cancer of the urinary tract. A distinction must be made between markers of exposure and markers of effect (injury), which are discussed elsewhere (see Chapter 4). Three issues must be considered: (1) What properties constitute a useful marker of exposure? (2) What techniques are available to measure these markers? (3) How can a marker of exposure be correlated with the dose that reaches a target tissue?

Because the goal in using a marker of exposure is to detect exposure as early as possible, the marker should be highly sensitive and should be a stable indicator. The criterion of sensitivity means that the entity being measured (e.g., presence of a metabolite) should be detectable at a concentration below that at which any overt injury occurs. If that is not the case and the marker is not detectable until the exposure causes toxicity, the marker is not useful in helping to prevent injury. Once again, establishing the relevant concentrations is difficult when the risk of disease is low.

TABLE 3-5 Cancers of the Urinary Tract Associated with Environmental Agents

Renal parenchyma
    Renal adenoma and adenocarcinoma
        Aromatic amines
        Nitroso compounds
        Hydrazines
        Alkylating agents
        Anticancer drugs
        Cadmium and lead
    Squamous-cell carcinoma
        Various chemical carcinogens
        Chronic infections
Renal Pelvis or Ureter
    Transitional-cell carcinoma
        Papillary tumors
            Various industrial carcinogens
    Urothelial carcinoma
        Balkan nephropathy
        Analgesics
Bladder
    Transitional-cell carcinoma
        Cigarette-smoking
        Coffee-drinking
        Aromatic amines (2-naphthylamine, xenylamine)
    Squamous-cell cancer
        Chronic irritation, such as presence of urinary calculi
        Chronic urinary tract infection
        Neurogenic bladder
        Parasitic infection
    Adenocarcinoma
        Bladder extrophy
        Persistent urachus

The criterion of stability refers to the chemical properties of the marker or metabolite; markers of exposure are generally terminal metabolites that are chemically unreactive (or not very reactive) and thus do not spontaneously decompose to other substances. If the marker is stable, it is assumed that it is formed in amounts that are proportional to exposure dose. That assumption depends on the kinetic properties of the enzymatic pathway responsible for generation of the metabolite. For example, if the enzyme that generates the terminal metabolite saturates at a relatively low substrate concentration, the amount of metabolite formed will not be proportional to exposure at normal doses. The K_m of the critical enzyme therefore must be higher than substrate concentrations that are typically encountered. Most terminal metabolites are highly polar molecules that are readily excreted from the body. That provides a marker source (e.g., urine or feces) that makes the detection procedure noninvasive. The terminal metabolites that can often be used as markers and that are relevant to chemicals that can produce renal injury include *N*-acetylcysteine *S*-conjugates, sulfoxides, sulfones, thioethers, glucuronide conjugates, and sulfate conjugates. For most parent chemicals of interest, the metabolites are recovered in urine. It must be kept in mind that some xenobiotics, such as pesticides, can be retained within the body in fat; if they can mimic or alter hormonal factors, the tiny amounts that leach into circulation can have profound effects and be difficult to detect in the form of excreted metabolites.

Many analytic techniques for detection and quantitation of terminal metabolites of drugs are available. The choice of method depends on the sensitivity required and on whether the analysis is being performed in laboratory animals or in humans. Radiolabeled drugs can be used and recovery of labeled metabolite can be determined as a sensitive marker of exposure, but this method is generally suitable only for laboratory animals.

A variety of chromatographic methods can be used to identify metabolites of xenobiotics. They include high-performance liquid chromatography (HPLC) and gas chromatography-mass spectrometry (GC-MS) and can provide unambiguous identification and measurement of metabolites by comparison with standards. The recent development of nuclear magnetic resonance (NMR) spectrometry techniques has expanded the ability to detect stable, terminal metabolites noninvasively, but NMR techniques are generally not very sensitive.

HPLC analysis of urinary metabolites obtained after exposure of rats to the hepatotoxic and nephrotoxic halogenated hydrocarbon trichloroethylene showed only the oxidation products generated by cytochrome P-450 (Dekant et al., 1986a). It has been demonstrated, however, that a small portion of the administered trichloroethylene is metabolized by the GSH conjugation pathway and that this minor pathway is associated with nephrotoxicity (Dekant et al., 1987a). Therefore, even though more than 90% of metabolites is recovered by HPLC and it might be sensitive enough to indicate exposure, it might not be sensitive enough to detect the metabolites that are critical indicators of potential nephrotoxicity.

For some of the halogenated hydrocarbons, many of which are potent and specific nephrotoxic agents, ^{19}F NMR spectroscopy has been used to detect terminal metabolites in urine and in experimental situations. An example of the latter was the detection of stoichiometric amounts of chlorofluoroacetate as a stable metabolite of the nephrotoxicant chlorofluoroethene (Dekant et al., 1987b). Although the studies were performed for analytic purposes in in vitro preparations, the basic method is easily applied to metabolites of halogenated hydrocarbons in rat and human urine. A recent study by Harris and Anders (1991a,b) demonstrated formation of aldehydes and glucuronides as terminal metabolites. The relevance for human exposure analysis and risk assessment is that these and many similar chlorofluorocarbons are being investigated as substitutes for the ozone-depleting chlorofluorocarbons now in use. Because of the potential commercial importance of the substitutes, there is notable potential human exposure to them.

As mentioned previously, HPLC or NMR might not be sensitive enough to detect metabolites of interest. GC-MS was used to detect and measure urinary metabolites of S-carboxymethyl-L-cysteine (CMC) in human urine (Hofmann et al., 1991). CMC is a model compound for characterizing the relative roles of the pertinent biotransformation pathways in humans. Many analogues of CMC, including several cysteine S-conjugates of halogenated hydrocarbons, are nephrotoxic and thus pose a health risk to humans. Study of human metabolism of CMC can suggest how the other nephrotoxic chemicals are handled.

A different approach to the use of biologic markers for nephrotoxicant exposure was developed by Woods and colleagues (Bowers et al., 1992). They determined that urinary porphyrin excretion patterns are altered in response to methyl mercury exposure and that the changes can correlate with both the magnitude of exposure and whether irreversible renal injury has occurred. That is, a molecule other than a terminal metabolite of the chemical under study is being used as a marker of exposure.

The HPLC, GC-MS, and NMR studies illustrate how the methods can be used for noninvasive measurement of stable terminal metabolites. They can provide two types of information: metabolite data can show which enzymatic pathways are involved in biotransformation of a xenobiotic, and a detected metabolite can be used as a marker of exposure to the xenobiotic. Interpretation of the first type of information can generally be straightforward because many terminal metabolites can be formed by only one pathway. In the latter case, however, it is not always simple to correlate urinary concentrations of a terminal metabolite with exposure; with the appropriate choice of metabolite, this can be achieved, and the metabolite can function as an early marker of exposure.

SUMMARY

For diseases of the urinary tract, the

most efficient program for determining the importance of occupational and environmental toxicants and carcinogens requires the identification of susceptible populations and the correlation of disease processes with the magnitudes and durations of exposure to the agents. Various factors modify human susceptibility to the effects of occupational and environmental nephrotoxicants and carcinogens.

Although much of the information on nephrotoxic chronic renal failure is circumstantial and comes from epidemiologic surveys that started with ESRD patients, for some agents the evidence is substantial. The most obvious group at risk consists of persons exposed to known or suspected nephrotoxicants in the workplace. Also at risk are people who live in regions of documented contamination. The possible link between a family history of renal disease and the development of renal failure might be an inherited susceptibility or a common geographic exposure. Altered nutrition and some coexisting diseases, including addictive behavior, are additional characteristics that indicate increased risk associated with nephrotoxicants.

Gender, race, and socioeconomic status provide tantalizing clues, but much more information needs to be collected than is currently available. Targeting populations at risk for future evaluation and followup is the most efficient strategy for the identification of patients early in the course of their toxic injury. This strategy might make it possible to introduce protective measures to reduce the progression of renal disease and to decrease the rate of entry of patients into ESRD programs.

Susceptibility factors for cancer can be either hereditary or acquired. For example, various hereditary conditions are associated with the development of progressive renal disease. Moreover, specific genes that predispose people to develop disease, such as tumor-suppressor genes, have been identified. These people inherit one defective copy and are at much greater risk for cancer than people who have two intact copies. Likewise, individual variations in the metabolic pathways play a large role in susceptibility to both urogenital cancer and nephrotoxicity.

The identification of specific markers of susceptibility and exposure is a daunting task. A susceptible person might be characterized by specific genotypic or phenotypic markers. Identifying these markers is likely to require a more complete understanding of the biochemical and physiologic properties of the kidney and lower urinary tract, as well as better insight into factors that control links between cellular growth, differentiation, proliferation, and malignant transformation. Consequently, these issues are addressed in detail in this report. The goal of identifying the markers is not to separate one population from another, but rather to identify exposures that can be tolerated by all.

4
BIOLOGIC MARKERS OF EFFECT

Biologic markers of effect are perhaps the most important set of markers. Markers of effect permit the early identification of adverse effects of toxic xenobiotic exposure. They constitute the missing link in the continuum of disease development, linking markers of exposure and susceptibility with onset of disease and offering the possibility of detecting disease at the very early stages of development. They offer the most potential for clinical intervention before irreversible effects have occurred. With proper use of such markers, carcinogenesis can be detected before clinically apparent tumors result, and nephrotoxicity can be identified while sufficient kidney function still remains.

This chapter focuses on relatively noninvasive measurements of early genitourinary consequences of human exposure to noxious chemical, biologic, or physical agents. It discusses available markers of renal effects, their limitations, and directions for future work, but it excludes in vitro procedures extensively used for evaluating cytotoxic effects in animals (which is covered in Chapter 5) or for studying mechanisms of nephrotoxicity. It also discusses markers of carcinogenesis that signal tumor development and address other disorders, such as interstitial cystitis.

Our objective here is to identify and evaluate markers of general or specific effects on renal function and integrity. Measurement of these markers is preferably minimally invasive, accurate, and suitable for screening populations at risk. A number of techniques have been developed to assess renal function, but these techniques generally lack the sensitivity and specificity necessary for detecting preclinical disease or are not applicable to large populations. For example, renal biopsy and radioisotopic procedures are not appropriate for screening, and classical clearance procedures are seldom reproducible to within $\pm 10\%$, even in the laboratory. Compromises have therefore been made with respect to convenience, sensitivity, and invasiveness. The conventional markers are reviewed from this perspective, and newer results from cellular and molecular studies are presented in light of their potential to provide new markers of effect that will be more sensitive to early nephrotoxicity and can be applied widely to at-risk populations. The conventional markers also can play an important role in validating new markers, particularly in people with compromised renal function who are exposed to additional potentially damaging events. The loss of a small number of tubular cells in a person with only minimal residual kidney function can lead to large changes in conventional markers whereas the same loss in a normal person would be undetectable. Such a change, now easily detectable in a compromised person, can be used to validate new markers against conventional markers.

EXTERNAL VISUALIZATION

Several techniques permit external visualization of the kidney and delineation of functional changes. Radiopaque contrast media and ultrasonic agents permit imaging of changes in blood or urine flow or in renal solute accumulation and secretion. Similar results are routinely obtained with gamma-emitting radioisotopes. In addition to the risks associated with radiation exposure, these tests do not reflect subtle changes in renal function and are therefore limited in their applicability to population screening.

A visualization technique that might hold promise and that has been applied to small animals is magnetic resonance imaging (MRI). Acara et al. (1991) observed conspicuously hyperintense regions in renal papillae in rats and related them to early hydronephrotic changes. Further improvements in MRI might make the technique more useful for screening human populations, but it is still time-consuming and expensive.

URINARY CLEARANCE MEASUREMENTS

Direct

Measurement of urinary clearances is not convenient for use in large populations at risk, but it continues to constitute a basic approach to evaluating renal function. Indeed, the introduction of the clearance concept into renal physiology by Moller et al. in 1929 provided for the first time a relatively simple and informative measure of renal function and was a critical contribution to the study of the kidney. Urinary clearance (C) is the virtual volume of plasma that is cleared per unit time by excretion into urine; it is expressed as $C_x = U_x V/P_x$, where $U_x V$ is the amount of substance x excreted per unit time and P_x is the plasma concentration of substance x. Depending on the solute chosen, clearances can provide information on glomerular and tubular function; they are useful under many conditions for evaluating functional integrity of the kidney in humans or animals.

The direct measurement of clearance, however, has disadvantages, especially for the study of large populations. It is clear from the formula $C_x = U_x V/P_x$ that it is essential to keep P_x constant during the clearance period or at least to be able to calculate the mean plasma concentration of the chosen solute. For exogenous solutes, such as inulin (often used to evaluate renal function), this requirement is best met by continuous intravenous infusion of 1 hour or longer, but it can at times be approximated by subcutaneous or intramuscular depot injection. In either case, blood has to be sampled repeatedly to ensure the required constancy of P. The procedure is simplified by reliance on an endogenous solute, such as creatinine, which is usually produced in the body at a constant rate and normally excreted almost entirely by glomerular filtration. Creatinine clearance can be a measure of glomerular filtration rate. Clearance of endogenous urea has also been used to evaluate glomerular function, but it suffers from the double disadvantage of less-constant plasma concentration than that of creatinine and substantial influence by urinary flow rate.

The need for accurately timed and complete urine collection presents a second difficulty in clearance procedures. Because of the cumulative errors arising from lack of constant plasma concentrations and from inaccurate timing or incomplete urine collection (and because of analytic errors), clearance determinations are seldom reproducible to within ±10%, even in a laboratory setting. Such low reproducibility might be adequate for many clinical purposes, but it interferes with attempts to demonstrate early changes in the ability of the kidney to excrete a given solute. In addition, early effects of nephrotoxicants might result in diminution of glomerular reserve capacity without changes in overall clearance (as described in Chapter 2). The glomerular reserve capacity reflects the ability to increase the glomerular filtration rate in response to stimuli. A failure to elicit such a response implies the presence of a subclinical disease process.

Total functional reserve capacity of the kidney can be measured by, for example, imposing a stress, such as ingestion of a high-protein meal. The increase in creatinine clearance during pregnancy can also reflect functional reserve capacity. For tubular secretion or reabsorption, total capacity is evaluated at saturating levels of a test solute, in which case secretion is expressed as excreted load minus filtered load, whereas reabsorption is calculated as filtered load minus excreted load. These techniques require determination of the glomerular filtration rate (GFR) as well as prolonged infusion of large amounts of solute and therefore are not suitable for routine screening.

Direct clearance measurement takes place primarily in the laboratory or special clinic. It can be simplified in some cases by use of radioisotopes, but this is not appropriate for general screening. It would be useful if convenient and sensitive chemical procedures could be developed for such compounds as chromium ethylenediaminetetraacetic acid (Cr-EDTA) and iodothalamate, both good solutes for measurement of GFR but so far used only in the radiolabeled form.

Indirect

A number of indirect measures of

renal clearance have been proposed. They are generally simple but lack sensitivity and precision, as can be illustrated by reference to the formerly extensively used measurement of the rate of excretion of a chemical after injection of a standard dose of some substance. For instance, the dye phenol red (PSP) is excreted by tubular secretion, and the rate of its recovery in urine is therefore controlled by independent variables, such as renal plasma flow, the integrity of tubular transport mechanisms, and a linear dependence of the rate of transport on plasma concentration of the dye at the dose used. Measurement of fractional excretion of a standard dose of PSP over 15 or 30 min can yield little information on specific aspects of renal function.

A more commonly used indirect method measures the declining slope of plasma concentration of an injected solute after a suitable equilibration period. An example is the estimation of GFR with iodothalamate.

Another indirect measurement of urinary clearance is based on the assumption that creatinine is filtered and excreted at a constant rate. Even in the presence of diurnal variations in GFR, normalizing urinary concentrations of a specified substance by the urinary concentration of creatinine permits conclusions to be drawn from spot urine samples and avoids the difficulty of obtaining 24-hour collections. Any deviation of a solute-to-creatinine concentration ratio from normal is equated to a corresponding change in the clearance of this solute. Normalization can also be achieved on the basis of total urine concentration as reflected in its specific gravity, with the empirically derived mean value of 1.018 as a reference point. Because GFR undergoes diurnal variation, spot samples should be collected at the same time each day. It must be kept in mind that specific lesions cannot always be blamed for a change in fractional excretion. For instance, the fractional excretion of filtered sodium normally falls below 1%. If GFR becomes depressed for any reason, maintenance of normal sodium excretion in the face of a reduced filtered load necessitates an increased fractional excretion.

In selected cases, clearance values can be computed indirectly from analysis of serum. The solutes most commonly studied in this manner are creatinine and urea. Creatinine is synthesized and excreted at an approximately constant rate. It follows that its steady-state concentration in plasma will vary indirectly with GFR. The agreement between measured creatinine clearance and that predicted from plasma creatinine concentration shows considerable variation. The predicted creatinine clearance is based on the formula $C_{cr} = [(140 - \text{age in years})(\text{weight in kg})]/72 S_{cr}$, where C is in milliliters per minute and S in milligrams per deciliter (Cockroft and Gault, 1976), for men. For women, 85% of the value is chosen. Both may be expressed in terms of body weight or body surface area. Blood urea has been similarly used as an indicator of glomerular function, but variations in the rate of urea synthesis and tubular reabsorption render it unsuitable for quantitative purposes. Attempts have also been made to use plasma concentrations of beta$_2$-microglobulin to measure GFR; this is further considered in the next section. All these methods are of qualitative, rather than quantitative, use.

URINALYSIS

Toxic injury to the kidney can involve any of the tissue compartments of the urinary tract. Ultimately, the products of such injury—whether sloughed necrotic renal epithelial cells themselves or the products of the functional, inflammatory, and reparative responses that the injury provokes—find their way into the urine. Simple urinalysis remains perhaps the most useful screening test for the detection of renal injury in large populations. Changes in urinary excretion of numerous solutes have been suggested as markers of exposure and effect. Bowers et al. (1992), for instance, showed that urinary porphyrin can serve as a marker of mercury accumulation and nephrotoxicity. It is unlikely, however, that porphyrin analysis provides more information than some of the simpler tests routinely performed.

Total Urinary Solute Concentration

In almost all cases of nephrotoxic injury, an early and reliable marker is the failure of the kidney to excrete concentrated urine. The failure is caused by the inability of the damaged kidney to generate and maintain a high osmolar concentration in the renal papilla. The two routine tests of this capacity are the urinary specific gravity (SG) and osmolality.

Urinary SG correlates well with the osmolar concentration with some exceptions. Normally, urinary SG exceeds 1.020 after overnight water deprivation, and values below 1.015 indicate a concentration defect. Care must be taken to correct the determined values for abnormally high glucose or protein concentrations. SG can be determined directly with a hydrometer. Indirect, easy-to-use measurements are gaining acceptance. The most common of these is based on the change in ionization of a polyelectrolyte as a function of the ionic strength of urine; the resulting change in pH is noted with a dipstick.

Osmolality can be measured with vapor-pressure osmometry or freezing-point depression. In the United States, subjects ingesting a normal diet excrete urine of up to around 900-1,200 mOsmol/L. By comparison, the osmolality of plasma is 280 mOsmol/L. In chronic renal disease or after acute toxic injury, the ability to concentrate urine maximally is impaired; and with severe injury, the maximal urine osmolality approaches that of plasma.

Dipsticks

Dipsticks have found extensive use for evaluation of urinary pH and the presence of proteins, glucose, hemoglobin, and bile pigments.

- *pH*. This is not greatly affected by most renal diseases, and the ability to acidify urine maximally is preserved even in severe forms of renal dysfunction. However, some forms of renal injury are associated with defects in acid excretion. Such so-called tubular acidosis can be induced by some xenobiotics, by solvents and metals, and by obstruction of the urinary tract. Urine excreted in cases of lead, cadmium, and toluene toxicity might be alkaline. The samples should be collected after overnight water deprivation to avoid the alkaline pH produced by stimulation of gastric HCl production.
- *Proteinuria* (discussed in detail later). The colorimetric reagent strip of most dipsticks uses tetrabromphenol blue for the semiquantitative detection of protein. The test is more sensitive for albumin than for other urinary proteins, and its sensitivity is greatly diminished at high pH.
- *Glucosuria*. Like proteinuria, glucosuria generally accompanies acute renal injury after exposure to nephrotoxicants. The glucose oxidase reagent is highly specific for glucose.
- *Hematuria*. Presence of hemoglobin (or large amounts of myoglobin) is readily demonstrated. The presence of red blood cells in the urine is abnormal. It does not necessarily reflect renal lesions themselves, inasmuch as diseases in any portion of the urinary tract can lead to this finding.
- *Bile pigments*. Nephrotoxicants often damage liver and red-cell membranes, so bilirubin and urobilinogen are important indicators of liver injury. Because bilirubin is unstable and light-sensitive, samples should be stored in the dark.

Microscopic Analysis

Little information is usually gained from microscopic examination of urinary sediment if chemical analysis had negative results. In large screening studies, microscopic examination of urine should be reserved for urine samples whose chemical screens were positive. First morning specimens, which are highly concentrated, as opposed to those passed later in the day, are not useful for this purpose, because cells rapidly deteriorate in hypertonic bladder urine. Therefore, freshly voided samples should be used only after partial hydration, which will result in a lower urine specific gravity. Preservatives can be added to ensure the integrity of cellular elements. This preservation will be especially important when more sophisticated techniques—such as immunocytology, in situ hybridization, and polymerase chain reaction (PCR), are introduced to examine these specimens (see the next major section, on markers of cytotoxicity). Formaldehyde-releasing tablets are now available as a preservative, but all preservatives might interfere with routine tests. For example, false-positive results for protein and glucose have been observed after addition of the preservatives thymol and formalin, respectively.

Preparation of the sediment should be standardized. Most commonly, 15 ml of urine is spun at 2000 g for 5 min. The supernatant is partially decanted and the sediment resuspended in 1 ml of supernatant. In such a preparation, the normal concentration of red cells and white cells is 1-2 cells per high-power field. The presence of casts, which can be formed especially at high salt concentrations in acidic urine, reflects injury to the glomerular capillary wall (in the case of red-cell casts) or tubules; the presence of granular and hyaline casts probably depends on the combined effects of tubular epithelial damage, inflammatory reactions, and abnormal protein excretion and aggregation.

Study of urinary cells is discussed in greater detail in the next major section. Normal urine specimens can contain epithelial cells and cells of the pelvic, ureteral, bladder, and urethral epithelium that cannot be distinguished by routine light microscopy. Some inflammatory cells can also found. A recent study by Mandal (1986) failed to identify tubule cells in normal urine with electron microscopy. Under pathologic conditions, cell identification requires a variety of techniques to identify phenotypic markers. Granular casts, consisting of contents of disintegrated cells, and Tamm-Horsfall protein (see the next section) are commonly found in the urine from patients with acute renal failure (ARF), and acute or chronic tubulointerstitial disease. Muddy-brown casts, coarsely granular casts, red cells and casts, and white cells are all associated with ARF. Their presence correlates with the degree of damage, so mild cytotoxicity is accompanied by few urinary abnormalities. However, major renal impairment can be present in acute interstitial nephritis with little or no abnormality of the urinary sediment. Acute interstitial nephritis with prominent peripheral eosinophilia can lead to the presence of eosinophils in urine.

Proteinuria

Under normal circumstances, the glomerular capillary wall provides a barrier to the filtration of protein. Permeation of the glomerular basement membrane by proteins is influenced by the size, charge, and rigidity of individual protein molecules. Small uncharged molecules, such as horseradish peroxidase, pass freely into the urinary space. Negatively charged molecules of intermediate size, such as albumin, are impeded at the endothelial-cell surface and the proximal layers of the glomerular basement membrane. Uncharged or positively charged molecules, such as myeloperoxidase, permeate the glomerular basement membrane and are retarded at the slit diaphragm of epithelial cells. The net effect of those properties is to exclude relatively large, negatively charged, and rigid proteins from passage into the urinary space. Thus, although albumin and gamma globulin account for the preponderance of serum protein in normal humans (5.0 and 1.5 g/dL, respectively), less than 60 mg and 20 mg, respectively, is excreted in the urine daily. Serum proteins that weigh less than 20,000 daltons (low-molecular-weight proteins) and peptide hormones contained in plasma pass readily into the urinary space, but almost the entire filtered load is later reclaimed in the proximal tubule.

Measurement of Urinary Protein

The urinary dipstick is the most common device for semiquantitative measurement of urinary protein. Although it can detect protein concentrations as low as about 10-15 mg/dL, it is relatively insensitive and unreliable at concentrations below 30 mg/dL, yielding negative or trace-positive results in more than half such samples tested (Rennie and Keen, 1967). Although the test is sensitive to small quantities of negatively charged proteins, such as albumin, it is insensitive to positively charged proteins, such as some immunoglobulin light chains, and to Tamm-Horsfall protein. Low-molecular-weight proteins (which are less negatively charged than albumin), even when present in abnormal amounts, do not usually reach a concentration high enough to give a clearly positive test result.

Urinary protein can also be estimated with precipitation. Urinary protein is precipitated either by adding 5% sulfosalicylic acid, 10% trichloroacetic acid, or concentrated nitric acid to the urine or by heating the urine and adding 5% glacial acetic acid. Those methods detect positively charged proteins, as well as albumin, but they have the same limitations as the dipstick test in quantifying total protein; that is, small amounts of protein might not be detected in dilute urine, and normal amounts of protein might be detected in concentrated urine.

For mass screening purposes, precipitation tests offer little advantage over the dipstick method and are more difficult to perform. Neither dipstick tests nor precipitation tests are useful in testing for abnormal low-molecular-weight proteinuria.

Neither the dipstick methods now in use nor the usual qualitative tests for acid precipitation of protein are sensitive enough to function as reliable screening tests for the detection of borderline or low concentrations of proteinuria (200-500 mg/g of creatinine) in random urine samples. For example, the sample sensitivity of two dipstick methods was less than 50% in consecutively acquired specimens, and the negative predictive value, or the ability to establish the absence of proteinuria, was only 64-69% (Allen et al., 1991). Quantitative assessment of protein excretion with measurement of the protein-to-creatinine ratio has been recommended to supplement the dipstick in screening for proteinuria in cases in which misclassification would lead to serious problems (Shaw et al., 1983). Interpretation of borderline positive results for protein in screening examinations is further clouded by the fact that functional proteinuria—elicited by upright posture, by exercise, and even by excitement—is frequently encountered, particularly in adolescents and young adults (Houser, 1987; Houser et al., 1986). Woolhandler et al. (1989) thought that screening healthy, asymptomatic young adults for proteinuria was not helpful, in that population-based studies showed that less than 1.5% of patients with a positive result for protein on the dipstick test had a serious and treatable disorder of the urinary tract. However, the yield might well be much higher in selected populations, such as hospitalized patients, older subjects, or those exposed to environmental hazards.

Albuminuria

Albumin is a 69-kilodalton polyanionic macromolecule with a Stokes radius of 3.6 nm, and a pI of 4.8. In urine, it is considered a high-molecular-weight protein. Albumin is synthesized in the liver and has a relatively high concentration in the plasma (about 4 g/dL). Because this high concentration of albumin is maintained within the vascular space, it generates so-called "oncotic pressure", which prevents fluid from moving into the tissues and causing edema. Albumin is thought to exist almost entirely in monomeric form in biologic fluids. Some albumin is normally filtered by the glomeruli with low-molecular-weight proteins. Filtered protein is primarily reabsorbed in the proximal tubules by brush-border enzymes. Albumin has less affinity for the brush-border enzymes than does the low-molecular-weight protein $beta_2$-microglobulin. Absorption of $beta_2$-microglobulin averages 99.7%, but that of albumin only 90%. In normal people, the fractional excretion of albumin is about 90 times higher than that of $beta_2$-microglobulin (Peterson et al., 1969). Isolated albuminuria without increased excretion of low-molecular-weight proteins results from change in the glomerular filtration of high-molecular-weight proteins and thus has been termed *glomerular proteinuria*.

Various mechanisms have been proposed to explain increases in glomerular filtration of albumin. In advanced renal disease, when GFR is 25-30 mL/min, the remaining functional nephrons might have increased perfusion, which could disrupt the cellular integrity of capillary membranes and lead to increased albuminuria (Hostetter et al., 1982). Reduction in the fixed anionic charge of the glomerular filtration barrier enables greater leakage of serum albumin across the glomerular basement membrane. Nephrotic syndrome induced by puromycin aminonucleoside is associated with reduction of glomerular anionic sites (Michael et al., 1970). Conversely, if the serum protein itself is modified to be less negatively charged, it might result in increased filtration, as has been proposed for glycosylated albumin in diabetes (Ghiggeri et al., 1985). In general, then, the clinical finding of isolated urinary excretion of high-molecular-weight proteins, particularly albumin, is seen as a manifestation of disturbances of the glomerular filtration barrier.

Increased excretion of albumin is usually the result of glomerular injury but sometimes is the consequence of reversible hemodynamic changes. Exercise, fever, infusions of epinephrine or norepinephrine, emotional stress, prolonged assumption of the lordotic position, and congestive heart failure are often accompanied by mild to moderate albuminuria. Those stresses amplify the excretion of albumin when albuminuria is already present.

Proteinuria in excess of 1 g/g of creatinine, or 1 g/day in an adult, is almost always indicative of glomerular injury. Excretion of large amounts of abnormal, positively charged immunoglobulin light chains, as in multiple myeloma, constitutes the chief exception to this rule. Although, with the foregoing exception, albumin is always the predominant constituent, proteins of higher molecular weight are excreted in small amounts in proportion to the degree of injury to the glomerular barrier. Thus, in minimal-change childhood nephrosis, in which glomeruli appear normal with light microscopy, virtually no proteins larger than albumin appear in urine, but in inflammatory or infiltrative glomerular diseases, the proteinuria is "nonselective," so urine contains variable amounts of high-molecular-weight plasma proteins, such as immunoglobulin G.

Day-to-day variation of 35-50% in urinary albumin excretion is well documented (Giampetro and Clerico, 1990). Normalizing urinary albumin with creatinine (discussed earlier) can increase the sensitivity and specificity of the test, but it requires a second assay, which increases cost and imprecision. Besides the physiologic factors noted above, various disease states and a number of nephrotoxic agents can increase albuminuria as well.

Microalbuminuria

The standard detectability limits for screening albuminuria are far above the normal range; this decreases the rate of false-positive results and thereby approaches a specificity of 100% but substantially reduces the sensitivity (Steffes et al., 1989). The hallmark of overt diabetic nephropathy is dipstick-positive proteinuria, as seen with "Albustix" (Ames Company). However, by the time the specified albuminuria (250-300 mg/L of urine or 250-300 mg/24 hours) is present, the progression to end-stage diabetic renal failure might be inexorable. In the patient with Type I or insulin-dependent diabetes mellitus (IDDM), once persistent proteinuria of this degree develops, the progression to ESRD is predictable (Konen et al., 1990; Mathiesen et al., 1984). Before clinical (dipstick-positive) proteinuria appears, there is a period of increased albumin excretion not detectable by dipstick (Mogensen, 1987). The small amount of albumin excreted can be conveniently detected with radioimmune assays. This "microalbuminuria" is said by some to be a marker of the likelihood of diabetic nephropathy. However, its prognostic utility in individual cases is uncertain, in that many patients with diabetes and microalbuminuria do not develop dipstick-positive proteinuria even after many years. Microalbuminuria in the absence of hypertension and decreased creatinine clearance does not accurately elucidate the severity of the underlying glomerular lesion in patients with type I diabetes (Chavers et al., 1989).

There has been disagreement regarding the albumin excretion that constitutes microalbuminuria. In 1985, a consensus conference agreed to define microalbuminuria as a urinary excretion rate (UAER) of greater than 20 µg/min but no higher than to 200 µg/min in an overnight or 24-hour sample (Mogensen et al., 1985-1986). A UAER of 20-200 µg/min is approximately equivalent to 30-300 mg/24 hr or 3-30 mg/mmol creatinine.

Low-Molecular-Weight Proteinuria

Virtually all filtered low-molecular-weight proteins are reabsorbed in convoluted and straight portions of the proximal tubule. Brush-border receptors have a considerably higher affinity for low-molecular-weight proteins than for albumin (which is why the concentration of these proteins in urine is so low), and a high absorptive capacity exists in relation to the normal filtered load for all low-molecular-weight proteins that have been tested (Maack et al., 1985). Nevertheless, with an increase in filtered load, urinary excretion generally increases well before the tubular maximum for absorption is reached (Waldmann et al., 1972).

Because low-molecular-weight proteins are reabsorbed solely and virtually completely in the proximal tubule, their appearance in urine in increased amounts is generally taken to indicate a reduction in proximal tubular function and an early sign of proximal tubular damage. But low-molecular-weight proteinuria can have other causes that lead to increases in the plasma concentration and glomerular filtration of the proteins. An increase in beta$_2$-microglobulin occurs in lymphoproliferative malignancy. An increase in the plasma concentration can also be secondary to renal insufficiency; functional competition for absorption at brush-border sites by charged molecules, such as lysine, arginine, and aminoglycosides; and nonrenal febrile diseases.

One of the important low-molecular-weight proteins in urine is beta$_2$-microglobulin, whose molecular weight is 11 800. For accurate assay results, urine must be made alkaline to prevent hydrolysis of the protein. The protein was isolated in 1968 by Beggard and Bearn from the urine of patients with two conditions characterized primarily by proximal tubular damage: Wilson's disease and chronic cadmium poisoning (Beggard and Bearn, 1968). Urinary beta$_2$-microglobulin has been used often to detect proximal tubular injury in clinical and experimental settings and to test for toxic effects of environmental exposures and antibiotic and chemotherapeutic agents.

Normal adults produce 150-200 mg of beta$_2$-microglobulin a day. It is cleared from the plasma almost completely by the kidney with a half-life of about 2 hours in persons with normal renal function. Serum concentration averages 2.0 mg/ml (range, 1.1-2.7 mg/ml), and the normal urinary excretion is less than 370 µg per day (Schardijn and Statius van Eps, 1987). An increase in beta$_2$-microglobulin production occurs whenever nucleated-cell turnover increases and results, even in the absence of renal malfunction, in raised serum concentration and urinary excretion. When serum beta$_2$-microglobulin exceeds 5 mg/ml, its urinary excretion is invariably high, which means that the threshold for maximal tubular reabsorption has been exceeded (Bernard et al., 1988).

The same receptor on the brush-border membrane is likely to be involved in the tubular reabsorption of beta$_2$-microglobulin and of albumin; however, the brush-border membrane has far less affinity for albumin than for beta$_2$-microglobulin and other low-molecular-weight proteins. Proximal tubular injury, therefore, results in a disproportionate increase in the excretion of low-molecular-weight proteins like beta$_2$-microglobulin, compared with that of albumin. In contrast, an increase in albuminuria without concomitant increase in urinary excretion of low-molecular-weight proteins signifies an increase in the glomerular filtration of albumin.

The urinary excretion of beta$_2$-microglobulin varies throughout the day in a circadian rhythm, the maximum occurring around midday to early afternoon and the minimum around 4 a.m. (Koopman et al., 1987). This day-night variability must be taken into account when single, spot samples of urine are measured in clinical or epidemiologic surveys.

The normal range of excretion of beta$_2$-microglobulin by adults can be expressed as micrograms of beta$_2$-microglobulin per milligram of creatinine. In adults, the ratio of beta$_2$-microglobulin clearance to creatinine clearance does not change with age (Evrin and Wibell, 1972). Urinary excretion of beta$_2$-microglobulin does not depend on body mass, sex, water load, or moderate physical activity, but it has been shown to increase with furosemide-induced diuresis (Guarnieri et al, 1979; Poortmans and Jeanloz, 1976; Wibel and Karlsson, 1976).

Because most heavy metals accumulate in the kidney, especially in the proximal tubules, beta$_2$-microglobulin excretion has been studied extensively as an early indicator of renal toxicity due to industrial exposure. Cadmium is a common pollutant in industrialized countries. Frank (glomerular) proteinuria has been recognized as a sign of renal dysfunction associated with chronic cadmium toxicity since 1948 (Friberg, 1948), but the earliest sign of cadmium toxicity is tubular proteinuria. The incidence and degree of beta$_2$-microglobulinuria increase with the duration and intensity of exposure to airborne cadmium and with increasing urinary cadmium. Cadmium workers can have beta$_2$-microglobulin excretion 100-1,000 times higher than normal people who have not been exposed. Inhabitants of areas exposed to cadmium in Sweden, Japan, and the United Kingdom had an increase in urinary excretion of beta$_2$-microglobulin related to duration of exposure, dietary cadmium, and urinary cadmium content (Elinder et al., 1985).

Another low-molecular-weight protein excreted in urine is retinol-binding protein (molecular weight, 21400). It is more stable than beta$_2$-microglobulin at acidic pH in urine, and its excretion has therefore been suggested as a practical and reliable test of tubular protein reabsorption.

Finally, alpha$_1$-microglobulin has been proposed for evaluation of tubular function (Guder and Hofmann, 1991). In healthy adults, alpha$_1$-microglobulin is present in the urine at approximately 20-100 times the normal concentration of beta$_2$-microglobulin and of retinol-binding protein; this circumstance might facilitate its measurement. Like albumin excretion, its excretion increases after exercise, so exercise proteinuria might be tubular, as well as glomerular (Murakami and Kawakami, 1990). The diurnal variation in alpha$_1$-microglobulin excretion is similar to that of beta$_2$-microglobulin, but the changes are less marked than those of albumin. Excretion is usually lower at night than during the daytime.

The usefulness of alpha$_1$-microglobulin was compared with that of urinary N-acetyl-beta-D-glucosaminidase (NAG) and albumin in 409 urine samples sent to a clinical chemistry laboratory for screening. The results were compared with the usual test-strip procedure (for pH, protein, leukocytes, blood, and glucose). Two-thirds of the samples had an abnormal result in at least one test. In this series, alpha$_1$-microglobulin excretion was rarely increased without a simultaneous increase in urinary NAG activity (Hofmann and Guder, 1980).

Alpha$_1$-microglobulin excretion might well be useful in monitoring early signs of renal toxicity after industrial exposures. Its ultimate utility in this field remains to be established.

Brush-Border Proteins

An early event in most forms of ischemic or toxic injury to the kidney is sloughing of the brush border of proximal tubules, particularly in the straight portion (S3) in the outer medulla and in the medullary rays of the cortex (Cheung and Swaminathan, 1989). Appearance of specific brush-border proteins in the urine might therefore be a sensitive early test for renal damage, and monoclonal antibodies that react specifically with such proteins have been developed (Birk et al., 1991). Examples of such tests include those for adenosine deaminase-binding protein (Birk et al., 1991), Na$^+$-D-glucose cotransporter protein (Birk et al., 1991), intravillus brush-border proteins (Birk et al., 1991), and the intestinal type of alkaline phosphatase (Verpooten et al., 1989). Those techniques, as yet experimental and therefore expensive, have great promise because of their inherent specificity and sensitivity. They have yet to be applied to large populations in screening tests, and their practicality in comparison with other measures of proximal tubular damage, such as enzyme excretion, remains to be determined.

Tamm-Horsfall Protein

The so-called Tamm-Horsfall protein (THP) has been found to be the most abundant protein in normal human urine (Lynn et al., 1982) and the predominant protein in hyaline casts (Fletcher et al., 1970; McQueen, 1962; Rutecki et al., 1971). In humans, the daily urinary excretion of THP ranges from 20 to 200 mg with an average of about 2 mg/hr, or about 50 mg/day. THP Is synthesized and secreted into the urine by cells that line the thick ascending limb of Henle's loop. This "urinary mucoprotein" was first described by Morner in 1895, but its function is still not clear (Kumar and Muchmore, 1990). Tamm and Horsfall described in 1950 the salt precipitation of a normal human urinary glycoprotein that inhibited myxovirus-induced hemagglutination (Tamm and Horsfall, 1950). The more recently described "uromodulin" (Muchmore and Decker, 1985) is identical in amino acid sequence with THP and has immunosuppressive properties.

THP is an 85-kD, 616-amino-acid mature protein with a 24-amino-acid leader sequence (Fletcher et al., 1970; Hession et al., 1987; Kumar and Muchmore, 1990). It is homologous with the human low-density lipoprotein receptor, epidermal growth factor (EGF), and GP-2, the major component of zymogen granule membranes of the exocrine pancreas (Hoope and Rindler, 1991). It is rich in cysteine and has numerous intrachain disulfide bonds. THP is about 30% carbohydrate by weight and about 1% lipid (Fletcher et al., 1970). The predominant carbohydrate-protein bonds are N (as opposed to O) linkages (Fletcher et al., 1970; Muchmore and Decker, 1987; Muchmore et al., 1987). Hydrophobicity plotting shows that THP does not have a classic transmembrane hydrophobic region; it appears to attach to the lipid bilayer via phosphatidyl inositol. It is likely to be released from the membrane by proteolytic or phospholipolytic cleavage (Rindler et al., 1990). THP has a low isoelectric point and has a tendency to form gels in aggregates of several million daltons. Gel formation is increased by the presence of albumin, acidic pH, high THP concentration, sodium and calcium ions, and possibly radiocontrast media (Kumar and Muchmore, 1990). Methods for determining THP in human urine include salt precipitation (as first noted by Tamm and Horsfall), radioimmunoassay, radial immunodiffusion, electroimmunoassay, and lectin adherence.

THP has been found in the kidneys of numerous placental mammals but not marsupials or reptiles (Wallace and Nairn, 1971). The rate of THP excretion can differ in a given individual over the course of a day by as much as a factor of 18 (Hession et al., 1987). The pattern of excretion varies from person to person, with no obvious circadian rhythm. Eating and recumbency do not appear to influence THP excretion. Radioactive-tracer experiments indicate that THP is formed de novo in the renal tubules, rather than transported there; it has been identified in urine produced by isolated perfused kidneys (Cornelius et al., 1965). Most studies have localized THP in humans to the ascending part of Henle's loop, the early distal convoluted tubule, and tubular casts (Hoyer et al., 1974; Lewis et al., 1972; Pollak and Arbel, 1969; Schenk et al., 1971; Wallace and Nairn, 1971). There is conflicting evidence as to its presence in macula densa cells (Hoyer and Seiler, 1979; Sikri et al., 1981). In normal humans, THP is characteristically absent from glomeruli, proximal tubules, and the interstitium. It appears to be produced by the aforementioned tubular cells, transported to the tubular lumen, and, as noted above, attached to the lipid bilayer via phosphatidyl inositol. The daily excretion of THP is about equal to its content in the kidneys, so the total amount of THP must be synthesized de novo each day (Schoel and Pfleiderer, 1987).

Localization of THP to the distal tubule and the gel-forming tendency of THP have led to speculation that it is partly responsible for the impermeability of the thick ascending limb to water (Kumar and Muchmore, 1990). In patients with chronic renal failure, THP excretion is diminished. Large amounts of the protein might be present in hyaline casts; this could be related to its distal secretion into relatively acidic tubular fluid. Changes in renal THP have also been extensively found in patients with glomerular nephritis, the nephrotic syndrome, and in renal-allograft recipients.

Although THP is the major protein in normal human urine, measures of its excretion have unknown utility as indexes of renal function or markers of renal injury.

MARKERS OF CYTOTOXICITY AND CELLULAR RESPONSE

Toxic agents have been implicated in the etiology of diseases in every major histopathologic category of glomerular and tubulointerstitial nephritis. For that reason, any marker of renal-cell damage or inflammation must be considered a possible indication of toxic insult. Toxic injury can affect the cells in any major compartment: glomeruli, tubules, interstitium, and vessels. A single agent can damage cells in more than one compartment. Furthermore, the structural and functional interdependence of cells in different compartments can complicate or confound the identification of primary targets of injury. Injurious effects can be acute, chronic, or both. Both acute and chronic damage can result from direct toxicity and from immunologic responses (allergic, autoimmune, and inflammatory) triggered by a toxicant. In addition, nonimmunologic factors, such as hemodynamic stress, can contribute to cell damage after exposure to toxicants. Furthermore, the cellular sites of injury can be different for acute and chronic effects of a single toxicant and can change during the course of a chronic disease, especially if immunologic factors become important.

Mechanisms of injury can be evaluated in vitro with tissue cultures, isolated kidney or nephron components, and cell-free extracts, but it is absolutely essential that nephrotoxicity be studied in vivo to elucidate the cell interactions and systemic factors that can trigger the response to injury or perpetuate a chronic disease. Studies of mercury nephropathy in rats, especially by the group of Druet (Dubey et al., 1991; Pelletier et al., 1987; Pelletier et al., 1988), documented this well. The rat model of mercury exposure is a paradigm of the potentially complex natural history of toxic injury to the kidney. The complex, long-term effects of exposure to such drugs as adriamycin, cyclosporin, and puromycin have also been studied successfully in animal models by several laboratories. Those investigations provide valuable insights into the possible relationships between acute and chronic effects of nephrotoxicants, and might provide important clues to cell-specific markers of renal injury.

Abnormal changes within the kidney can involve endogenous cells, marrow-derived cells, and connective-tissue matrix. The endogenous cells of the kidney that can be affected by toxicants include those of glomeruli, tubules, interstitium, and vessels. Despite the microanatomic and functional heterogeneity of the cells those compartments, responses to injury can be divided into a few major categories: hypertrophy, proliferation, degeneration (detachment, vacuolization, lysis, necrosis, and apoptosis), altered metabolism (synthesis, secretion, and uptake), surface-membrane remodeling (changes in shape, charge, antigen, and receptor expression), regeneration, inclusion formation, alterations in cytoskeleton or organelle composition, and activity.

A substantial population of marrow-derived cells can be found in the glomeruli, interstitium, and perivascular tissue of normal kidneys. In pathologic states, those blood-borne cells can increase in number through infiltration and local proliferation, differentiate into functionally or phenotypically distinct cells, and become metabolically activated.

Abnormal functions of endogenous cells and marrow-derived cells can lead to alterations in the extracellular matrix of the kidney. The direct action of a toxicant and the accumulation of immune reactants, coagulation factors, etc., can also cause pathologic changes that affect the structure of the matrix (thickening, thinning, splitting, duplication, and disruption), its biochemical composition (inclusions and abnormal constituents), and its physical properties (charge and permeability).

Urinary Cells

Conventional light-microscopic examination of cells in urinary sediments does not permit precise distinctions between types of epithelial cells and leukocytes. Therefore, the cellular composition of urinary sediment has not proved useful as a marker of renal injury. That is unfortunate, in that accurate analysis of urinary cells is a potentially powerful tool for identifying the site, severity, and mechanism of renal damage.

Some recent innovations hold promise, but further technologic advances and clinical studies will be required before they can be applied in routine screening of urine samples. The use of monoclonal antibodies and immunohistochemical staining should make it possible to distinguish among epithelial cells arising from individual nephron segments or structures of the lower urinary tract. Similarly, cell-specific antibodies should allow precise and detailed characterization of macrophages and lymphocytes that reach the urine. The cytocentrifuge should prove valuable for preparing cells for both conventional and histochemical staining (Schumann, 1986). Even more important might be the development of techniques for preparing urinary cells that were developed in the course of successful studies that used urinary cells in the diagnosis and management of bladder cancer. These techniques, when coupled with fast image-analysis systems, offer the potential for automated analysis to detect cells from specific regions of the kidney. Transmission electron microscopy could also aid in the identification of epithelial cells (Mandal, 1986; Mandal and Bennett, 1988). Methods of tissue culture might provide clinically useful information about the metabolism of cells isolated from urine (Racusen et al., 1991). Techniques of in situ hybridization have been widely disseminated in recent years for analyzing changes in gene expression that accompany tissue damage. Those methods of molecular biology have been applied successfully to the investigation of mechanisms of renal injury in tissue samples and should prove similarly useful in detecting abnormal gene expression or synthetic capacity among cells in urinary sediments.

Hematuria is an important and often early marker of renal injury. Analysis of the red cells in urine (number, structure, and cast formation) can sometimes provide clues about the nature and site of the lesions (Kohler et al., 1991; Pollock et al., 1989). However, despite persistent attempts to identify criteria that might increase the diagnostic value of red cells in urine, their presence remains a relatively nonspecific indication of injury.

Markers of Injury or Inflammation in Renal-Tissue Samples

With the widespread use of the kidney biopsy for diagnosis, a systematic approach to the evaluation of renal-tissue samples has been developed that facilitates the detection of changes in kidney architecture. Histologic abnormalities can be recognized and analyzed with considerable sophistication, rigor, and reliability. Conventional methods of light and electron microscopy permit ready identification of deviations from the normal appearance, configuration, and number of endogenous cells. Immunofluorescence and immunohistochemical staining procedures, which constitute a powerful adjunct for the microscopic examination of renal tissue, are now used routinely to identify antibodies, complement components, coagulation factors, mononuclear cells, and basement-membrane constituents. With the recent rapid dissemination of highly specific monoclonal antibodies and genetic probes, it is becoming possible to obtain detailed information on the metabolism of renal cells from appropriate studies of tissue specimens.

Conventional methods of histology have been especially useful in the identification and classification of glomerular damage. A major limitation has been that early (preclinical) stages of human glomerular damage have not been available for study. Histologic aspects of tubular injury have not been as comprehensively or formally described. That deficiency is especially unfortunate, inasmuch as tubular cells are most commonly the primary site of toxic injury. Optimal preservation of tubular cell structure requires techniques of perfusion fixation that can be applied in vivo to animals only. As a result, mild damage to the luminal membrane, slight alterations in tubular diameter, and other subtle features of tubular injury can be difficult to assess or even be completely overlooked in human tissue samples. Furthermore, samples obtained by biopsy often fail to provide adequate representation of the highly specialized segments of the tubular system. Finally, the interstitial cells that lie near the tubular cells remain mysterious. Until the microanatomic heterogeneity, functional specialization, and developmental origins of normal interstitial cells are better understood, it will be difficult to identify subtle deviations from normal conditions. Inflammation of the interstitium is usually easy to recognize, and immunohistochemical staining methods have made the mononuclear cell composition of interstitial infiltrates amenable to sophisticated analysis.

Markers of renal-cell injury have been studied more analytically in animals than in humans (see Table 4-1[1]). The animal models cover a wide array of pathogenic mechanisms, including immune complex disease, specific antibody-mediated injury, heavy-metal exposure, drug toxicity, ischemia, surgical reduction of renal mass, hypertension, diabetes, and systemic autoimmunity. In addition to methods of microscopy, cell abnormalities associated with renal injury in animal models have been evaluated with tissue culture, isolated whole-kidney perfusion, tubular microperfusion, micropuncture, cell fractionation and extraction, and the study of kidney-slice preparations and isolated membrane vesicles. The biologic markers that have been linked to cell injury on the basis of studies of renal tissue can be divided, for the purposes of review, into several categories:

- Immunologic factors:
  -- Humoral—antibodies and antibody fragments; components of complement cascade, and coagulation factors.
  -- Cellular—lymphocytes, mononuclear phagocytes, and other marrow-derived effectors (eosinophils, basophils, neutrophils, and platelets).
- Lymphokines.
- Major histocompatibility antigens.
- Growth factors and cytokines: platelet-derived growth factor, epidermal growth factor, transforming growth factor (TGF), tumor-necrosis factor, interleukin-1, etc.
- Lipid mediators: prostaglandins, thromboxanes, leukotrienes, and platelet activating factor.
- Extracellular-matrix components:
  -- Collagens.

---

[1] All tables referenced in this chapter can be found at the end of the chapter.

Biologic Markers of Effect

-- Procollagen.
-- Laminin.
-- Fibronectin.
- Adhesion molecules.
- Reactive oxygen and nitrogen species.
- Transcription factors and proto-oncogenes: c-*myc*, c-*fos*, c-*jun*, c-Ha-*ras*, c-Ki-*ras*, and *Egr*-1.
- Tubule antigens, Tamm-Horsfall protein, brush-border molecules, and cystatin.
- Heat shock proteins.
- Endothelin.

Some methods permit precise identification of cell sites of response to injury. Proliferation, evaluated microscopically according to the uptake of appropriately labeled nucleic acid precursors, has been localized in individual cells or histologic compartments of the kidney in various animal models. Special tracers and staining methods have been used to recognize and localize alterations in charge distribution of cell and basement membranes. Theoretically, with appropriate probes (e.g., monoclonal antibodies and genetic probes) it should be possible to identify with high precision the cell sources of synthesis or the sites of accumulation of most of the molecules that reflect metabolic alterations produced by injury. In practice, however, relatively few markers have been unequivocally linked to cell- or site-specific injury. Technical problems are considerable. For example, it has been very difficult to distinguish the secretory activities of mesangial cells of the glomerulus from those of marrow-derived cells that are near and might be numerous in hypercellular glomeruli. In addition, metabolic responses to injury might be similar in cells in different compartments of the kidney and in exogenous and endogenous cells.

Potential sources of biologic markers of cell injury or inflammation in renal tissue samples are listed in Table 4-2. It can be seen, for instance, that cytokine production has been attributed to mesangial cells, macrophages, and tubular cells; the expression of major-histocompatibility-complex class II antigens is also a potential of all three cell types. A wide variety of hormone-like proteins and biologic response modifiers can be produced in the kidney in response to cell damage and tissue inflammation. Although marrow-derived mononuclear cells are a major and well-studied source of those molecules, many can also be produced by cells intrinsic to the kidney. The table summarizes a large and rapidly growing literature on possible intrarenal sources of some biologic markers of cell injury that might have clinical value. Glomerular sites of synthesis of individual molecules have been relatively easy to identify, because glomeruli are readily separated from other renal components for extraction, tissue culture, and other analytic procedures. Tubular, interstitial, and vascular cells have been less amenable to detailed investigation. Recent advances in tissue-culture methods, however, hold promise of more complete characterization of the metabolic and synthetic capacity of specialized renal cells outside the glomerulus.

Markers of Renal-Tissue Injury in Urine

Some of the potential markers of cell injury identified through studies of renal tissue have been detected in urine, but many of the findings have been isolated, inconclusive, or fragmentary. Urinary concentrations of some cytokines, arachidonic acid metabolites, adhesion molecules, growth factors, and connective-tissue matrix components have been measured in animal models and human diseases (see Table 4-1). That urinary concentrations of the markers are usually correlated with at least one histopathologic or pathophysiologic index of disease severity indicates the potential value of urine assays in detecting or monitoring for nephrotoxicity.

Results of studies with EGF suggest that its urinary excretion can be especially useful as a marker of effect (see Chapter 6). The origin of urinary EGF seems to be the kidney. That renal clearance of EGF exceeds that of inulin in rats and mice indicates renal secretion (Nielson et al. 1989). Removal of other sources of EGF does not affect its renal excretion, and removal of one kidney reduces urinary EGF excretion by half (Olsen et al., 1984). With the finding of its mRNA in the kidney, results of those studies strongly suggest that urinary EGF is derived from the kidney. Several recent studies in experimental renal failure and human renal disease demonstrate that both the renal production and the renal excretion of EGF fall after renal injury. Cisplatin (Safirstein et al., 1989) and aminoglycoside nephrotoxicity (Verstrepen et al., 1991) reduce renal production of the mRNA associated with the EGF precursor molecule prepro-EGF and reduce EGF excretion. Ischemia (Safirstein et al., 1990) markedly reduces prepro-EGF mRNA and excretion, as does unilateral or bilateral ureteral obstruction (Storch et al., 1992). In those studies, the decline in EGF excretion correlated very closely with the degree of renal impairment. The renal excretion of EGF falls in human diabetic renal disease (Mathiesen et al., 1988) and correlates well with the degree of renal damage. Urinary EGF excretion is low in patients soon after receipt of a renal allograft (Kvist and Nexo, 1989) and rises as renal function is restored. In experimental models of polycystic renal disease, prepro-EGF mRNA is reduced (Gattone et al., 1990). The assay of urinary EGF is simple, and urine samples can be collected under routine conditions and stored under refrigeration for long periods.

Urinary Enzymes

The modern era of using urinary enzymes in the investigation and diagnosis of renal injury or disease was initiated by Rosalki and Wilkinson, who reported increased lactate dehydrogenase (LDH) activity in the urine of patients with renal disease (Rosalki and Wilkinson, 1959). Acceptance of urinary enzymes, in contrast with serum enzymes, as a diagnostic tool has been slow; in his sentinel review of the application of urinary enzymes in evaluating both nephrotoxicity and renal disease, Price (1982) concluded that "the principal reason for the slow development of this field [urinary enzymes] is the difficulty involved in the assay of enzymes in a fluid which varies in volume, composition, and which is a hostile environment for many enzymes." In the 1990s, the difficulty might be rephrased: the major barriers to widespread acceptance of urinary enzymes as diagnostic markers include uncertainty about the exact location of the enzyme in the nephron and about how it reaches the urine. Questions also remain about pathologic correlates of urinary enzyme activity, its relationship with exposure dose and percentage of tissue destruction, and its therapeutic or prognostic importance.

Dubach, who has contributed much to the understanding of the diagnostic application of urinary enzymes, has concluded that "empiricism" dominates the field and that only by focusing on the definition of the cellular location of the enzyme can this application gain respect (Dubach and LeHir, 1984). This focus and the characterization of the mechanisms by which enzymes gain access to urine, are the principal tasks of the basic investigator. For the clinical investigator, the problem is "clarifying pathophysiological mechanisms for increased excretion of urinary enzymes." Dubach and LeHir (1984) discourage the use of urinary enzymes in screening for renal disease, because drugs, diagnostic procedures, and co-existing systemic diseases (e.g., sepsis and hyperthyroidism) and myocardial infarction can markedly influence urinary enzyme activity. However, enzymuria has proved useful in screening for selected circumstances of occupational exposure and should be used to evaluate new drugs or procedures with potential renal effects. A recent interest has been the early diagnosis of renal-transplant rejection and the differentiation of upper from lower urinary tract infections.

Theoretical and Diagnostic Importance

The theoretical basis for recommending the use of urinary enzymes in assessing renal injury is well reasoned. Enzymes are uniquely distributed along the course of the nephron; at least 10 separate segments have been defined (Guder and Ross, 1984). Furthermore, enzyme location in renal cells is restricted to specific subcellular components, thus providing further detail about injury site and potential mechanism. Access of plasma enzymes to the glomerular filtrate is limited by the permeability of the glomerular membrane.

In discussing the assessment of renal injury with urinary enzymes, Plummer et al. (1985) summarized the various possibilities that might explain how enzymes reach the urine. In addition to their low rate of filtration, the processes include the normal shedding from tubular cell surfaces and the release of enzymes that occurs with cell injury or death. Other cells in urine might contribute (i.e., bacteria, red cells, white cells, and lymphocytes), as might other components of the genitourinary system (i.e., bladder, glandular secretions, semen, and tumors). Under experimental conditions, contamination from food, feces, etc. might also contribute enzymatic activity to urine.

Well over 100 urinary enzymes have been investigated for diagnostic use in various states of renal injury, although only a few have gained notable acceptance for routine clinical use. To be clinically useful, the baseline excretion rates of a urinary enzyme must be low and allow recognition of an increase without excess background noise. Recent advances in technology have improved both detection and the reproducibility of measurement, including the application of automation for processing large numbers of samples, such as might arise from field screening projects. Substantiation of the theoretical basis of urinary enzymes as a renal diagnostic tool has not come easily. To understand the unresolved problems, it is best to start by defining the criteria by which urinary enzymes should be judged if they are to be accepted as diagnostic markers.

Recently, Guder and Hofmann (1991) stipulated the following criteria for the diagnostic use of tubular enzymes or antigens. They divided the items to be evaluated into technical and biologic groups. The following five items were to be considered in judging the technical aspects of specific urinary enzymes as diagnostic tools: precision, standardization (accuracy), interferences, technical performance (automation), and cost. Six biologic items were to be considered: origin (in or outside the nephron), intracellular location, mechanisms of release into urine, stability in urine at 37°C, sampling conditions, and the influence of other factors (diet, blood pressure, and biologic variation).

The two major applications of urinary enzymes are monitoring for subtle renal dysfunction and clarifying mechanisms of nephrotoxicity. Only a few enzymes have been generally accepted as valuable urinary markers, including NAG, alanine aminopeptidase (AAP), and LDH. Others, such as intestinal alkaline phosphatase, are emerging, but they lack the wide application and reporting enjoyed by the first three. NAG has now been defined to have its highest activity in the straight (S3) location of the proximal tubule of humans, with less activity in the collecting-duct portion of the distal nephron (Schmid et al., 1986). That observation, coupled with the refinement of colorimetric assays, makes it one of the most useful and best studied of the diagnostic urinary enzymes. Both enzymes are stable in the frozen state. For acute changes—such as those induced by drugs, transplant rejection, and acute renal injury—the enzymes offer an excellent diagnostic aid, although baseline values are often required to ensure proper interpretation of the findings.

Disadvantages as Markers

Urine is readily available and convenient to sample, but it constitutes a harsh environment for most enzymes because its low pH, slight buffering capacity, and the presence of many inorganic and organic compounds, which can act as either inhibitors or activators. To circumvent those problems, urine is stored at 4°C until collection is complete, and it is dialyzed or diluted before analysis whenever possible.

Unless a convenient reference standard, such as creatinine, is available, accurate timing and volume measurements of urine specimens for analysis are essential. The debate continues about whether the proper method for expressing urinary enzyme excretion is concentration in units per liter, with or without correction for creatinine, or rate of excretion in units per unit time. Because of diurnal variation, especially for NAG and AAP, the use of a concentration expression demands that similar collection intervals during the day be used if comparisons are to be useful.

Another source of difficulty in the use of urinary enzymes arises from lack of information on several factors: What is the influence of aging on the excretion of urinary enzymes? Does it follow the same pattern as creatinine? Does sex have any influence? Another deficiency of urinary enzyme studies is the absence of pathophysiologic correlations. In addition, the overlap between normal ranges and values recorded in patients with stable chronic renal disease provides poor specificity and sensitivity for the diagnosis of renal disease. Finally, identification of location in the nephron has been lacking, although this is changing with the advent of specific isoenzymes such as intestinal alkaline phosphatase (Verpooten et al., 1989).

MARKERS OF NEOPLASIA

General Considerations

By the time many cancers are detected on the basis of symptoms, it is already too late for effective intervention. Although conventional therapy might prolong life, most patients with metastatic disease eventually die of it. It is clear, therefore, that reducing cancer death rates will depend on prevention. Elimination of exposure can be effective, but it is of no use for those already exposed; such an approach cannot achieve significant cancer reduction except over long periods. Cancer usually has a roughly 20-year latent period in humans, and recent findings strongly suggest that carcinogenesis can be detected through biologic markers and inhibited by treatment with chemopreventive agents. Markers of effect play a crucial role in this process by identifying people who have demonstrable intermediate end points short of clinical disease. Detecting people in a more easily treatable preclinical phase shifts the emphasis from treatment to prevention.

Modern research is identifying hundreds of potential markers. If a marker is to be useful in the clinical setting, its accuracy must be sufficient to allow treatment decisions. The "gold standard" is the 5-year prospective study, but when the additional costs of developing a laboratory test with quality control suitable for widespread implementation are included, it is clear that only a very small number of markers can be subjected to the final test of the prospective study. Strategies will need to be developed that can quickly identify markers worthy of clinical study.

Proliferation

A hallmark of cancer initiation, promotion and progression is an increase in cell proliferation. Increased cell proliferation can promote the carcinogenic process itself by increasing the likelihood of inheritable changes that promote carcinogenesis. That is, cells are more likely to mutate during cell division and acquire invasive and metastatic potential. The frequency of mitotic figures and the presence of abnormal mitotic figures are good indicators of malignancy and are used in pathologic grading of tumors. But an increase in the number of proliferating cells is not always diagnostic of malignancy because many benign conditions—including hyperplasia, repair, inflammation, and stones—can also increase cell division. In addition, although tumor size and progression are correlated, bulky tumors do not necessarily invade and metastasize. For instance, many benign prostatic glands are larger than primary prostatic cancers, and many low-volume, high-grade prostatic cancers metastasize (Norming et al., 1989, 1992; Wheeless et al., 1991, 1993).

Tumors commonly have abnormal cell divisions that result in abnormal numbers of chromosomes, and therefore abnormal amounts of DNA, but it is difficult to estimate these with precision in hyperdiploid tumors when there are overlapping populations of cells. The hyperdiploid fraction is the most reliable predictor of bladder-cancer behavior and is accurate for determining proliferating cells in diploid tumors (Wheeless et al., 1991, 1993). The biologic characteristics of the tumor are not the only source of error in the determination of the fraction of cells in the synthetic phase of the cell cycle. Others include individual laboratory accuracy, variation in mathematical modeling programs, and percentage of apoptotic and necrotic cells that can influence dye binding. Tritiated-thymidine uptake assays that use autoradiographic techniques (Steel, 1977) to determine DNA synthesis are generally considered the "gold standard" for cell proliferation, and assays that use 5-bromodeoxyuridine incorporation are the next best (van Weerden et al., 1993; Waldman et al., 1993), but both are too complex for routine clinical studies. More recently, in vivo labeling has been used with iododeoxyuridine; this reflects the biology of invasion and metastasis more accurately.

Figure 4-1 depicts some of the most common proliferation-associated antigens and markers used to estimate proliferative rates and dissect the cell cycle. Until recently, proliferating-cell nuclear antigen, PCNA (Galand and Degraef, 1989; Raska et al., 1989; Thaete et al, 1989; van Dierendonck et al., 1991;

Biologic Markers of Effect

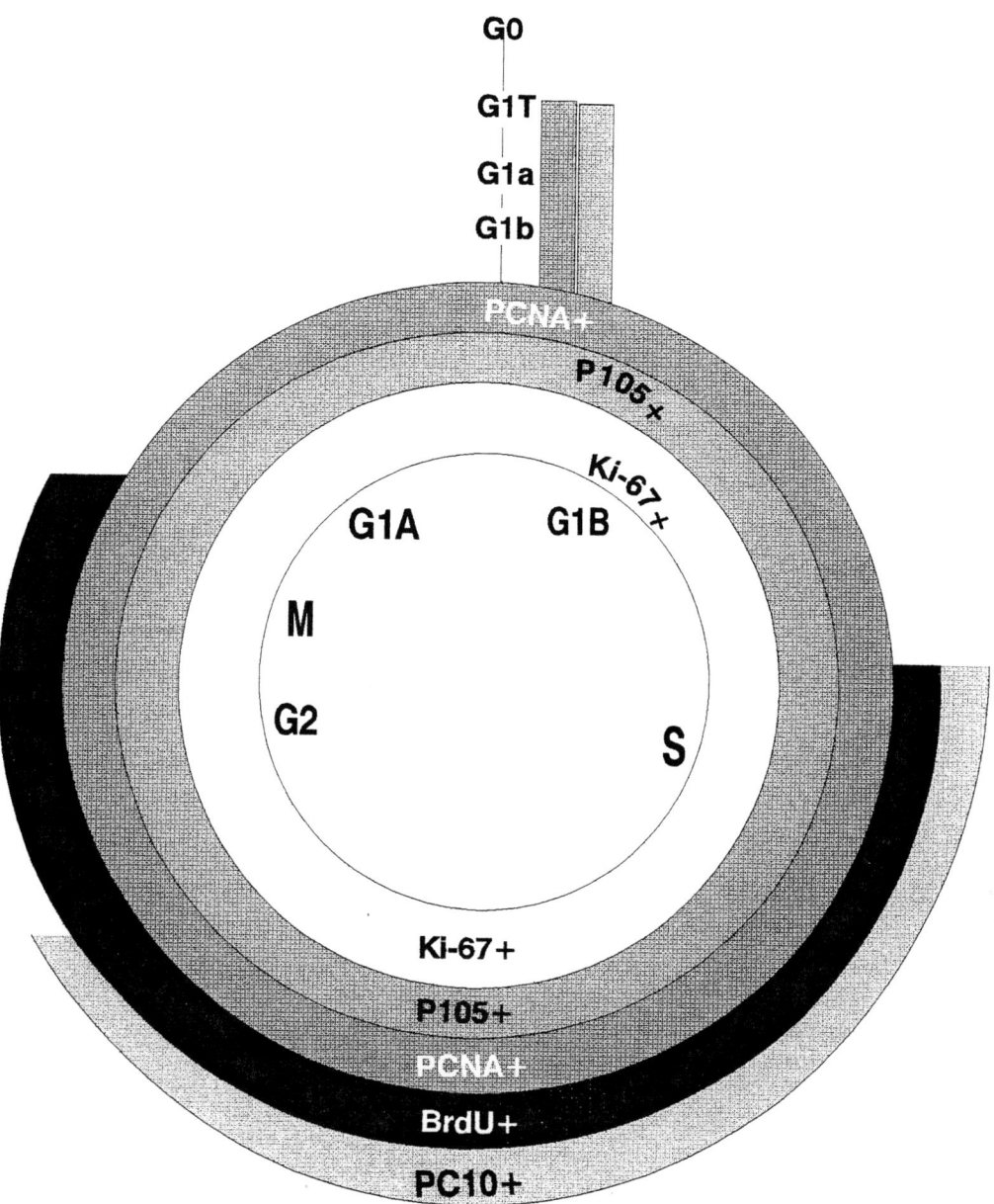

FIGURE 4-1. Markers of the cell cycle. Superimposed on the diagram of the cell cycle are indications of where different proliferation markers are positive. PCNA and p105 are positive in all cells that are not in G0, including the G1 transition phase. Ki-67 is positive in all but the transition area; BrdU and PC10 are positive during much more restricted portions of the cycle. Use of different markers can lead to different estimates of the fraction of proliferating cells.

Waldman et al., 1993) and p105 (Clevenger et al., 1987a,b) were the only markers useful for retrospective studies with paraffin tissue blocks. Cyclin, an auxiliary protein of DNA polymerase-δ, is perhaps the best-characterized marker of cycling cells, but even here there is no universal agreement (McCormick and Hall, 1992). Newer antibodies to recognize PCNA and other antibodies to proteins that mark the S phase of the cell cycle are under investigation (see below). PC10 (Figure 4-1) reportedly reacts only with cells in the S phase (Landberg and Roos, 1991) but might have a longer duration in the cell-cycle phase than previously thought (Scott et al., 1991). Ki-67 closely correlates with BrdUrd labeling, and the new anti-Ki-67 MIB1-3 antibodies (Key et al., 1993) are now available, allowing retrospective studies of paraffin-embedded tissue. Evaluation of data in the literature is difficult. Assay control standards have not consistently been used to assess the quality of each assay. The use of different assay methods by different investigators can result in discrepancies. And, different vendors offer different antibodies that might recognize different sites of a protein or even different epitopes. Clearly, the one feature of DNA ploidy analysis that consistently correlates with cancer survival involves cell-proliferative rates. Biologic markers to assess populations of cells in each stage of the cell cycle accurately should prove beneficial in evaluating the efficacy of chemotherapy and predicting tumor aggressiveness. Systematic studies with standard controls of well-established proliferative rates need to be performed to determine which technically feasible proliferative markers are clinically useful in predicting metastases or survival.

Renal-Cell Carcinoma

Several secondary chromosomal abnormalities occur in nonpapillary renal-cell carcinoma. Various aberrations, including deletions and duplications, have been reported in chromosomes 3, 5, 7, 8, 9, 14, and 20 and the Y chromosome. Those aberrations, used as markers, are presumed to be associated primarily with tumor progression and might reflect the marked genetic instability of tumors. Trisomy of chromosome 5 has been observed in about 15% of the cases. Abnormalities of chromosome 5, including segment loss and translocation, have also been associated with renal-cell carcinoma and the familial adenomatosis gene in sporadic colon cancers. The c-*fos* gene on chromosome 14 has also been observed to be aberrant in nonpapillary renal-cell carcinoma. Another common abnormality has been reported in chromosome 17, where deletions have been documented in high-stage nonpapillary renal-cell carcinoma. Chromosome 17 contains the important suppressor gene *p53*, which is associated with oncogenesis. Table 4-3 contains a list of various potential biologic markers of renal cancer.

Bladder Cancer

Conventional Markers in Bladder Cancer

The urinary epithelium consists mainly of a single type of cell, and exfoliation of these cells into urine provides a noninvasive source of both cellular and soluble markers. The cardinal sign of bladder cancer has been hematuria with or without irritative bladder symptoms. The hematuria can either be gross or microscopic. Most studies have shown that a majority of high-grade tumors have already progressed to metastasis by the time gross hematuria is evident, so its usefulness as a marker is slight.

Dipstick Testing

The sensitivity of dipstick and microscopic analysis for the detection of microhematuria is controversial, and the results afforded by the two methods have not always correlated. It has been estimated that a single-dipstick method for detecting blood in urine has a sensitivity of about 37% for detecting bladder cancer. A major consideration in the use of microhematuria as a marker of bladder cancer is that the microhematuria is intermittent and that in some cases neovascularization of the lesion must have occurred to a marked degree before the patient presents with gross hematuria. It can, however, be detected as long as 8 years before the detection of a tumor. Because of the lack of sensitivity of a single-dipstick test, Messing and colleagues (1990, 1992) introduced sequential dipstick analysis by the patient at home for 14 consecutive days. This method is inexpensive and has been estimated to detect about 80% of bladder cancers, although compliance is a relatively poor 45%. About 21% of the patient population aged 50 and older who smoke will have microscopic hematuria on at least one analysis, so many patients might require more extensive and high-cost testing to rule out the presence of bladder cancer. In retesting patients who have initially tested negative, it was found that 6% of patients were positive for hematuria on the second screen, and 1% of these patients were found to have bladder cancer. These results suggest the need for methods of detecting bladder cancer that have a higher degree of sensitivity and specificity.

Cytologic Analysis

The routine evaluation of Papanicolaou cytology for bladder cancer has recently been reviewed (Farrow, 1990). Papanicolaou cytology under most circumstances has a sensitivity between 20 and 85%, depending on stage and grade. Sample type—i.e., bladder wash or voided urine—also affects the sensitivity and specificity of the test. Papanicolaou cytology has been found to be particularly useful for monitoring patients who have established high-grade disease but has been generally ineffective in screening programs to detect bladder cancer in xenobiotic-exposed cohorts who are at risk for developing malignancy (Cartwright, 1986). Nearly 50% of such patients have been missed in conventional screening and have progressed to invasive disease. With an incidence of bladder cancer in the general population of 3/10,000, many persons need to be screened to detect one bladder-cancer case. Therefore, attempts have been made to automate urinary cytology; although some success has been achieved (Azzopardi and Evans, 1971; Bauer, 1988), it is still hard to distinguish atypia or dysplastic changes associated with inflammatory conditions from changes associated with malignant disease.

Karyotypic changes

Low-grade papillary tumors have been associated with changes in chromosome 9, and high-grade carcinomas and carcinoma in situ appear to develop along a second pathway that often involves alterations in suppressor oncogenes, such as *p53*. Whether those alterations are primary and initiate carcinogenesis or secondary and result from genetic instability remains to be determined. The genetic mutations in the *p53* gene on chromosome 17 are frequent and varied. Whether specific mutations are associated with various types of carcinogenic exposure is a point for future investigation, which might shed light on the underlying mechanisms of carcinogenesis.

Markers in Occupationally Exposed Cohorts

Several cohorts are being monitored nationally and internationally for bladder cancer with markers of varied effectiveness (Davies et al., 1988; di Sant' Agnese, 1988; di Sant' Agnese and de Mesy Jensen, 1987; Doctor et al., 1986; Eaton et al., 1988). The markers include elements of conventional urinary cytology, the addition of DNA ploidy (to determine the presence of an excessive number of chromosome sets) and various forms of differentiation markers, growth factors and their receptors, and oncogenes and their protein products.

Table 4-4 outlines a number of markers that should be useful for bladder cancer. The most widely used marker in the genitourinary tract is quantitation of DNA ploidy, which has been applied to routine cytology with either flow cytometry or image analysis. The significance of cells that are atypical in appearance has been a major issue for cytologists particularly with respect to low-grade malignancies of the genitourinary tract. Experiments with acridine orange dye, Hoechst dye, and Feulgen staining have demonstrated that with the addition of tests of DNA ploidy some degree of increased sensitivity is gained in bladder-cancer detection.

Studies of bladder-field disease have identified several cellular markers that become abnormal at different times in the development of disease, including altered ploidy as a late event (Rao et al., 1993). A major group of potential markers is monoclonal antibodies to tumor-related antigens.

Prostatic Cancer

Prostatic cancer is the second-most common cancer of men; about 200,000 cases of prostatic cancer will be diagnosed in the United States in 1994. That is substantially more than in previous years and is almost certainly due to the introduction of prostate-specific antigen (PSA) testing. PSA is a protease specific to prostatic tissue that is secreted into the serum and constitutes a marker of effect. Its introduction has highlighted a number of important issues with regard to biologic markers of effect, including health-care costs, its effective use, its sensitivity and specificity, and the relationship between cryptic and active disease. Current research into the molecular and cellular biology of the prostate and of growth regulation are expected to result in new tests with higher specificity and sensitivity that should permit individual risk assessment and differentiation of indolent and active disease. The introduction of PSA without completing a prospective long-term followup study illustrates one of the major problems in prostatic cancer research in that such a study would have required 15 years for completion. Because the discovery of organ-confined disease is often life-saving, there is a reluctance to defer the application of a new technology until the results of a long-term study are available.

The complex organization of the prostatic architecture reflects the multiplicity of potential cellular interactions critical to cellular homeostasis. Any of the cellular types can abort normal regulatory control and directly or indirectly participate in tumorigenesis. Hormonal growth control has been the primary focus of investigational research and clinical therapy. However, in today's clinical setting, the tumor cells that progress and the eventual demise of the patient are hormone-independent and can reflect a reawakening of mechanisms that were active during embryogenesis. Hidden in the histopathologic architecture of the gland are neuroendocrine cells that are adjacent to basement membrane cells and associated stromal cells that support the functional epithelial cells. These neuroendocrine cells are the participants in supporting the biologic media so important to human fertility and reproductive success. The scientific challenge is to determine which cells are the primary targets of xenobiotics and which genetic factors are implicated in the deregulation of adult paracrine growth control and the emergence of autocrine independence.

Investigations at the cellular and molecular levels have begun to identify structural and biochemical changes associated with the emergence of the malignant phenotype. At the structural level, a series of graded dysplastic changes is now recognized as being premalignant. These changes are known as prostatic intraepithelial neoplasia (PIN). Investigations of molecular biologic markers using immunohistochemistry and quantitative morphology have shown a strong correlation between PIN and other recognized biologic markers of malignancy, such as aneuploidy. Quantification at the single-cell level and a definition of biochemical "cell talk" promise to define further the biochemical changes responsible for the histopathologic alterations and the selection of the deadly hormone-resistant clones. Each of the histopathologic manifestations of increased cellular proliferation, invasion, and metastasis reflects subtle quantitative alterations.

Serum Prostate-Specific Antigen

Biologic markers of disease have been detected primarily in the serum and are nonspecific for malignancy, reflecting quantitative alterations in normal cellular constituents. Until recently, acid phosphatase, an enzyme normally expressed in prostatic epithelial cells, was the principal serum biologic marker. Because of its poor specificity for detecting early disease, more recent efforts have focused on the evaluation of PSA (Wang et al., 1979). PSA is a glycoprotein found in the seminal plasma. Antibodies to this 30-kD molecule were produced by Wang et al. (1979). PSA is found only in prostatic tissues, benign prostatic hypertrophy (BPH), and prostatic cancer. The gene coding for PSA has been sequenced and is on chromosome 19. Genetic deletions or mutations of this chromosome will result in clonal cells that do not predominantly express the protein so important for the liquefaction of seminal ejaculate. PSA production is limited to the epithelial cells, and its quantification within these cells establishes its potential as a cellular biologic marker for prostatic cancer. The chymotrypsin-like activity of the protein cleaves the seminal vesicle-specific antigen produced by the seminal vesicle cells. This interaction establishes the rationale for the theoretical development of perhaps more specific and sensitive assays in the future (see Chapter 6). It should be kept in mind that some cancers seem to have deleted the gene for PSA production, which might mean that they do not produce PSA or contain less of it than normal or BPH cells. However, PSA is still usually increased because it is not normally released into the serum unless the cell membranes are altered or damaged. Serum PSA can be increased in BPH or prostatic infection, in addition to cancer.

Because PSA is increased in patients with BPH and inflammatory conditions of the prostate, attempts have been made to improve the sensitivity and specificity of this assay for the diagnosis of prostatic cancer. Various thresholds were established before implementation of the assay in wide scale clinical testing. This has raised multiple issues concerning uncritical reliance on the PSA test in clinical medicine. The use of this assay has stimulated interesting questions related to technology transfer and assay evaluation; optimal methods for interpreting test results in light of PSA's poor specificity (age-adjusted PSA, PSA density, PSA velocity); the detection of clinically inactive or latent prostatic cancer; adaptation as a screening test without clinical trials to establish efficacy; and the cost-benefit relationship in the management of patients with prostatic cancer.

Prostatic Intraepithelial Neoplasia

A strong association between the presence of prostatic intraepithelial neoplasia (PIN) and high-grade prostatic cancer suggests that PIN can be a precursor of prostatic cancer. Using PIN tissue, McNeal and Bostwick (1986) have proposed a group of potential biologic markers of prostatic cancer (Table 4-5). Assessment of any of them awaits the development of more sensitive methods for detecting alterations in serum, prostatic secretions, or urine. Fine-needle aspiration of cells is also possible with needles smaller than those commonly used in venipuncture to obtain blood. A grading scale based on the distribution of vimentin and various keratins might prove helpful.

Morphometric analysis of PIN has revealed graded alterations in nuclear size, roundness, membrane ruffling, and nuclear chromatin similar to those in frank cancer. The greatest utility of such tissue might be as a resource to develop genotypic changes correlated with progression. One potential strong advantage of the use of such tissue is that some phenotypic changes that can predict progression are seen in otherwise normal cells of the field; this could allow detection of a cancer in the prostate, even if the lesion itself cannot be sampled. That feature might be applicable to the design of programs to screen persons at risk with fine needle aspiration and to use more specific core biopsies for confirmation.

Cytokeratins are used mainly to identify the origins of cells in complex epithelia. Research on cytokeratins has focused on delineating the integrity of the basement membrane adjacent to the prostatic epithelial cells in normal tissue, benign lesions, premalignant lesions, PIN, and malignant lesions. Studies with monoclonal antibodies, such as MH-903 (Shah et al., 1991), revealed disruption of the basement membrane in 39 of 40 cancer lesions with a graded increase in frequency noted in PIN.

DNA Ploidy and Prostatic Cancer

DNA ploidy can be determined with such techniques as flow cytometry and image analysis. Image analysis facilitates rare-event detection, correlates with cytologic characteristics, and tends to be more precise in biochemical quantification of epitopes (complex antigenic molecules). Flow cytometry is particularly useful for studying suspensions of cells with recognized homogeneity—i.e., blood cells as opposed to mixtures of stromal and epithelial cells found in the prostate. The lack of standardization of sample collection and processing and signal analysis continues to hinder the usefulness of these approaches to defining ploidy, but further refinement of ploidy studies and combination of ploidy studies with antibodies to various proliferation markers are possible.

A major limitation of marker research

in prostatic cancer is the requirement for long-term longitudinal followup of patients. Archival specimens that are available might or might not have been appropriately fixed to maintain the quantitative or qualitative characteristic of the reacting epitope. Aneuploidy reflects genetic instability and marked tumor heterogeneity. Fluorescent in situ hybridization studies should improve the sensitivity of ploidy analysis and lead to a redefinition of ploidy changes in the future. The identification of markers of specific events in cellular proliferation, invasion, and metastasis should supplant the rather crude measure of DNA ploidy based on flow cytometry or image analysis.

Cytoskeletal and Nuclear-Matrix Proteins

Recent evidence suggests that the actin filaments of the cytoskeleton and nuclear matrix might be useful as early markers of tumorigenesis (Pienta et al., 1989; Partin et al., 1993; Rao et al., 1990, 1991, 1993). Recently, Partin et al. (1993) identified a nuclear-matrix protein uniquely present in all of 14 prostatic cancers but absent from normal prostates and benign-prostatic-hypertension (BPH) tissues. Characterization of the protein revealed an isoelectric point of 6.58 and a molecular weight of 56000.

Tumor-Suppressor Genes

Prostatic cancer can be latent (or occult) or progressive (or biologically active). Latent disease is identified with great frequency at autopsy. It is not known whether the biologically active form develops along a separate oncogenic pathway or occurs as an additional mutation in the low-grade form of the disease.

In addition to oncogenes and growth factors, cancer research has focused on the tumor-suppressor genes. Several tumor-suppressor genes have been identified in prostatic cancer, including *Rb* and *p53*. Various methods are available for detecting tumor-suppressor genes. The importance of *p53* in relation to growth control was delineated by Gumerlock and DeVere White (Chi et al., 1994), who demonstrated a biologic correlation between *p53* mutations and concentrations of IL-6, a cytokine known to be important in cellular proliferation. Growing evidence suggests that changes in the *p53* expression occur late in tumorigenesis, are associated with local invasions, and reflect marked tumor heterogeneity. It remains to be determined whether the marker is a stronger marker for predicting tumor invasion and metastasis than is DNA ploidy or tumor grade. Data on *Rb* are scant, compared with data on *p53*.

Growth Factors

- *Nerve growth factor.* One of the exciting advances in prostatic cancer is the recognition of the potential importance of neuroendocrine cells, the associated growth factors, and their potential neuroendocrine function. Cohen and Ellwein (1991) pointed out the importance of neuroendocrine cells as a component of the ducts and acini. They reported that half of all clinically manifest tumors show neuroendocrine differentiation, and suggested that the cells are of greater prognostic value than conventional histopathologic grading systems. Cultured stromal and epithelial cells from prostatic cancer contain a protein-like nerve growth factor (NGF) tissue. The NGF-like protein is localized to the stroma in normal prostatic tissue, which has undergone benign hypertrophy, and cancerous tissue. However, NGF receptors reside primarily on the epithelial component of the prostate. The neuroendocrine cells in the prostate can be stained with chromogranin A; in benign hypertrophy, these cells also stain positive for the androgen receptor. Cohen hypothesized that the cells lacking the androgen receptor might be important in the evaluation of prostatic cells that lack neuroendocrine regulation (Cohen et al., 1991a).

- *Epidermal growth factor.* The DU145 prostate cell line has epidermal growth factor (EGF) and EGF receptor, which provide an autocrine loop for growth control of these cells (Connolly and Rose, 1990). An anti-EGF-receptor antibody reduces growth in these cells. However, no increase in urinary EGF excretion was found in males with prostatic or bladder cancer. Mattila and others found that adenocarcinoma did not stain for EGF in 15 of 20 cases (Mattila et al., 1988).

- *Insulin-like growth factor.* There might be a complex interaction between the insulin-like growth factors (IGF-1, IGF-2) and EGF receptors (Connolly and Rose, 1992). In vitro results with the DU145 human prostatic-cancer cell line suggest an interrelationship between the response mediated by IGF receptors and the autocrine EGF response. Results of a second set of experiments involving a rat transplantable prostatic-cancer line that metastasizes to bone reveal the presence of the IGF-1 receptor and the absence of the IGF-2 receptor (Polychronakos et al., 1991). Prostatic epithelial cells in tissue cultures were found to respond to IGF-1 and not IGF-2 or insulin. The effect of IGF-1 might be modulated by a 24-kilodalton IGF-binding protein (Cohen et al., 1991a). Recent studies suggest that fibroblasts from BPH express increased mRNA for IGF-2. The importance of IGF-2 remains to be determined. The quantity and presence of IGF-binding protein is potentially an important and rate-limiting factor.

- *Transforming growth factor alpha.* Transforming growth factor alpha (TGF alpha) has been identified in the premalignant expression in four of 27 benign and 18 of 34 malignant biopsies. Positive staining has also been identified in premalignant lesions and within the stromal elements. TGF alpha might enhance the detection of premalignant lesions (Lloyd et al., 1992). In tissue sections from the prostate, TGF alpha is preferentially expressed in prostatic-cancer cells in comparison with EGFR, which stains primarily the basal cells in both BPH and prostatic cancer.

MARKERS OF INTERSTITIAL CYSTITIS

Several causes of interstitial cystitis (IC) have been suggested. The various suggestions are not necessarily mutually exclusive, and the syndrome could have multiple etiologies. Each proposed etiology tends to be associated with a marker. One difficulty has been the inadequacy of diagnostic criteria. Spontaneous remission is common in the disease, and it has been difficult to define specific markers or to monitor the effects of treatment.

The suggested etiologies of IC were recently reviewed by Messing (1989). Table 4-6 lists the proposed etiologies, briefly summarizes the evidence for and against each, and indicates the potential markers that might be useful. The suggested etiologies include infection, obstruction of vascular and lymphatic channels, neuropathy, endocrinopathy, inflammation, autoimmunity, deficiencies in the bladder protective mechanisms, and the presence of toxic substances in urine.

Several recent studies have linked bladder damage with various mechanisms of inflammation. Elgebaly and co-workers (1992) found that high concentrations of neutrophil chemotactic factors were released after bladder injury by an acidic solution (pH 4.5) in a rabbit model and that this preceded an inflammatory response with high concentrations of neutrophils in the bladder tissue. They hypothesized an early step in an inflammatory response that eventually produces IC. Of great interest was the finding that 11 patients with IC had high urinary concentrations of neutrophil chemotactic factors; but these were not found in patients with diseases other than IC or in interstitial-cystitis patients treated with the anti-inflammatory agent dimethyl sulfoxide (DMSO).

A recent study suggested a further link among inflammation, symptoms, and altered innervation in IC. A population of interstitial-cystitis patients—selected to have a consistent set of symptoms, including mastocytosis—showed nerve content in submucosa and muscle 4 times greater than that in controls; much of the additional nerve content was ascribable to nonadrenergic noncholinergic fibers (Hohenfellner et al., 1992). Recent advances in neurochemistry with in situ identification of neuropeptides and other neurotransmitters have made it possible to readily identify the functional system to which nerves belong (Hohenfellner et al., 1992). The authors (Hohenfellner et al., 1992) point out the existence of links between the nervous system and inflammatory tissue reactions, which are referred to as neurogenic inflammation, in that increased sympathetic outflow can lead to increased local mast-cell populations and pain; they report similar findings in autoimmune diseases, such as rheumatoid arthritis and inflammatory bowel disease.

Although the presence of mast cells in the detrusor muscle has been a variable feature considered by some to be an important diagnostic factor and by others to have no importance (Yun et al., 1992), an interesting study showed that urinary histamine concentrations after hydrodistention more than doubled in interstitial-cystitis patients; in patients without this diagnosis, there was no significant increase. (There was no significant difference in urinary histamine concentration before hydrodistention.) It is not clear how histamine released from mast cells, which are found in the muscularis well below the urothelial surface, would be able to reach the urine.

In contrast, functional characteristics of peripheral lymphocytes examined with flow cytometry did not support an autoimmune etiology (Miller et al., 1992). There was no discernible difference between normal and interstitial-cystitis patients with respect to the distribution of lymphocyte immunophenotypes before stimulation, after incubation with and without autologous urine and with and without mitogenic stimulation, or in activation of T cells. The investigators concluded that no primary immunologic disorder was evident in patients with IC and that urine, which might leak into the bladder if the urothelium were leaky, does not stimulate the immune system. However, studies of lymphocytes in bladder tissue showed that CD4-positive lymphocytes (helper T cells) predominated over CD8-positive lymphocytes (suppressor/cytotoxic T cells) in the lamina propria in IC and other forms of cystitis; in the urothelium, CD8-positive lymphocytes predominated in IC and CD4-positive lymphocytes predominated in bacterial and mechanical cystitis (MacDermott et al., 1991). That suggested that the urothelium was not involved in the inflammatory process in IC, and the authors therefore suggested that the initiating factor did not originate in the bladder lumen. Those findings seem to contradict findings of a slightly earlier study that produced distinct evidence of persistent inflammatory reactions in both peripheral and bladder lymphocytes (Harrington et al., 1990).

A recent study of markers of cellular activation in urothelial cells showed that urothelial cells from interstitial-cystitis patients expressed increased HLA-DR but not increased ICAM-1 or IL-1à (Liebert et al., 1993). Urothelial cells in culture respond to cytokine stimulation in the same way as most epithelial cells—that is, ICAM-1 and HLA-DR increase simultaneously. The authors concluded that the simultaneous increase in ICAM-1 and HLA-DR, although not peculiar to IC, might be a useful marker of this condition. The marker might indicate an unusual type of activation of the urothelial cells in IC.

Further evidence has also accumulated to support the theory that the urothelium becomes deficient in the protective layer of glycosaminoglycan. An extensive investigation of urinary concentrations of macromolecular uronate (a measure of glycosaminoglycan and proteoglycan content) as a marker of IC showed that, as a group, subjects with the disorder had markedly less urinary macromolecular uronate than did normal controls. The median values in the two populations differed by a factor of 2. As suggested by earlier work with the acid-damaged rat-bladder model, the decrease is consistent with binding of glycosaminoglycan to damaged bladder epithelium, rather than with a specific effect of IC. In preliminary studies with a relatively unselected patient group at two centers, 58% sensitivity and 79% specificity were obtained with a diagnostic threshold of macromolecular uronate of 40 nmol/mL (Nickel et al., 1993). An investigation of the ultrastructure of the extensive cell-surface coating of the bladder-surface macromolecules with an antibody against bladder mucin to stabilize the layer showed no discernible difference in the depth or ultrastructure of the layer coating the umbrella-cell layer of urothelium. Any differences must therefore be in composition and would not affect structure.

Whether or not the layer is abnormal, the bladder epithelium in IC was again shown to be much more permeable than normal. In an extensive study of 31 normal persons and 56 with IC, a highly significant difference in bladder-surface permeability to urea was demonstrated between interstitial-cystitis patients and control urology patients (Fowler et al., 1988). In contradiction of an earlier report, no evidence of increased permeability to Tamm-Horsfall protein was found in the bladders of interstitial-cystitis patients. It should be noted that permeability to high- and low-molecular-weight substances might be quite different. A potential link between changes in permeability to low-molecular-weight substances and inflammation was suggested by results of a recent study that showed in a mouse model of autoimmune cystitis that permeability differences were similar to those reported by Parsons in humans (Parsons et al., 1991).

SUMMARY

The ideal cytologic marker of nephrotoxicity would represent a nonspecific, if not universal, cellular response to injury. Such a marker should be produced in the kidney and secreted into urine in a readily detectable form. Substantial changes in urinary concentration of the marker should correlate well with pathophysiologic or histopathologic manifestations of renal injury. The marker should be expressed soon after injury is sustained. Persistence of marker expression would increase its clinical value. The marker should be relatively easy to measure. It should be stable in storage, resistant to degradation, and unambiguously identifiable. In the search for suitable markers, more emphasis should be placed on the response to injury of tubules and interstitium, because these compartments appear to be the major sites of susceptibility to toxic injury. Finally, extensive use must be made of appropriate animal models to evaluate potential markers because complex responses of intact organisms (vis-a-vis) isolated systems) can influence response to injury, inflammation, and repair in vivo.

Several markers that fulfill many of the above criteria have already been identified. Urinalysis and clearance measurements will continue to provide important functional markers of renal injury. Of particular use in that regard would be a nonisotopic technique for analyzing iodothalamate or Cr-EDTA, both markers of glomerular filtration rate.

There is no "ideal" marker, i.e., a single marker able to provide all the information necessary to identify people at risk. The main priority for research should be the identification of strong markers that define preclinical, potentially dangerous disease. Addressing disease at this point minimizes costs, morbidity, and mortality and is now possible because of the availability of biologic markers. To achieve this, the following should be emphasized.

- Interdisciplinary research will be needed in which clinical medicine, basic science, and statistics are combined. None of these fields is capable of proceeding alone.
- Because strong markers are necessary for accurate clinical decision-making, funding of marker research should require that small studies demonstrating a high degree of separation of populations of interest be undertaken early in the marker-development process.
- Multiple marker profiles are likely to be necessary to provide all the information needed for individual risk assessment.
- Attention needs to be given to quality control and suitability of the method to use in a routine environment before large-scale clinical trials with a marker are attempted.

TABLE 4-1 Potential Markers of Cell Injury or Inflammation in the Kidney[a]

Category	Detected As		Disease	Reference[b]	Species	Kidney Tissue Site[b]			
	RNA	Product				Glomeruli	Tubules	Other	Urine
CYTOKINES									
Tumor-necrosis factor		x	Purine amino-nucleoside nephrosis	Diamond and Pesek, 1991	Mouse	x			
		x	Immune-complex glomerulo-nephritis	Noble et al., 1990	Rat	x			x
		x	Anti-glomerular-basement-membrane nephritis	Tipping et al., 1991a	Rabbit	x			
	x	x	Lupus nephritis	Boswell et al., 1988	Mouse	x	x		
		x	Polycystic kidneys	Gardner et al., 1991	Human	x		Cyst fluid	
Interleukin-1		x	Purine amino-nucleoside	Diamond and Pesek, 1991	Rat	x			
	x	x	Immune-complex glomerulo-nephritis	Noble et al., 1990 Werber et al., 1987	Rat				x

Cytokine		Disease	Reference	Species		Notes
	x	Anti-glomerular-basement-membrane nephritis	Matsumoto and Hatano, 1989	Rat	x	
	x	Lupus nephritis	Tipping et al., 1991b	Rabbit		
	x		Boswell et al., 1988	Mouse	x	
	x	Crescentic glomerulonephritis	Matsumoto et al., 1988	Human	x	
Interleukin-2	x	Polycystic kidneys	Gardner et al., 1991	Human		Cyst fluid
	x	Polycystic kidneys	Gardner et al., 1991	Human		Cyst fluid
Interleukin-6	x	Mesangial proliferative glomerulonephritis	Horii et al., 1989	Human	x	
	x	Interstitial nephritis	Fukatsu et al., 1991	Human	x	
Monocyte chemotactic peptide	x	Renal ischemia	Safirstein et al., 1991	Rat	x	

TABLE 4-1, continued

	Detected As					Kidney Tissue Site[b]			
Category	RNA	Product	Disease	Reference[b]	Species	Glomeruli	Tubules	Other	Urine
GROWTH FACTORS									
Platelet-derived growth factor	x	x	Mesangial proliferative glomerulonephritis	Gesualdo et al., 1991	Human	x			
	x	x		Iida et al., 1991	Mouse	x			
		x		Yoshimura et al., 1991	Rat	x			
	x	x	Partial nephrectomy	Floege et al., 1992a	Rat	x			
	x		Ureteral obstruction	Fries and Collins, 1992	Mouse			Ureter	
	x		Lupus nephritis	Nakamura et al., 1992	Mouse				
Insulin-like growth factor I	x	x	Partial nephrectomy	Lajara et al., 1989	Rat		x		
		x		Stiles et al., 1985					

Growth factor		Condition	Reference	Species		
Transforming growth factor beta	x	Diabetes	Flyvbjerg et al., 1990	Rat		x
	x	Antiglomerular-basement-membrane nephritis	Noh et al., 1991	Rabbit	x	
	x	Mesangial proliferative glomerulonephritis	Okuda et al., 1990	Rat	x	
	x	Purine aminonucleoside nephrosis	Jones et al., 1991	Rat		
Epidermal growth factor	x	Aminoglycoside nephrotoxicity	Nonclercq et al., 1991	Rat		x, x^c
	x	Cisplatin nephrotoxicity	Safirstein et al., 1989	Rat		x, x^c
	x	Renal ischemia	Safirstein et al., 1989	Rat		x, x^c
	x	Diabetes	Mathiesen et al., 1988	Human		x^c
		Ureteral obstruction	Storch et al., 1992	Rat		x

TABLE 4-1, continued

Category	Detected As		Disease	Reference[b]	Species	Kidney Tissue Site[b]			
	RNA	Product				Glomeruli	Tubules	Other	Urine
Epidermal growth factor (continued)	x	x	Polycystic kidneys	Gattone et al., 1990	Mouse		x		
MATRIX COMPONENTS									
Collagens	x	x	Purine aminonucleoside nephrosis	Jones et al., 1991	Rat	x	x	Interstitium	
	x	x	Cyclosporin A nephrotoxicity	Wolff et al., 1990	Mouse		x	Interstitium	
	x	x	Lupus nephritis	Nakamura et al., 1991	Mouse	x		Interstitium	
	x	x	Mesangial proliferative glomerulonephritis	Floege et al., 1991	Rat	x			
	x	x	Antiglomerular-basement-membrane nephritis	Merritt et al., 1990	Rabbit				

Component			Disease	Reference	Species		Location
	x	x	Partial nephrectomy	Floege et al., 1992b	Rat	x	
		x	Proliferative nephritis	Yoshioka et al., 1989	Human	x	
	x	x	Polycystic kidneys	Ebihara et al., 1981	Mouse		Around cysts
	x	x	Purine aminonucleoside nephrosis	Jones et al., 1991	Rat	x	Interstitium
Laminin	x	x	Lupus nephritis	Nakamura et al., 1991	Mouse	x	
	x	x	Mesangial proliferative glomerulonephritis	Floege et al., 1991	Rat	x	
	x	x	Polycystic kidneys	Ebihara et al., 1981	Mouse		Around cysts
	x	x	Purine aminonucleoside nephrosis	Jones et al., 1991	Rat	x	Interstitium
Fibronectin		x	Partial nephrectomy	Floege et al., 1992b	Rat	x	

TABLE 4-1, continued

Category	Detected As		Disease	Reference[b]	Species	Kidney Tissue Site[b]				
	RNA	Product				Glomeruli	Tubules	Other	Urine
Fibronectin (continued)		x	Mesangial proliferative glomerulonephritis	Okuda et al., 1990	Rat	x			
Proteoglycans		x	Purine aminonucleoside nephrosis	Jones et al., 1991	Rat	x[c]	x		
	x		Lupus nephritis	Nakamura et al., 1991	Mouse	x			
		x	Partial nephrectomy	Floege et al., 1992b	Rat	x			
		x	Mesangial proliferative glomerulonephritis	Okuda et al., 1990	Rat	x			
		x	Polycystic kidneys	Ebihara et al., 1981	Mouse			Around cysts	
PROCOAGULANT ACTIVITY		x	Antiglomerular basement-membrane nephritis	Wiggins et al., 1985	Rabbit	x			x

	Condition	Reference	Species			Platelets	
x	Mercury chloride nephropathy	Kanfer et al., 1987	Rat	x			
x	Proliferative glomerulo-nephritis	Tipping et al., 1988	Human				x
x	Acute serum-sickness glomerulo-nephritis	Tipping et al., 1987	Rabbit	x			

LIPID MEDIATORS

Prostanoids (cyclo-oxygenase products)

	Condition	Reference	Species				
x	Mesangial proliferative glomerulo-nephritis	Gesualdo et al., 1992; Lianos et al., 1991; Stahl et al., 1990	Rat	$x(TxB_2)^d$			
x	Membranous glomerulo-nephritis	Rahman et al., 1988a	Rat	$x(TxB_2)^d$			
x	Lupus nephritis	Kelley et al., 1986; Patrono et al., 1985	Mouse; Human				$x(PGE_2, TxB_2)^{d,e}$

TABLE 4-1, continued

Category	Detected As		Disease	Reference[b]	Species	Kidney Tissue Site[b]				
	RNA	Product				Glomeruli	Tubules	Other	Urine
Leukotrienes (lipoxygenase products)		x	Antiglomerular basement-membrane nephritis	Fauler et al., 1989 Lefkowith et al., 1992 Lianos et al., 1985 Rahman et al., 1988b	Rat	x(LTB$_4$,12 HETE)			x(LTB$_4$)
	x		Mesangial proliferative nephritis	Oberle et al., 1992	Rat	x(LTB$_4$,12 HETE)			
		x	Membranous glomerulonephritis	Rahman et al., 1988b	Rat	x(LTB$_4$)			
	x		Lupus nephritis	Spurney et al., 1991	Mouse	x(LTB$_4$, LTC$_4$)		Platelets	
Platelet-activating factor	x		Membranous glomerulonephritis	Noris et al., 1991	Human				x

Mediator		Disease	Reference	Species	
REACTIVE SPECIES OF OXYGEN AND NITROGEN					
Oxygen (O$_2^-$, H$_2$O$_2$, OH)	x	Antiglomerular basement-membrane nephritis	Bertani et al., 1987; Lianos and Zanglis, 1990	Rat	x
	x	Immune-complex glomerulonephritis	Cook et al., 1987	Rat	x
	x	Mesangial proliferative nephritis	Oberle et al., 1992	Rat	x
	x	Antiglomerular basement-membrane nephritis	Boyce et al., 1989	Rabbit	x
Nitrogen (NO$_2^-$, NO$_2$, NO)	x	Antiglomerular basement-membrane nephritis	Cattell et al., 1990	Rat	x
MAJOR HISTOCOMPATIBILITY ANTIGENS	x	Immune-complex glomerulonephritis	Noble et al., 1990	Rat	x

TABLE 4-1, continued

Category	Detected As		Disease	Reference[b]	Species	Kidney Tissue Site[b]				
	RNA	Product				Glomeruli	Tubules	Other	Urine
MAJOR HISTO-COMPATIBILITY ANTIGENS (continued)		x	Puromycin aminonucleoside nephrosis	Schreiner et al., 1984	Rat	x			
		x	Antiglomerular-basement-membrane nephritis	Schreiner et al., 1984	Rat	x			
		x	Mercury chloride nephropathy	Aten et al., 1988; Madrenas et al., 1991	Rat		x	Interstitial cells	
	x	x	Lupus nephritis	Boucher et al., 1986; Halloran et al., 1988	Mouse		x	Interstitial cells	
		x		Wuthrich et al., 1989	Human		x		
	x	x	Renal ischemia	Shoskes et al., 1990	Mouse		x	Interstitial cells	

	Condition	Reference	Species		Interstitial cells
x	Proliferative glomerulonephritis	Müller et al., 1987	Human	x	
x	Ischemia	Ouellette et al., 1990	Rat	x	
x		Rosenberg and Paller, 1971	Rat	x	
x		Safirstein et al., 1990	Mouse	x	
x	Polycystic kidneys	Cowley et al., 1987	Mouse	x	
x	Folic acid toxicity	Cowley et al., 1987	Mouse	x	
		Cowley et al., 1989	Mouse		
		Kujubu et al., 1991			
x	Partial nephrectomy	Aulitzky et al., 1992	Mouse		

TRANSCRIPTION FACTORS

TABLE 4-1, continued

Category	Detected As		Disease	Reference[b]	Species	Kidney Tissue Site[b]			
	RNA	Product				Glomeruli	Tubules	Other	Urine
MISCELLANEOUS									
Complement C5b-9	x		Membranous glomerulo-nephritis	Pruchno et al., 1989	Human	x			x
		x		Schulze et al., 1991	Rat	x			x
Endothelin	x		Proliferative glomerulo-nephritis	Ohta et al., 1991	Human		x		x
Vimentin	x		1,2-Dichlorovinyl cysteine toxicity	Wallin et al., 1992	Rat		x		
		x	Daunomycin nephrosis	Gröne, et al., 1987	Rat		x		
		x	Mercury chloride toxicity	Gröne, et al., 1987	Rat		x		
Intercellular adhesion molecule 1	x	x	Lupus nephritis	Wuthrich et al., 1990	Mouse	x	x	Vessels	
Cystatin		x	Isoproterenol treatment	Cohen et al., 1990	Rat		x		

			Condition	Reference	Species			
		x	Sodium chromate nephrotoxicity	Colle et al., 1984; Esnard et al., 1988	Mouse			x
Heat-shock proteins								
x	x		Mercury chloride toxicity	Goering et al., 1992	Rat		x	
	x		Ischemia	Emami et al., 1991; Sawczuk et al., 1987	Rat		x	
	x		Heat stress	Emami et al., 1991	Rat		x	
Clusterin								
x	x		Prostatic involution	Bandyk et al., 1990; Buttyan et al., 1989	Rat		Prostatic epithelium	x
x	x		Ureteral obstruction	Buttyan et al., 1989	Rat	x		
x	x		Ischemia	Rosenberg and Paller, 1971	Rat			
	x		Gentamycin toxicity	Aulitzky et al., 1992	Rat			x

TABLE 4-1 (continued)

[a] This table lists a wide variety of biologic markers of tissue damage or inflammation that potentially serve as clinically useful markers of renal injury to the kidney. Important changes in gene expression or protein synthesis have been linked to renal histopathology or pathophysiology. For most markers, disease processes were found to stimulate increased gene expression or protein synthesis. In other cases, renal disease was associated with reduced amounts of the marker. Some reports documented links between abnormal markers and renal damage; in a few instances, abnormal intrarenal sites of marker synthesis were linked to injury. It should be noted that for some growth factors and hormones, abnormal local receptor expression or inhibitor production has also been found to accompany injury or repair. At present, the literature on receptor and inhibitor production is relatively sparse and fragmentary. A careful review in the not-too-distant future could well reveal other potentially useful markers among those molecules.

[b] The review has been limited to studies in which samples were obtained directly from animals or human patients with renal damage or disease. Reports of studies to analyze the action of nephrotoxicants in vitro (tissue culture, isolated perfusion, etc.) have not been included. In many of the studies listed here, observations were confined to isolated glomeruli or cells and extracts prepared from them; tubules and interstitium were not examined at all. In some investigations that focused on tubules, only cortical tubules were evaluated. For that reason, blank spaces under the heading "Kidney Tissue Site" reflect the absence of information, rather than a determination that production of the marker in question has not localized to a particular renal-cell compartment.

[c] With renal injury, normal concentrations were reduced, not increased.

[d] Tx = Thromboxane.

[e] PG = Prostaglandin.

Biologic Markers of Effect

TABLE 4-2 Potential Sources of Biologic Markers of Cell Injury or Inflammation in Renal-Tissue Samples

Microanatomic Compartment	Molecules*
Glomerular mesangium	PDGF, IL-1, IL-6, IL-8, PAF, IGF-1, TNFα, MCP-1, TGFβ, endothelin, prostaglandins, O_2 and NO metabolites, MHC antigens, adhesion molecules
Tubular epithelium	TNFα, endothelin, MHC antigens, EGF, uromodulin, adhesion molecules
Endothelium	PDGF, PAF, MHC antigens, endothelin, adhesion molecules
Bone-marrow-derived cells (glomerular mesangium, interstitium, infiltrates)	PDGF, IL-1, IL-6, TNFα, TGFβ, PDGF, PAF, IGF-1, MHC antigens, eicosanoids, O_2 and NO metabolites

*Abbreviations: MHC = major histocompatibility; PAF = platelet-activating factor; PDGF = platelet-derived growth factor; IL = interleukin; IGF = insulin-like growth factor; TNF = tumor-necrosis factor; TGF = transforming growth factor; EGF = epidermal growth factor; MCP = monocyte chemotactic peptide

A wide variety of hormone-like proteins and biologic-response modifiers can be produced in the kidney as a response to cell damage and tissue inflammation. Although bone-marrow-derived mononuclear cells are a major and well-studied source of those molecules, many might be produced by intrinsic kidney cells as well. This table summarizes a large (and rapidly growing) literature documenting possible intrarenal sources of some biologic markers of cell injury that might have clinical value. Glomerular sites of synthesis of individual molecules have been relatively easy to identify because glomeruli are readily separated from other kidney tissue components for extractions, tissue culture, and other analytic procedures. Tubular, interstitial, and vascular cells have been less amenable to detailed investigation, although recent advances in tissue-culture methods suggest that the metabolic and synthetic capacity of the specialized kidney cells outside the glomerulus will soon be more completely characterized.

TABLE 4-3 Potential Biologic Markers of Renal Cancer

Marker	References
Clinical and pathologic markers	
Nuclear grade	Fuhrman et al., 1982; Medeiros et al., 1988; Madeiros and Weiss, 1990; Selli et al., 1983
Tumor size	Fuhrman et al., 1982; Medeiros et al., 1988; Madeiros and Weiss, 1990; Selli et al., 1983
Tumor stage	Fuhrman et al., 1982; Libertino et al., 1987; McNichols et al., 1981; Medeiros et al., 1988; Robson et al., 1969; Skinner et al., 1971; Waters and Richie, 1979
Morphometric markers	
Nuclear abnormalities	Tosi et al., 1986
Quantitative morphometry	Tarnowski et al., 1993; Unger et al., 1993; vanden Houte et al., 1991
Proliferative markers	
BrDU	Rew et al., 1991
DNA ploidy	Baisch et al., 1985, 1990; Bringuier et al., 1993; el-Naggar, et al., 1994; Grignon et al., 1989; Kloppel et al., 1986; Kumar and Kumar et al., 1993; Lanigan et al., 1993; Leyh et al., 1992; Ljungberg et al., 1956; Oosterwijk et al., 1988; Rainwater et al., 1987, 1991; van Leeuwen et al., 1987; Veloso et al., 1992; Yoshida et al., 1986
Ki-67	Baretton, et al., 1991; Chow et al., 1993; de Riese et al., 1993; Kaiser et al., 1991
PCNA	Delahunt et al., 1993; Hiaish et al., 1993; Lipponen et al., 1994; Tanioka et al., 1993
Ploidy S-phase	al-Abadi and Nagel 1988; Baretton et al., 1991; Ellis et al., 1992; Kieseweter et al., 1987; Larsson et al., 1993; Masters et al., 1992; Rew et al., 1991; Tanioka et al., 1993; Yu et al., 1991, 1993

Biologic Markers of Effect

Table 4-3, continued

Marker	References
Growth factors	
EGF and EGF receptor	Chow et al., 1993; Fuse et al., 1992; Ishikura et al., 1991; Kaiser et al., 1991; Kimball et al., 1984; Kotake and Kinouchi 1994; le Coutre et al., 1992; Narayan and Roy, 1992; Tokito et al., 1991; Volm et al., 1993; Weidner et al., 1990
Endothelin-2 (ET-2)	Ohkubo et al., 1990; Onda et al., 1991; Shinmi et al., 1993; Tokito et al., 1991; Yorimitsu et al., 1992
fgf	Fujimoto et al., 1991; Herrera, 1991
tgf-alpha and tgf-beta	Gomella et al., 1989; Herrera et al., 1991; Kimball et al., 1984; Kotake and Kinouchi, 1994; le Coutre et al., 1992; Mydlo et al., 1989; Petrides et al., 1990; Thalacker and Nilsen-Hamilton, 1987
tnf	Dosquet et al., 1994; Muraki and Nakazano, 1992; Wade et al., 1989
IFN	Hoogstraten et al., 1982; Nagata, 1993
IGF	Martinerie and Perbal, 1991; Zumkeller et al., 1993
p170-glycoprotein	Eguchi et al., 1992; Rochlitz et al., 1992
p-glycoprotein	Inoue et al., 1989; Lizard et al., 1992; Moll et al., 1993; Moriyama et al., 1991; Nakagawa et al., 1994; Sato et al., 1990; Volm et al., 1993; Yamazaki et al., 1988, 1991
Oncogenes	
cl 4-2	Boldog et al., 1991
c-*etsl*	Wemert et al., 1992
erb (c-*erbB*-2)	Herrera, 1991; Lipponen et al., 1994; Lunec et al., 1992

Table 4-3, continued

Marker	References
fos, abl	Peter, 1991; Sunderman et al., 1990
myc	Drabkin et al., 1985; Gemmill et al., 1989; Kinouchi et al., 1989; Kotake and Kinouchi, 1994; Martinerie et al., 1992; Weidner et al., 1990; Yao et al., 1988
neu	Rotter et al., 1992; Toyoshima, 1990
ras (p21)	Karthaus et al., 1987; Kumar et al., 1991; Mannens et al., 1987; Nanus et al., 1989; Ohgaki et al., 1991; Trapman, 1992; Volm et al., 1993; Waber et al., 1993; Weiss et al., 1991
src	Nanus et al., 1991
Tumor-suppressor genes	
p53	Bot et al., 1994; Brooks et al., 1993; Haber and Buckler 1992; Lipponen et al., 1994; Suzuki and Tamura 1993; Suzuki et al., 1992; Trapman, 1992; Weghorst et al., 1994
Rb	Brooks et al., 1993; Horikawa and Oshimura, 1991; Sabatier et al., 1989; Strong, 1993; Trapman; 1992; Waber et al., 1993
Cytogenetics	
CA50	Hershman et al., 1987
CEA	Akaza et al., 1987; Blouin et al., 1989; Cisternino et al., 1986; Lanzafame et al., 1986; Martin et al., 1976; Popper et al., 1987; Uchida et al., 1986

Table 4-3, continued

Marker	References
Chromosome 3	Anglard et al., 1991; Boldog et al., 1991; Brauch et al., 1990; Dietrick and Droz, 1992; Erlandsson et al., 1990, 1991a,b; Fournet et al., 1992; Fuzesi and Cober 1992; Glenn et al., 1991; Hoogstraten et al., 1982; Kohno et al., 1993; Kovacs et al., 1989a, 1991a; Ljungberg et al., 1991; Presti et al., 1991, 1993; Sugao et al., 1992; Szymanski et al., 1993; Teyssier and Ferre, 1990; van der Hout et al., 1993; Yamakawa et al., 1991
Chromosome 5	Hino et al., 1993; Jordan et al., 1989; Kovacs et al., 1987; Morita et al., 1991; Presti et al., 1991
Chromosome 7	Dal Cin et al., 1992; Emanuel et al., 1992; Hughson et al., 1993; Kovacs et al., 1991b; Limon et al., 1990; Ljungberg et al., 1991; Maloney et al., 1991; Miles et al., 1988; Nordenson et al., 1988; Sugao et al., 1992; Weaver et al., 1988; Yoshida et al., 1988
Chromosome 16	Green et al., 1994; Grundy et al., 1994; Huff et al., 1992; Maw et al., 1992
Chromosome 17	Anglard et al., 1991; Bergerheim et al., 1989; Haugen et al., 1990; Kovacs, 1989; Kovacs et al., 1991b; Ogawa et al., 1992; Presti et al., 1991; Reiter et al., 1993
Chromosome Y	Dal Cin et al., 1989; Jordan et al., 1989; Kovacs et al., 1991b; Walker et al., 1992
Chromosome 12	Annab et al., 1992; Parshad et al., 1992
Chromosome 14	Boehm et al., 1988; Davis et al., 1991; Finver et al., 1989; Fournet et al., 1992; Royer-Pokora et al., 1989
Chromosome 18	Casalone et al., 1992; Grollino et al., 1993; Presti et al., 1991
Chromosome 20	Kovacs et al., 1989b
Chromosome 21	Nanus et al., 1989

Biologic Markers in Urinary Toxicology

Table 4-3, continued

Marker	References
Cytokeratins	Blouin et al., 1989; Dierick et al., 1991; Droz et al., 1990; Kaiser et al., 1991; Kumar and Kumar, 1993; Kumar et al., 1988; Popper et al., 1987; Schroder et al., 1992
Double minutes	Moriyama-Gonda et al., 1991
Exo-1	Klingel et al., 1992
Fibronectin	Kloppel et al., 1986
Glutathione *S*-transferase	Grignon et al., 1994; Kurokawa et al., 1990; Sabatier et al., 1989; Volm et al., 1993
IL-6	Koo et al., 1992; Miki et al., 1989; Sabatier et al., 1989; Takenawa et al., 1991
leu-7	Schroder et al., 1992
PDGF	Ambrus et al., 1992
Pepsinogen II (PG II)	
Polyamines	Balitskaia et al., 1992; Koide, 1992
Tissue plasminogen activator (TPA)	Blouin et al., 1989; Gohji et al., 1990; Hata, 1989; Popper et al., 1987; Yamazaki et al., 1988, 1991

Tumor-associated antigens

C219	Rochlitz et al., 1992
Uromonoclonals	Bander et al., 1989; Blouin et al., 1989; Cohen et al., 1988; Gu et al., 1991; Hashimura et al., 1989; Kerr et al., 1990; Yagoda and Bander, 1989

Serum markers

CRP	Blay et al., 1992
Erythropoietin	Da Silva et al., 1990; Rigatti et al., 1990
fgf	Fujimoto et al., 1991
gamma-Enolase	Takashi et al., 1988

Table 4-3, continued

Marker	References
IGF-I	Pollak et al., 1989
IL-6	Blay et al., 1992

Biologic Markers in Urinary Toxicology

TABLE 4-4 Potential Biologic Markers of Bladder Cancer

Marker	References
Clinical and pathologic markers	
Dysplasia	Berman et al., 1991
Host inflammatory response	Levin et al., 1991; Nouri et al., 1991; Tomita et al., 1990
Morphologic variants of cancer	Christopher et al., 1991; Fossa, 1992
Tumor volume	Herr, 1992; Lee et al., 1993
Tumor stage	Escudero Barrilero et al., 1991
Tumor grade, WHO	Jordan et al., 1987
Morphometric markers	
Chromatin abnormalities	Van Velthoven et al., 1994
Nuclear abnormalities	Lipponen and Eskelinen, 1990; Lipponen et al., 1991a, 1992; Nielsen et al., 1988; Sanchez-Fernandez de Sevilla et al., 1992; van der Poel et al., 1991
Nucleolar abnormalities	Cairns et al., 1989; Hitmair et al., 1994; Neilsen et al., 1988
Proliferative markers	Cohen et al., 1993
BrDU	Hattori et al., 1988; Waldman et al., 1993
Ki-67	Mulder et al., 1992
p105	Horowitz et al., 1990
PCNA	Malmstrom et al., 1992; Waldman et al., 1993
Ploidy S-phase fraction	Lipponen et al., 1991b; Wheeless et al., 1993
Ploidy analysis and genetic markers	Trapman, 1992
Chromosome 1	Borland et al., 1992; Hopman et al., 1991
Chromosome 3	Kao et al., 1919; Klingelhutz et al., 1992
Chromosome 5	Borland et al., 1992

TABLE 4-4, continued

Marker	References
Chromosome 7	Berrozpe et al., 1990; Borland et al., 1992; Hopman et al., 1991; Waldman et al., 1991
Chromosome 9	Borland et al., 1992; Hopman et al., 1991; Tsai et al., 1990
Chromosome 11	Borland et al., 1992; Fearon et al., 1985; Hopman et al., 1991; Kao et al., 1919; Proctor et al., 1991; Tsai et al., 1990
Chromosome 17	Borland et al., 1992; Olumi et al., 1990; Tsai et al., 1990
DNA ploidy	Bonner et al., 1993; Gerber et al., 1991; Hemstreet et al., 1991; Koss and Czerniak; 1992; Montironi et al., 1987; van der Poel et al., 1991; West et al., 1987; Wheeless et al., 1993
erb-B2	Coombs et al., 1991; Gardiner et al., 1992; Lunec et al., 1992; Moriyama et al., 1991b; Serio, 1991; Wright et al., 1991
Loss of heterozygosity	Brewster et al., 1994; Cairns et al., 1991; Ishikawa et al., 1991; Oka et al., 1991
myc	Kotake et al., 1990
neu (p185)	Rao et al., 1993; Wood et al., 1991
p53	Borland et al., 1992; Fujimoto et al., 1992; Lunec et al., 1992; Morkve and Hostmark, 1991; Oka et al., 1991; Sidransky et al., 1991; Trapman, 1992; Wright et al., 1991
ras (p21)	Borland et al., 1992; Enomoto et al., 1990; Grimmond et al., 1992; Miao et al., 1991; Nagata et al., 1990; Pratt, et al., 1992; Saito, 1992; Theodorescu et al., 1991; Wang, 1991
Rb	Borland et al., 1992; Cairns et al., 1991; Goodrich et al., 1992; Horowitz et al., 1990; Ishikawa et al., 1991; Trapman, 1992
src	Fanning et al., 1992

TABLE 4-4, continued

Marker	References
Growth factors	
e-cadherin	Bringuier et al., 1993
EGF and EGF receptor	Fuse et al., 1992, Harney et al., 1991a,b; Ishikura et al., 1991; Kageyama et al., 1991; Messing et al., 1987; Neal et al., 1989, 1990; Rao et al., 1993; Smith et al., 1989; Theodorescu et al., 1991; Wood et al., 1992
fgf	Ravery et al., 1992; Valles et al., 1990
P-glyprotein	Kageyama et al., 1991; Moriyama et al., 1991a
Plasminogen activator	Hasui et al., 1992; Hiti et al., 1990; See, 1992
tgf-beta and tgf-alpha	Hiti et al., 1990; Kawamata et al., 1992
tnf	Hitti et al., 1990
Differentiation	
amf	Javadpour and Guirguis, 1992; Nabi et al., 1992
Cytokeratin	Basta et al., 1988; el-Mohamady et al., 1991; Helmy et al., 1991; Konchuba et al., 1992; Schaafsma et al., 1990; Sumi et al., 1990
ECP	Lose and Frandsen, 1989
EMA	Ring et al., 1990
F-actin	Rao et al., 1991, 1993
G-actin	Rao et al., 1993
PNA, WGA	Orntoft et al., 1988
Blood-group antigens	
ABH	Das and Glashan, 1988; Das et al., 1986; Limas, 1990; Malmstrom et al., 1991; Orntoft et al., 1988; Sanders et al., 1991; Tichy et al., 1991; Yamada et al., 1991

Biologic Markers of Effect

TABLE 4-4, continued

Marker	References
Due ABC 3	Decken et al., 1992
HLA	Cordon-Cardo et al., 1991; Eryigit and Kirkali, 1990; Levin et al., 1991, 1992; Nouri et al., 1990; Tomita et al., 1990
Lewis-X	Limas, 1991; Matsusako et al., 1991
Lewis antigens	Fradet et al., 1990a; Langkilde et al., 1991a,b; Limas, 1991; Matsusako et al., 1991
Tumor-associated antigens	
10D1	Hijazi et al., 1989
12F6	Hijazi et al., 1989
19A211	Cordon-Cardo et al., 1992; Fradet et al., 1990a
2A6	Messing et al., 1984
2E1	Messing et al., 1984
3-48-2, 48-1,3-50-3	Summerhayes et al., 1985
3-71-1,94-3	Summerhayes et al., 1985
3C6	Summerhayes et al., 1985
3G2-C6	Lin et al., 1988; Young et al., 1985
4-72-2	Summerhayes et al., 1985
486P 3/12	Arndt et al., 1987; Huland et al., 1990, 1991
6D1	Hijazi et al., 1989
7C12	Hijazi et al., 1989
8-30-3,771-,2-94-2	Summerhayes et al., 1985
9A7	Messing et al., 1984
AN43	Liebert et al., 1989
BB369	Liebert et al., 1989

TABLE 4-4, continued

Marker	References
BIUH4, BIUH6, BIUH9	Guo, 1992
BL 2-10D1	Longin et al., 1989
BLCA-8	Lose and Frandsen, 1989; Walker et al., 1989
C3	Young et al., 1985
CA50	Morote et al., 1990
CEA	Boileau et al., 1987; Morote et al., 1990; Piana et al., 1991
DD23	Bonner et al., 1995; Grossman et al., 1992
G4,E7	Chopin et al., 1985
GF 26.7.3	Baricordi et al., 1985
HBA4, HBE3	Masuko et al., 1984
HBE10	Masuko et al., 1984
HBF2	Masuko et al., 1984
HBG9	Masuko et al., 1984
HBH8	Masuko et al., 1984
J143	Fradet et al., 1984, 1986
M344	Bonner et al., 1993; Cordon-Cardo et al., 1992; Fradet et al., 1987; Rao et al., 1993
Mano 4/4	Arndt et al., 1987
OM5	Fradet et al., 1984, 1986, 1990b
P7 A 5-4	Ben-Aissa et al., 1985
RBS-31, RBS-85, RBA-1, HBP-1	Masuko et al., 1989
SK 4H-12	Ben-Aissa et al., 1985
T16	Bretton et al., 1989; Fradet et al., 1984, 1986, 1990b
T23	Fradet et al., 1984, 1986

TABLE 4-4, continued

Marker	References
T43	Fradet et al., 1984, 1986, 1990b
T87	Fradet et al., 1984, 1986
T110	Fradet et al., 1984, 1986
T138	Fradet et al., 1984, 1986, 1990b
Thomsen-Friedenrich	Yamada et al., 1991
Tu-MARK-BTA	Yogi et al., 1991
urine TPA	Carbin et al., 1989; Morote et al., 1990

TABLE 4-5 Potential Biologic Markers of Prostatic Cancer

Marker	References
Clinical and pathologic markers	
Atypical adenomatous hyperplasia	Bostwick et al., 1993
Hyperplasia metaplasia atrophy	Morote et al., 1986
Host inflammatory response	van Weerden et al., 1993
Morphologic variants of cancer	Davies et al., 1988; Lloyd et al., 1992; Partin et al., 1989
Prostatic intraepithelial neoplasia	Montironi et al., 1990
Tumor grade, (Gleason grade, nuclear grade)	Abrahamsson et al., 1987
Tumor stage	Azzopardi and Evans, 1971; Thompson, 1990
Tumor volume	Helpap, 1988
Morphometric markers	
Chromatin abnormalities	Shah et al., 1987
Nuclear abnormalities	Shah et al., 1987; Umbas et al., 1992
Nucleolar abnormalities	Wenk et al., 1977
Nucleolar organizer region	McNeal et al., 1988a
Proliferative markers	
BrDU	Abrahamsson et al., 1989; Ercole et al., 1987
Ki-67	Ercole et al., 1987; Shah et al., 1991; Thompson, 1990
PCNA	Sherwood et al., 1990
Ploidy S phase	Gittes, 1987; Grignon and Wright, (in press); Sporn, 1992; Stege et al., 1992
Thymidine labeling	Lovern et al., 1975
Growth factors	
egfr	Eaton et al., 1988; McManus et al., 1976; Sacks et al., 1975

Biologic Markers of Effect

TABLE 4-5, continued

Marker	References
fgf	Schiebler et al., 1992; Visakorpi et al., 1991
IGF, NGF, PDGF, KGF	Visakorpi et al., 1991
tgf	Fjellestad-Paulsen et al., 1988; Ito et al., 1988; Mellon et al., 1992; Schiebler et al., 1992; Tsukamoto et al., 1988
tnf	Fruehauf and Sinha, 1992
Oncogenes	
erb (c-*erb*B-2)	McManus et al., 1976
myc	Schiebler et al., 1992; Voeller et al., 1991
neu (p185)	Bostwick et al., 1994; Kuhn et al., 1993; Latil et al., 1994; Sadasivan et al., 1993; Veltri et al., 1994; Zhau et al., 1994
ras (p21)	Amico et al., 1991; Morote Robles and Ruibal Morell, 1987; Schiebler et al., 1992; Yoshiki et al., 1987
src, fos, abl	Visakorpi et al., 1991
Tumor-suppressor genes	
p53	Amico et al., 1991; Epstein and Lieberman, 1985; McManus et al., 1976; Voeller et al., 1991
Rb	Amico et al., 1991
Neuroendocrine differentiation	
ACTH	Bauer, 1988; Eble and Epstein, 1990; Fox et al., 1993; Fukutani et al., 1983; Hagood et al., 1991; Nemoto et al., 1990; Park et al., 1987; Purnell et al., 1984; Rubenstein et al., 1988; Sellwood et al., 1969; Tarle and Rados, 1991; Vuitch and Mendelsohn, 1981
ADH	Doctor et al., 1986; Guinan et al., 1989; Newmark et al., 1973
Aldosterone	Clar-Blanch et al., 1992

Biologic Markers in Urinary Toxicology

TABLE 4-5, continued

Marker	References
alpha-HCG	Eble and Epstein, 1990
Argentaffin	Turbat-Herrera et al., 1988; Weaver et al., 1992
beta-Endorphins	Eble and Epstein, 1990; Nemoto et al., 1990
beta-HCG	Eble and Epstein, 1990; Maddy et al., 1989; Nabi et al., 1992; Sesterhenn et al., 1991; Sukumar et al., 1991; Tawfic et al., 1993; Webster et al., 1959
BOM	Helpap, 1980; Purnell et al., 1984
Bombesin (GRP)	Maddy et al., 1989
Calcitonin	Bauer, 1988; Eble and Epstein, 1990; Eskelinen et al., 1991; Fox et al., 1993; Fuse et al., 1991; Heim et al., 1977; Helpap, 1980; Maddy et al., 1989; Purnell et al., 1984; Sarkar et al., 1992
CGRP	Fuse et al., 1991
Chromogranin	Bostwick et al., 1993; Maddy et al., 1989
Corticotropin	Fox et al., 1993
Enkephalin	Eble and Epstein, 1990
Glucagon	Eble and Epstein, 1990; Maddy et al., 1989; Purnell et al., 1984
Glucocorticoids	Clar-Blanch et al., 1992
HIAA	Radjaipour et al., 1994
Lipofuscin	Weaver et al., 1992
Neuron-specific enolase	Maddy et al., 1989; Shalitin et al., 1991
Parathormone	Fuse et al., 1991; Purnell et al., 1984
Prolactin	Maddy et al., 1989

Biologic Markers of Effect

TABLE 4-5, continued

Marker	References
Serotonin	Broder et al., 1977; Eble and Epstein, 1990; Eskelinen et al., 1991; Helpap, 1980; Maddy et al., 1989; Purnell et al., 1984; Sassine and Schulman, 1992
Somatostatin	Eaton, et al., 1988; Eble and Epstein, 1990; Eskelinen et al., 1991; Maddy et al., 1989; Purnell et al., 1984
TSH	Eble and Epstein, 1990; Maddy et al., 1989
Other proteins	
amf	Coombes et al., 1974; Mahadevia et al., 1983; Watanabe et al., 1988
CA50	di Sant' Agnese and de Mesy Jensen, 1987;
Cadherins	Delaere et al., 1988; Scrivner et al., 1991
Cathepsin B	Buck et al., 1992; Guenette et al., 1994; Hasnain et al., 1992
CEA	di Sant' Agnese and de Mesy Jensen, 1987; Milani et al., 1986; Morote et al., 1990; Papapetrou et al., 1980
Cytokeratins	Boag and Young, 1992; Jarrett et al., 1964; Partin et al., 1993; Perlman and Epstein, 1990; Shulkes et al., 1991
Glutathione S-transferase	Capella et al., 1981
Inhibin	Fekete et al., 1989
leu-7	Grasso, 1952
nmp	Kleer et al., 1993
5'-Nucleotidase	Pretlow et al., 1994
Ornithine decarboxylase	Wu et al., 1994
PD-41	Bostwick, 1990
Pepsinogen II (PG II)	Grasso, 1952

Biologic Markers in Urinary Toxicology

TABLE 4-5, continued

Marker	References
Polyamines	Doctor et al., 1986; Manteuffel-Cymbrorwska, 1993; Ryzlak et al., 1992
Tissue inhibitor metalloproteinases (TIMP I-III)	Baker et al., 1994; Knox et al., 1993; Stearns and Wang, 1994; Stetler-Stevenson et al., 1993
Tissue plasminogen activator (TPA)	Grasso, 1952
Type IV collagenase	Epstein and Woodruff, 1986
Blood-group antigens	
A and B	Bussemakers et al., 1991
Lewis-X antigen	Bussemakers et al., 1991
Serum markers	
gamma-Seminoprotein	Akimoto et al., 1988; Anonymous, 1985; McNeal et al., 1988b; Molland, 1978; Oesterling, 1991; Tetu et al., 1989; Trapman, 1992; Wang and Kawaguchi, 1987
p21 (not ras)	Beckett et al., 1993
PAP	Bostwick, 1989; Cantrell et al., 1981; Gleason, 1990; McNeal et al., 1988b; Miki et al., 1980; Oesterling, 1991; Piana et al., 1991; Sade and Barrack, 1991; Steiner and Barrack, 1992; van Dalen et al., 1988; Weber and Rohner, 1987; Wise et al., 1965
PSA	Bookstein et al., 1993; Cantrell et al., 1981; di Sant' Agnese, 1988, 1992; Gleason, 1990; Miki et al., 1980; Nagle et al., 1991; Ohashi et al., 1987; Rojas-Corona et al., 1987; Sherwood et al., 1991; Shinoda et al., 1988; Stamey et al., 1987; Steiner and Barrack, 1992; Tinari et al., 1993; van Dalen et al., 1988; Yogi et al., 1991

TABLE 4-6 Etiology, Evidence, and Potential Markers of Interstitial Cystitis

Etiology	Evidence Pro	Con	Potential Markers
Infection	Finding of fastidious organisms	Failure to respond to antimicrobials; no findings of positive markers	Antibodies, genome, or immunohistochemical detection of proteins of infectious agent
Obstruction of vascular and lymphatic channels	Fibrosis, submucosal edema; onsets related to pelvic infection or operations	Experimental models do not produce IC; surgical treatments for IC will produce blockage	Fibrosis; submucosal edema
Neuropathy	Pain, focal neural inflammation; response to amitriptyline; afferent fibers or neurotransmitters can control inflammatory response directly or through mast cells	Perineural inflammation not found in other studies; clinical therapy highly unpredictable	Focal inflammation in and around intramural and perivesical nerve bundles; neurotransmitter levels and receptors; mast cells
Endocrinopathy	Association with oophorectomy, sex bias, small numbers in premenopausal and postmenopausal women	Symptomatic response to hormonal therapy unpredictable	Endocrine receptors and levels
Psychoneurosis	Neurotic behavior pattern	Behavior seems to be a result of disease	--

TABLE 4-6, continued

Etiology	Evidence Pro	Evidence Con	Potential Markers
Inflammation	Similarities to systemic lupus and other autoimmune disorders; chronic inflammation in pathology; some responses to steroids, NSAIDS, and immunosuppressants; positive FANA at times; Ig deposits in bladder	Ig deposits nonspecific in origin; inflammation in bladder frequently not present; phenotypes of inflammatory cells in bladder; variable response to immunosuppressant therapy; relief on diversion	Immune mediators; immune-cell typing; specific antibody levels
Deficiencies in bladder lining	"Leaky" epithelium in IC, also produced in normal subjects by protamine and reversed by heparin; decreased GAG excretion in IC patients; increased GAG produced by hydro-distention; ultrastructure shows loose urothelium	GAG treatment not always effective; deficiency never demonstrated directly	Decreased GAG in urine; altered GAG on umbrella cells
Toxic substances in urine	Diversion relieves symptoms; increased serum anti-THP IgG; cytotoxic urine	Some evidence suggests defect is from inside bladder	Antibodies vs. urinary constituents

5
BIOLOGIC MARKERS IN EXTRAPOLATION

Extrapolation is common in many scientific disciplines—so common that it can easily go unrecognized. Extrapolation is concerned with the translation or transfer of relationships (e.g., clinical measures and mathematical variables) observed in one setting to another setting. When we extrapolate from experimental systems to humans, we assume that some cause-effect relationships are the same in humans as in the experimental systems. The more we understand about the variables in an experimental situation, the more we will understand about the validity of such extrapolation to human situations. In the context of biologic markers and urinary toxicity, our goal is to gain a better understanding of the overall relationship between exposure and disease by examining those markers. The relationships among markers that can be observed and tested in experimental systems might be extrapolatable from those systems to situations of concern with respect to human health risks, such as occupational or environmental exposures to urinary toxicants.

The main purpose of extrapolation in any context is to make it possible to predict. In the clinical setting, the observation of particular symptoms in a patient leads clinicians to conclude that the patient has a particular disease or will soon manifest other symptoms. In that case, the clinicians are extrapolating from their experience with some patients to a new patient. Extrapolation is required to support prediction and the design of a suitable treatment.

Extrapolation from epidemiologic studies is commonly used to predict risks to other cohorts. Every epidemiologic study is restricted to some population. To extrapolate from a study population to other potentially affected populations, one must consider the differences between the study population and the potential target populations. Variables related to and affecting the development of the health effect under investigation must be considered; differences between the study and target populations with respect to those variables will influence how the extrapolation is completed by influencing or modifying the relationships between variables and the health effect.

Consider the situation of a new chemical that is proposed for use or that will be a byproduct of some new operation. Direct evidence that the chemical causes adverse effects in humans is lacking. Some important considerations include the potential for adverse health effects in humans exposed to the chemical and, if adverse effects do result from exposure, the magnitude of the effects after exposures of different severities. It is clear that predictions are required. However, the basis for the predictions cannot be the previous human experience; there is no previous experience. It might be possible to extrapolate, but the model from which the extrapolation is made will of necessity be a nonhuman model.

In the scenario just described, the need for extrapolations from experimental systems to humans is apparent. Many other scenarios, both clinical and "population-based," will require the prediction of human responses from data obtained in nonhuman test systems, especially in light of the thousands of chemicals that are produced, used, and released into the environment as byproducts of our way of life. Indeed, it is the desire to be predictive that drives the need to develop and apply good experimental systems. Such systems have at least four advantages: they allow predictions of human health effects and the magnitude of those effects before human exposure occurs or before adverse effects are manifested in exposed populations; they can be altered to clarify aspects of the process leading from exposure to adverse health effects when similar experimentation in humans would be unethical; they can be designed to eliminate many factors that confound the determination of cause-effect relationships in epidemiologic studies; and they can suggest directions for epidemiologic investigation by providing the hypotheses that epidemiologic studies might be able to test.

Previous chapters have focused on markers of susceptibility, exposure, and effect (particularly early effect) and their value in clinical situations; the sooner a disease state or precursor of a disease state can be identified, the greater the chance of successful therapy or treatment. This chapter focuses on the prediction of effects, not in an individual patient but rather in a (hypothetical) population of humans potentially exposed to a supposed toxicant. In this context, one is concerned about maintaining the health of the population by predicting whether an activity or an exposure is likely to produce harmful consequences in that population—often without previous observations of humans exposed at the magnitudes of interest. The objective is to learn how to tie chemical exposure under various scenarios to the dose or amount of the chemical that reaches the body, to the amount that is absorbed and distributed to target tissues, and ultimately to the effect.

How does one make such predictions? The discipline of risk assessment addresses that question. Human-health risk assessment is a complex, multifaceted process that relies on data and scientific principles from many disciplines to determine whether a chemical is toxic and the likelihood of manifestation of detrimental effects under specific conditions. Risk assessment draws on a variety of methods and models to examine and evaluate information about the toxicity of a chemical. A useful and well-described system for looking at toxicity information is the risk-assessment paradigm originally depicted by the National Research Council in the 1983 publication *Risk Assessment in the Federal Government: Managing the Process*. That system for organizing and analyzing risk information follows an understandable series of steps portraying qualitative and quantitative aspects and typically includes some or all of the following: hazard identification, dose-response assessment, exposure assessment, and risk characterization (NRC, 1983). The result is a characterization of the potential adverse health effects of human exposures to a chemical.

The characterization of potential adverse human health effects requires extrapolation. That typical of many risk assessments includes extrapolation from animal test species to humans, from large exposure to small exposure, and from one route of exposure to another. Risk assessments are not infallible, and the relative accuracy of a risk assessment depends not only on the scope and quality of the scientific data but also on the reliability of the methods and the validity of the models used. The degree of confidence in a risk assessment depends on how well data and model quality are validated and on the extent to which uncertainty is quantified.

As described in Chapter 2, various aromatic amines are recognized human bladder carcinogens (NRC, 1981). Consider, for example, an assessment of human bladder-cancer risk associated with dermal exposure to one of those aromatic amines—4,4'-methylene-bis(2-chloroaniline), or MOCA. For such an assessment, we might be required to extrapolate relationships that were observed in dogs exposed to MOCA in their diet (Stula et al., 1977) at doses far greater than the human exposures of interest. The conclusion that the occurrence of bladder cancers in dogs exposed to MOCA implies a bladder-cancer risk for humans assumes that the qualitative relationship between MOCA exposure and bladder cancer can be extrapolated from dogs to humans. Such cross-species extrapolation is typical of the hazard-identification component of risk assessment, that is, the determination of the existence of a cause-effect relationship between chemical exposure and adverse health effect.

Mathematical extrapolation is particularly relevant to risk assessment of a most useful kind, i.e., quantitative risk assessment. Quantitative risk assessment is a means of providing a measure of the risk of some harm as a result of a specific exposure to some substance or activity (Almeder and Humber, 1987), and mathematical extrapolation can be conceived of as the transfer of the quantitative relationships estimated in one scenario to another scenario, whether those scenarios differ in species (animal-to-human extrapolation), in magnitude of exposure (dose extrapolation), or in route of exposure (route-to-route extrapolation). Again, it is assumed, lacking data to the contrary, that a relationship observed in one scenario is valid in the other. Risk can be stated as the magnitude of exposure that is estimated to be without substantial likelihood of harmful effect, or stated as the probability of occurrence of harmful effect.

The remainder of this chapter concerns answers to two questions: What is the basis for concluding that extrapolation is reasonable? Can it be said that one approach to extrapolation is better than another approach? It is in relation to the second question that the relevance and utility of biologic markers in risk assessment become apparent.

BASIS OF EXTRAPOLATION

Animal Studies

A fundamental principle of toxicology is that results of animal studies can be applied to humans. The scientific basis for assuming that animals are good surrogates for humans and therefore a suitable basis for extrapolation to humans is overwhelming. If one considers that the genetic makeup of a mouse or a rat is more than 95% and of a monkey is more than 99% identical with that of a human, it is reasonable to assume that these animals in particular and mammals, in general, will react to infectious agents and chemical stressors much as humans will. Among mammals, most of the host defense mechanisms (barrier and immune) and metabolic (anabolic and catabolic) systems are similar. In particular, the urinary systems of most mammals are very similar. Although specific, often subtle, differences between humans and other animals with respect to renal function have been demonstrated, the vast majority of human renal responses to xenobiotics mimic what has been observed in other species. Biologic markers of renal transport, concentrating, and metabolic functions of the kidney are reproduced in many species, including humans, although quantitative differences have been demonstrated. Therefore, it is reasonable to use animal models for extrapolation to humans unless specific information on specific chemicals dictates otherwise.

Many epidemiologic investigations have, in fact, been suggested as a result of animal studies, and the epidemiologic findings have tended to support the results of the animal studies. For example, several of the current epidemiologic studies of heavy-metal toxicity, including small exposures, were initiated in response to urinary toxicity observed in a large number of animal studies and were undertaken specifically because of the likelihood that the human response would mimic that seen in animal test species. Mechanistic studies of chemically induced nephrotoxicity in animals influenced or stimulated epidemiologic studies of a variety of substances, including many of the halogenated hydrocarbons, such as chloroform, hexachlorobutadiene, and bromobenzene. Much of the epidemiologic investigation and mechanistic understanding of anesthetic, analgesic, and antibiotic nephropathy has been driven by observations of nephrotoxicity in animals. Classic examples of the direct application of animal studies to epidemiology of known nephrotoxicants are the use of antineoplastic and immunosuppressant drugs (e.g., cisplatin and cyclosporin, respectively).

Studies of carcinogenic responses in laboratory animals and in humans have revealed substantial correlations (Crump et al., 1989). The results suggest that there are reasonable approaches to extrapolating cancer responses observed in test species—approaches that appear to predict fairly well the responses in humans.

Identification of chemical hazards should include assimilation and evaluation of all relevant information. Appraisal of physical and chemical properties and structure-activity relationships can sometimes provide important indications of potential toxic characteristics. Markers of urinary function and chemical toxicity in experimental animals can be studied at various levels of tissue structure and organization (Table 5-1). This is in contrast with human studies, in which only noninvasive studies of renal function are possible. Markers identified through in vitro studies of systems that use animal and human cells or tissues in culture can often give insight into potential toxicity. However, because of the intricacy of the body, only whole-animal studies or observations in humans provide information on the operation of multiple cells, tissues, and organs under the influence of complicated feedback mechanisms. Animals are necessary in the study of chemical-induced toxicity and for the development and validation of markers because studies that involve modulation of cellular responses and tissue sampling cannot be performed in humans.

Traditionally, experimental animal studies have been of most value for identifying markers that can be used to predict target-organ effects, understand dose-response relationships, and study mechanisms of action. The studies include measurements of function, blood chemistry, urinalysis (including cytology), histopathology, and electron microscopy. Metabolism and transport peculiar to the kidney are often routine parts of such studies.

Animal models developed as surrogates for humans in the study of renal and urinary function should conform to some general principles, which are applicable to any organ system, although they are discussed here in the context of renal and urinary toxicology. They include the following:

- The animal model should be reproducible within and among laboratories. It should not be so complex that only a few laboratories could do the study.
- The model should be peculiar to the part of the urinary tract under consideration. This characteristic can be realized only with sophisticated procedures that permit study of discrete nephron segments.

- The model should be sensitive enough to differentiate normal from abnormal changes in structure or function.
- The model should be able to measure alterations in renal structure or function caused by exogenous agents.

Whole-animal studies are usually the first step in evaluating the potential toxicity of a given agent, and these studies involve assessment of urine and plasma for markers indicative of organ function and toxicity. Noninvasive or nondestructive studies can be followed by application of histopathologic techniques to determine markers of target-organ or tissue-site injury. In most cases, rodents suffice, but there might be instances when only a primate can properly represent the human situation. For example, the route of administration might be important if direct extrapolation to humans is likely. With nephrotoxicants as with other toxicants, acute, subchronic, and chronic exposures are used to determine potential toxicity. The determination of markers of urinary toxicity is generally easier with acute protocols than with subchronic or chronic exposures. Studies often involve single exposures of both sexes of at least two species, usually rodents. Depending on the results of the rodent studies and the questions being asked, one might decide to study the agent in higher mammals, such as dogs or monkeys. The use of subchronic or chronic exposure regimens is usually driven by the nature of the potential human exposure, the agent being studied, and the possibility of chronic toxicity, including carcinogenicity.

Renal Parenchymal Injury

The difficulties in diagnosing renal injury and predicting its health consequences are considerable. That is primarily because the kidney can undergo substantial chemically induced injury without any clinical manifestation; subtle injury can be negligible, given the considerable functional reserve of the kidney. For example, the single cross-sectional measurement of glomerular filtration rate (GFR) might show only severe acute or chronic renal damage, as discussed in Chapter 2. Most studies indicate that quantitative urinary-enzyme secretion patterns cannot reveal either the type or the severity of renal injury, and often they do not correlate with structural or functional changes, as discussed in greater detail in Chapter 4. The need, therefore, is for standard diagnostic criteria that are sensitive enough to serve as markers of renal damage in the presence of renal functional reserve.

Much of the nephrotoxicity that follows the administration of inert, relatively nontoxic chemicals is related to the formation of reactive electrophiles during their metabolism (Ford and Hook, 1984). It is thought that the electrophilic products can react covalently with various nucleophilic sites on renal macromolecules and, by some mechanism yet to be defined, lead to renal damage. Measurement of the reactive electrophiles or the covalently bonded compound with sensitive techniques might yield markers of renal damage (Harris et al., 1987; Omichinski et al., 1987; Reddy et al., 1984; Tyson and Mirsalis, 1985). However, those procedures require renal tissue; although they might yield useful markers of renal damage in experimental studies, they are not suitable for use in humans exposed to potentially electrophilic products of nontoxic chemicals.

Glomerular Filtration Rate

Evaluation of the blood-urea nitrogen concentration (BUN) is a common procedure for the indirect assessment of the GFR in experimental animals. As discussed in Chapter 4, it is unsuitable for quantitative purposes but might have utility in establishing the course of chronic renal failure in an experimental setting if renal damage is severe enough. Measurement of the serum creatinine concentration and urinary creatinine excretion, with calculation of the creatinine clearance, is generally preferred as an indicator of GFR. However, in some animal models, variable amounts of creatinine can be excreted via tubular secretion, and that reduces its utility as a marker of GFR. More subtle changes in GFR can be assessed by evaluating the clearance of various exogenous substances, such as inulin, EDTA, and iodothalamate.

Sensitive analytic procedures are available for measurement of those markers and GFR. Again, however, the extent of reduction of GFR in the face of a nephrotoxic insult might be hidden by the inherent renal reserve, and even measurements of GFR often are not sensitive enough to detect modest renal damage.

GFR can be assessed in either conscious or anesthetized animals. In both cases, the same markers can be used and their clearance determined with standard renal physiologic techniques. Both creatinine and inulin are used commonly to determine GFR. The use of anesthetized animals permits a more accurate determination of urinary flow than the use of conscious animals housed in metabolism cages. The use of anesthetized animals also permits the collection of precisely timed blood samples for determination of the marker under study. However, if conscious animals are used, GFR can be determined with reasonable accuracy with subcutaneous injection of the marker in a concentrated gelatin solution and collection of a single blood sample at the end of a 60-min urine-collection period. Some researchers have suggested that the anesthetic agents by themselves can alter renal function.

Tubular Function

Tubular dysfunction in experimental animals can be assessed through relatively simple and inexpensive tests, such as those for the measurement of glucosuria, enzymuria, and osmolality. Some are sensitive enough to detect relatively small effects on the kidney after acute administration of a nephrotoxicant, but caution must be exercised in predicting specific effects on transport processes or cell viability on the basis of the data obtained from these types of in vivo tests (Berndt, 1981). Some researchers (Daugaard et al., 1988; Dieperink et al., 1983) have used the renal clearance of lithium as a more subtle technique for evaluating renal damage that occurs during chronic studies; this noninasive method is applicable to humans (Thompson et al., 1984), as well as to animals. The loss of renal tubular function can also be assessed with test conditions that impose stresses on renal function, e.g., maximal urinary dilution or concentration or urinary acidification or alkalinization. Similarly, tests of maximal tubular reabsorption of glucose or maximal tubular secretion of *p*-aminohippurate (PAH) can be valuable in assessing tubular damage. They can also be applicable to humans and yield relatively sensitive markers of renal damage. However, these approaches require carefully controlled experimental studies and are not suitable for casual observations in the workplace.

Proteinuria

Proteinuria is the appearance of proteins in the urine after increase in the permeability of the glomerular membranes, reduction in tubular reabsorption of filtered proteins, shedding of specific constituents into urine as a consequence of cellular turnover or selective renal tubular damage, or a combination of the above. Glomerular or tubular damage can occur in the absence of a substantial reduction in GFR, so it has long been thought that the evaluation of proteinuria can be useful in detecting renal dysfunction at either the glomerular or tubular level. Although this topic is discussed in considerable detail in Chapter 4, a few comments concerning proteinuria as a marker of renal dysfunction are incorporated here for the sake of completeness.

Although one can measure total protein in urine, it is a relatively insensitive assessment of renal damage. Total-protein measurement also offers no insights into whether one is assessing damage to glomerular membranes or to tubular membranes. A more rational approach is the use of electrophoretic separation of single proteins to provide a comprehensive approach to chemically induced renal dysfunction.

Proteins of relatively high molecular weight (over 45,000 daltons) usually are retained in the vascular compartment by the various glomerular membranes. Those membranes also serve as charge discriminators and tend to retain negatively charged proteins in the plasma compartment as well. Proteins of low molecular weight pass the glomerular barrier with various degrees of efficiency and are later (more or less efficiently) taken up by proximal tubular cells. Indeed, the reabsorption of proteins that pass the glomerular membranes is very efficient; even slight decreases in tubular fractional reabsorption due to tissue damage increases the excretion of relatively low-molecular-weight proteins.

Electrophoretic patterns of urinary proteins can reveal glomerular damage, tubular damage, or both. Careful assessment of electrophoretic patterns can reveal selective glomerular damage (loss of glomerular polyanion) or unselective damage (glomerular hyperfiltration). Studies in experimental animals must be undertaken cautiously because in some species large variations in urinary protein excretion can lead to incorrect conclusions. For example, sex-, age-, and diet-related changes that occur in male rats are not related to glomerular damage (Neuhaus et al., 1981). Young male rats can show "tubular" proteinuria, whereas aging rats show "glomerular" proteinuria; the proteinuria in the first instance is essentially physiologic, and that in the second is attributable to spontaneous nephropathy, which can be controlled in part by reducing dietary protein.

Tamm-Horsfall Protein

Excretion of Tamm-Horsfall protein (discussed in detail in Chapter 4) is increased after damage to the distal part of the nephron and is decreased when the renal mass is reduced.

Enzymuria

Several investigators have used enzymuria as a marker of nephrotoxicity, but Dubach et al. (1989) have indicated that none of the enzymes studied experimentally satisfies all the criteria of a nephrotoxic response. Because renal enzymes are not distributed uniformly along the nephron, it might be possible to localize renal damage within the nephron on the basis of the pattern of enzymuria. The site selectivity of single enzymes is questionable. Other factors that complicate the use of enzymuria as a marker of renal dysfunction have been suggested. For example, early renal changes induced by chemicals might be less selective, in which case the predictive value of enzyme markers would be compromised. Many procedures for analyzing urinary enzymes are poor and, rather than pinpointing specific nephron sites, might give rise to nonspecific patterns that are difficult to interpret. Most urinary enzymes are stable only over a narrow pH range, and their activity in urine could be affected by inhibitors, some of which can alter urinary pH (Price, 1982). Thus, enzymuria studies in experimental animals must be carried out under very carefully controlled experimental conditions. Contamination of urine with food or microorganisms must be minimized, and urine must be collected in vessels that are then stored in ice (Berlyne, 1984). (See Chapter 4)

Monoclonal Antibodies

Other potential markers of renal damage are immunoreactive tissue constituents that are released into urine because of increase in cellular turnover or cell death. Those constituents can be detected immunochemically, and monoclonal antibodies have been produced against both rat (Tokoff-Rubin, 1986) and human (Mutti, 1989; Mutti et al., 1985) brush-border antigens. The selectivity of these markers for identifying specific nephron segments is still debated, although experimental results suggest that they can be useful. The earlier studies suggested that the BB-50 brush-border antigen was also localized in peritubular capillaries; Mutti et al. (1988) identified a monoclonal antibody that reacted with an antigen that was peculiar to the brush border—the so-called brush-border antigen. Monoclonal antibodies to the S3 segment of the nephron, where alkaline phosphatase is, also have been produced (Verpooten et al., 1989). They might prove useful as markers of the effects of chemicals, such as mercury, that act selectively on the straight part of the proximal tubule.

Bladder Toxicity

Xenobiotics

On the basis of results of long-term carcinogenicity studies of 358 xenobiotics (Barrett and Huff, 1991), the bladder is among the 10 most prevalent sites of cancer development in rodents. According to histopathologic findings after chronic exposure, 16 xenobiotics (4%) caused bladder tumors in at least one sex of either rats or mice. Histologic evaluation of rat bladders after various doses and durations of exposure to 4-butyl-(4-hydroxybutylnitrosamine) (BBN) and N-(4-[5-nitrofuryl]-2-thiazolyl)formamide (FANFT) demonstrated a series of changes in structure that are good models of the changes noted in human bladder cancer. A considerable amount of information based on those models and the results obtained with 2-acetylaminofluorene (2-AAF) is available on the potential of xenobiotics to influence the development of bladder cancer (Ito et al., 1989; Soloway and Hardeman, 1990; Staffa and Mehlman, 1980).

It should be noted, however, that there is not complete concordance across species, even for the genotoxic bladder carcinogens. That fact, by itself, makes extrapolation to humans difficult. The text that follows should be viewed in this light. For example, BBN produces tumors at a lower rate in mice than in rats, and at a lower rate in hamsters than in mice, and guinea pigs do not develop bladder tumors after exposure to BBN (Hirose et al., 1976). Furthermore, in the long-term carcinogenicity studies, 10 chemicals produced bladder tumors in rats, but only six were associated with tumors in mice (Barrett and Huff, 1991). It is also noteworthy that the first xenobiotic (or group of xenobiotics) shown to induce bladder cancer in humans, the aromatic amines (see Rubber et al., 1985, for discussion), readily induced bladder cancer in dogs but did not induce bladder tumors in rats until massive doses were given chronically by oral gavage (Hicks et al., 1982). In fact, it has been stated that the failure of the rat bladder to respond to the aromatic amines was one factor that led to the use of the maximum tolerated dose (MTD) in rodent bioassays (Wiseburger, 1992).

The capacity of chemicals and of various physical injuries to induce hyperplasia in rat bladders has been measured by histopathologic means (Ito et al., 1989) and on the basis of increased DNA replication and increased fresh and dry organ weight (Anderson, 1991; Cohen and Ellwein, 1991; Ito et al., 1989). When those techniques have been applied after the same treatments, they yielded similar results and appeared to be equally valid ways to measure cell division in the bladder. The use of the techniques in humans is questionable because they depend on using the isolated organ.

It has been reported that one can ascertain the potential of a chemical to induce bladder damage by determining its ability to enhance concanavalin A's agglutination of bladder cells from rats treated with the chemical (Kakizoe et al., 1981). That technique is reportedly capable of distinguishing between complete carcinogens and tumor promoters (R.L. Anderson, Procter and Gamble, unpublished material, 1987). Complete carcinogens increase cell agglutination directly. Tumor promoters do not produce the response when given alone, but they can sustain the response in animals that were first exposed to a known initiator carcinogen.

Extracellular Calcium

Studies with nitrilotriacetate (NTA), a nongenotoxic compound that causes bladder tumors in rats but not in mice, have demonstrated that high doses of this metal-chelating chemical cause an increase in urinary calcium and a coincident decrease in bladder-tissue calcium in rats (Anderson and Alden, 1989). That state is accompanied by the presence of crystalline calcium-sodium NTA in collected urine. Uncomplexed NTA in urine extracts calcium from the urothelial extracellular pool more rapidly than it can be replenished from the circulation. The removal of extracellular calcium reduces cell-cell contact in the urothelium and results in increased cell loss and increased urothelial replication. If this process is continued chronically, it can result in urothelial tumors (at a low rate). Only one other rat-bladder carcinogen, terephthalic acid, which forms calcium terephthalate crystals in urine, is known to cause bladder tumors by the mechanism demonstrated for NTA (Chin et al., 1981). Attempts to demonstrate a broader base for this mechanism with other treatments known to induce bladder tumors have not been successful (Anderson, 1991).

Zinc

There are few data to support the notion that tissue concentration of zinc is related to the development of bladder cancer. What data are available are the result of a single study with short-term exposure (four weeks) to BBN that demonstrated that the treatment caused increased bladder-tissue weight and tissue zinc without a change in any other mineral (Anderson et al., 1986a). It should be noted, however, that exposure to a carcinogenic dose of sodium saccharin increased the tissue concentration not only of zinc, but also of several other minerals (Anderson, 1985; Schoenig and Anderson, 1985). Also, physical injury to the bladder of sufficient magnitude to induce tumors failed to show that urothelial proliferation is always accompanied by increased tissue zinc (Anderson, 1991).

Monovalent-Cation Salts and Urinary pH

Mineral analyses of bladders from rats fed various doses of sodium saccharin (NaSacc) showed that NaSacc caused a dose-dependent increase in the concentration of several minerals in the bladders of male but not female rats; note also that NaSacc is far more carcinogenic to the male than the female rat (Schoenig and Anderson, 1985). It has been proposed that tumor-promoting effects of several sodium salts in the bladder might result from high urinary sodium, which increases sodium transport into urothelial cells and induces increased cell replication (Ito and Fukushima, 1989).

Studies with NaSacc and several other monovalent-cation salts that have been shown to induce bladder tumors in male rats have led to the hypothesis that increased urinary pH and monovalent-cation concentration can cause bladder epithelial hyperplasia and even induce bladder tumors (Cohen et al., 1991b; Ito and Fukushima, 1989).

Although potassium salts have the same response, divalent-cation salts do not induce the effects, at least in short-term experiments (Anderson et al., 1988; Cohen et al., 1991b; Hasegawa and Cohen, 1986; Ito and Fukushima, 1989).

NaSacc might act as a tumor promoter when ingested chronically by male rates after initiation with FANFT (Cohen et al., 1991b). The effect has not been noted in other species. For example, after initiation with 2-AAF, mice do not develop an increased number of bladder tumors in response to chronic ingestion of doses of NaSacc that act as potent promoters in rats (Fredrick et al., 1989).

Overall, the experimental evidence suggests that the male rat bladder is particularly sensitive to tumor development when heavily exposed to monovalent-cation salts of a variety of acids. The effect has been noted especially in studies in which the rats were exposed to an initiating dose of a known carcinogen before exposure to the salt being studied. The results show good concordance with the pathologic sequence noted in human cancer development, but the mechanism by which the tumors are induced is controversial.

Mechanical Distention

In contrast with the idea that monovalent-salt-induced bladder toxicity results from high urinary monovalent-cation excretion and increased urinary pH, an alternative mechanism has been suggested to account for the bladder tumorigenicity associated with, for example, NaSacc. High NaSacc ingestion and the loss of urine concentrating capacity that accompanies the reduction in functional renal tissue in aging male rats result in an increase in urinary volume. The increase in urinary volume leads to an increased demand for tissue growth. The relationship between increased urinary volume and demand for tissue growth suggests that chronic diuresis can be causally related to bladder carcinogenicity (Anderson, 1991). Acute, subchronic, and chronic studies demonstrate that, at least in male rats, the bladder responds to increases in urinary volume with tissue proliferation and urothelial hyperplasia and even with bladder tumorigenesis if the treatment results in continued diuresis. The bladder-distention model is further supported by studies showing that bladder distention caused by an injection of water into the bladder or by intraperitoneal injections of water or saline causes a wave of increased cell replication in the urothelium that is associated with diuresis (Herbertson et al., 1982; Koo et al., 1979; Martin, 1962).

However, the failure of furosemide-induced diuresis to promote bladder tumors in rats after initiation with BBN and the lack of data showing increased bladder tumors in the spontaneously diabetic rat make the diuresis model tenuous (Shibata et al., 1989). It has been reported that patients with bladder tumors have reduced fluid intake, less-frequent micturition, and increased urine concentration—findings that are inconsistent with the diuresis model (Braver et al., 1987). Results of epidemiologic studies on this point are conflicting.

Growth Factors

Urine is an important component in bladder tumorigenesis, at least in experimental animals. Yara et al. (1989) fractionated urine and showed that the promoting activity is attributable to epidermal growth factor.

Treatment

The development of highly predictive models of bladder carcinogenesis in rodents has been useful in evaluating the potential for treatments for bladder tumors. With the FANFT mouse model, good concordance with clinical experience has been observed for both systemic and intravesical treatments for bladder cancer (Soloway and Hardeman, 1990). It should be noted, however, that at least two agents used clinically—adriamycin and mitomycin C—demonstrate tumor-promoting activity when instilled intravesically in rats after initiation with BBN (Ito and Fukushima, 1989).

Renal-Tumor Formation

The mechanism of renal-tumor formation associated with chronic high-dose ingestion of the nongenotoxic agent NTA provides an approach to ascertaining the relevance of animal results to human risk associated with exposure (Anderson and Alden, 1989). NTA is a metal-chelating agent that has been used in laundry detergents in Canada for many years with continuous monitoring of its concentration in drinking water, which affords an opportunity to define human exposure (Anderson et al., 1985). The mean drinking-water concentration of NTA after several years of use was 2.5 μg/L. Thus, a 70-kg person consuming 2L of the average water each day would ingest ≈ 0.07 μg of NTA per kilogram of body weight per day (3.7 x 10^{-4} μmol/kg per day).

Results of several chronic-ingestion studies of NTA (Anderson et al., 1985) showed that ingestion rates greater than 1.35 μmole/kg per day were associated with renal proximal tubular-cell carcinomas. Extensive studies comparing biochemical and histologic changes induced by established carcinogenic doses of NTA demonstrated that carcinogenesis depended on the chronic delivery of a dose of NTA that induces cellular proliferation in the proximal tubules (Anderson and Alden, 1989). The biochemical studies demonstrated that the induction of renal tubular-cell proliferation depended on increased delivery to and accumulation of zinc in the renal tubular cells. Zinc delivery was accomplished through the formation of a zinc-NTA complex in the blood, which was cleared by the kidney, which reabsorbed and accumulated the zinc but not the NTA. The formation of sufficient zinc-NTA in the blood to increase zinc delivery to the renal tubules showed a distinct threshold with respect to blood NTA concentration. The threshold for sufficient zinc delivery coincided with the threshold for induction of histologic changes in the renal tubules and with a no-effect concentration in chronic-exposure studies. In the case of NTA, the threshold dose of NTA ($\approx$q00 μmole/kg per day) is more than 10^7 times the mean human exposure from drinking water in Canada after many years of use of NTA in laundry detergents.

TECHNIQUES USED IN RISK ASSESSMENT

Animal Studies

Renal function and nephrotoxicity in experimental animals can be studied at various levels of tissue structure and organization (Table 5-1), whereas in humans, studies of renal function are likely to rely on noninvasive studies whenever possible. Under certain circumstances, tissues obtained by biopsy may be available for analysis, but this is not likely to be a routine part of most studies. The first stage of any investigation on the nephrotoxicity of a xenobiotic should be in vivo studies. In the absence of knowledge about the potential toxicity of a chemical or the potential risk to human injury associated with a chemical, the first goals of studies in experimental animals are to determine whether toxicity occurs, and if so, whether it is organ-specific. If target-

TABLE 5-1 Experimental Model Systems for Study of Renal Metabolism, Renal Function, and Nephrotoxicity

Model System	Advantages	Limitations	Common Uses and Relation to Whole Animal
Whole animal	1. Intact organ structure 2. No possibility of in vitro artifacts 3. Ability to study interorgan metabolism 4. Study renal function	1. Minimal control over incubation conditions 2. Expensive 3. Interanimal variability 4. Inability to distinguish intrarenal and extrarenal effects 5. Precise knowledge of exposure condition and site of action unavailable	Target-organ specificity studies; interorgan metabolism studies
Isolated perfused kidney	1. Structurally identical with in vivo kidney 2. Study renal function 3. Can distinguish intrarenal and extrarenal effects 4. Somewhat better-defined incubation conditions	1. Short-term use (up to 2 h) 2. Expensive 3. Interanimal variability 4. Incomplete control over conditions	Ideal for studies of renal function on whole-tissue level without extrarenal factors
Kidney slices	1. Maintenance of renal-tissue structure 2. Ease of preparation 3. Improved definition of incubation conditions	1. Short-term use (up to 2 h) 2. Limited access to brush-border membrane 3. Poor cellular oxygenation 4. Presence of many cell types	Can be useful for studies of renal transport and metabolism; less potential for isolation artifacts vs. other in vitro systems because more of renal structure is maintained

TABLE 5-1, continued

Model System	Advantages	Limitations	Common Uses and Relation to Whole Animal
Isolated single nephron	1. Intact tubular structure 2. Precise definition of incubation condition 3. Ability to separate nephron cell types 4. High degree of cell-type homogeneity 5. Limited damage during isolation	1. Short-term use (up to 2 h) 2. Complicated microdissection techniques required 3. Low yield	Most useful for electrophysiologic studies because of maintenance of cell-to-cell contacts within a specific nephron region
Isolated tubules	1. Intact tubular structure 2. Relative ease of preparation 3. Precise definition of incubation conditions 4. Perform several manipulations with paired controls	1. Short-term use (up to 4 h) 2. Often limited access to brush-border membrane 3. Possible damage during isolation if enzymatic or chemical method used	Metabolism, toxicity, and transport easily measured under controlled conditions; good model in vivo if functional integrity (e.g., open lumen) is maintained; short-term model
Freshly isolated renal cells	1. Bidirectional exposure 2. Relative ease of preparation 3. Precise definition of incubation conditions 4. Ability to separate specific nephron cell types 5. Perform several manipulations with paired controls	1. Short-term use (up to 4 h) 2. Possible damage during isolation if enzymatic or chemical method used 3. Loss of polarization	Metabolism, toxicity, and transport easily measured under controlled conditions; good model in vivo if cell polarity is not relevant; short-term model

Renal-cell culture lines	1. Long-term use (weeks to months) 2. Precise definition of incubation conditions 3. Easily obtained and subcultured	1. Dedifferentiation 2. Often ill-defined origin	Can be useful model if phenotype of interest is maintained; long-term (>24 h) processes can be studied in very reproducible system
Primary renal-cell cultures	1. Closely related to in vivo cells 2. Precise definition of incubation conditions 3. Longer-term viability (up to 2 weeks) 4. Ability to separate specific nephron cell types 5. Maintenance of polarity	1. Difficult to maintain 2. Dedifferentiation can occur 3. Limited lifetime	Potentially best model if phenotype is maintained; long-term (several hours to several days) processes can be studied

organ specificity is found, more detailed studies, both in vivo and in vitro, can then be performed to elucidate mechanisms of toxicity and potential protective strategies.

Noninvasive or nondestructive studies of organ function can be used to demonstrate target-organ specificity. These studies generally involve measurement of enzyme activities in plasma or serum or measurement of urinary, plasma, or serum concentrations of metabolites or waste products that are secreted or excreted by the tissue of interest. Such measurements can be used to assess organ function without removal of tissues. Histopathologic techniques can then be applied to determine target-organ specificity of injury and to determine the site within the tissue at which injury occurs. In vivo measurements are important as first steps in assessment of xenobiotic-induced injury and as adjuncts to in vitro studies to validate in vitro techniques and to permit correlations between in vitro and in vivo measurements. Although in vivo techniques allow the study of organ function in intact animals, their use for routine screening of chemical toxicity or for more-detailed mechanistic studies is severely limited (Table 5-1), and detailed mechanistic information might not be available from them. The development of in vitro methods permits extended maintenance of cellular function outside an organism so that other types of measurements can be applied for the assessment of toxicity.

Whole-Kidney Preparations

A first step toward development of methods to study nephrotoxicity permits some control over exposure conditions. That can be accomplished by the use of an isolated perfused kidney. The advantages of this method are that intact tubular structure is maintained and that some control of exposure conditions is possible. Substantial advances in development of perfusion buffers that closely simulate renal arterial plasma have been made in recent years (Maack, 1986); these advances have enabled investigators to maintain tissue viability for longer periods. Examples of measurements that are possible with the isolated perfused kidney are those of renal clearance and metabolism. Release of renal enzymes, which is commonly measured in vivo to assess nephrotoxicity, can be conveniently determined without complications of extrarenal sequestration or metabolism. The principal advantage of the method is that the maintenance of intact tubular structure requires that limits be placed on incubation and exposure conditions; this limits the information that can be obtained. As with in vivo investigations, a single animal serves as a single experimental manipulation, so studies with the isolated perfused kidney require a large number of animals and are expensive. In addition, the target site in the kidney for a given toxic or pathologic condition might not be easily discernible.

Micropuncture techniques can be

used to examine responses in specific regions of the nephron (Burg and Knepper, 1986; Quamme and Dirks, 1986). This method has been used extensively in the study of nephrotoxicity. Information derived from these studies has been fundamental in the understanding of the consequences of both acute and chronic injury to the kidneys. It has the great advantage of allowing measurements to be made with the kidney in its proper environment but suffers from some of the limitations of time and expense as occurs with the isolated perfused kidney (Table 5-1).

In Vitro Studies

In vitro models for studies of renal function generally use physical means or digestive enzymes to separate cellular material from specific regions of the kidney. To obtain detailed information on biochemical mechanisms of action of toxic chemicals or on pathologic states, it is necessary to use living material from specific nephron regions. Physiologic and biochemical differences among mammals in the various kidney regions and nephron cell populations have important toxicologic implications and must therefore be considered in risk assessment of potentially nephrotoxic chemicals. The ability to localize nephrotoxic responses to specific cell populations is an important step in understanding biochemical mechanisms of action. Tissue slices, isolated nephron segments, isolated tubule fragments, and isolated cell suspensions can be obtained from the renal cortex or medulla. The degree to which nephron heterogeneity is taken into account determines the type of model that is required for any investigation.

Various experimental approaches can be used to separate cells from different nephron regions (Table 5-2) (Schlondorff, 1986). Because each cell population has distinct structural characteristics, visual separation with the aid of a dissecting microscope is possible. This allows preparation of nearly homogeneous material (Wilson et al., 1987). The low yield of material and the length of time required for separation this way generally make biochemical studies difficult. Biochemical or physiochemical properties of individual cell populations have been used to achieve purification and separation. Differences in cell-surface properties are useful in this regard. For example, the high glycoprotein content of the proximal tubular brush-border membrane gives proximal tubular cells a high density of negative charge, so electrophoretic techniques have been used to purify proximal tubules and isolated proximal tubular cells (Kreisberg et al., 1977a). Although high purity can often be obtained with electrophoresis, the technique requires expensive equipment, and the yield of material is often low.

A common method to separate a mixture of cell populations into its component cell types is density-gradient centrifugation. This method is relatively easy to use and rapid, so long pro-

TABLE 5-2 Procedures for Separation of Cells or Tubules from Specific Nephron Regions

Separation Method	Advantages	Limitations
Microdissection	1. Extremely high purity	1. Very low yield 2. Difficult to perform 3. Time consuming
Microdissection and immunologic reactivity	1. Extremely high purity 2. Cell type specificity	1. Very low yield 2. Difficult to perform 3. Time-consuming
Electrophoresis	1. High purity	1. Very low yield 2. Requires special equipment 3. Time-consuming
Density-gradient centrifugation (e.g., sucrose, Ficoll, Percoll)	1. Easy to perform 2. High yield 3. Relatively rapid	1. Enrichment rather than absolute purification

cedures with complicated equipment or tedious steps are not necessary, and cell viability is generally not sacrificed. One limitation of separation methods based on cell density is that enrichment, rather than absolute purification, is obtained because cell density is not a discrete property; rather, a given cell population generally contains cells with a range of densities.

Either tubule fragments or isolated cells can be used as starting material from which to obtain enriched cell populations. Both have several advantages, including the ability to pair the materials with control samples from a single animal to perform several manipulations and replications and the ability to specify and control incubation conditions. In comparing tubule fragments with isolated cells and in choosing a model system, several points should be considered (Table 5-1) (Lash, 1989). First, the lumens in many isolated-tubule preparations do not remain open during incubation; this can be problematic if transport or enzymatic processes on the brush-border membrane are of interest, and it restricts access of nutrients or toxicants to the intracellular milieu and thereby alters cellular response to exposure. Second, when tubules are fragmented during the isolation procedure—whether by mechanical, chemical, or enzymatic means—tubule length might become nonuniform and thereby introduce additional variability into the model. Isolated cell preparations, in contrast, generally yield more uniform materials; preparation methods are adjusted so that predominantly single cells in suspension, rather than multicellular aggregates or tubules, are found in the final material. Third, a given tubule fragment might contain cellular material derived from more than one nephron region, depending on how connective tissue is digested during the initial isolation procedure; this limitation could substantially diminish the effectiveness of later density-gradient centrifugation steps that are designed to yield nearly homogeneous material. With suspensions of isolated cells, in contrast, single cells are the starting material for any further purification steps.

Isolated proximal tubule fragments have been used extensively to characterize the basic transport and energetic properties of this nephron segment (Gullans et al., 1984a,b; Brazy et al., 1984). These studies demonstrated the role of plasma membrane transport and other energy-requiring processes in determining the rates of cellular oxygen consumption. Because the in vitro model contains segments of intact epithelium, one can study such processes as transepithelial transport and membrane potential, provided that lumens remain open during the course of experiments. Additional studies with proximal tubule fragments that are of more toxicologic interest have focused on metabolic and mitochondrial inhibitors (Aleo and Schnellmann, 1992a; Dickman and Mandel, 1990; Gullans et al., 1982; Weinberg et al., 1990) and on anoxia (Almeida et al., 1992; Jacobs et al., 1991; Portila et al., 1992). Additional studies on cytotoxic effects of specific nephrotoxicants, such as cisplatin (Brady et al., 1990) and mercuric chloride (Zalups et al., 1993) have also been performed.

Ficoll and Percoll density-gradient centrifugation procedures have been used to separate proximal and distal tubular epithelial cells from rat renal cortex (Cojocel et al., 1983; Gesek et al., 1987; Scholer and Edelman, 1979; Vinay et al., 1981). Ficoll density-gradient centrifugation has been used to enrich proximal tubular cells from rat renal cortex (Kreisberg et al., 1977b), and Percoll centrifugation has been used to separate isolated cells from rat proximal and distal tubules (Lash and Tokarz, 1989). Both Ficoll and Percoll form density gradients spontaneously when a centrifugal field is applied. Ficoll is a nonionic, synthetic polymer of sucrose, is supplied as a powder, and must be placed in solution to be used. Percoll consists of colloidal, polyvinylpyrolidine-coated silica beads and is supplied as a liquid. Because of its chemical properties and because it is already in liquid form, procedures using Percoll are generally easier to perform and have more uniform density gradients than those using Ficoll (Pertoft and Laurent, 1982). Both methods are applicable to other regions of the nephron: Chamberlin et al. (1984) used Percoll to obtain an enriched preparation of medullary thick ascending-limb tubules from rabbit kidney, and Eveloff et al. (1980) used Ficoll to obtain an enriched preparation of single medullary thick ascending-limb cells from rabbit kidney as well.

Two types of approaches can be used with Percoll: The use of discontinuous and continuous gradients. Although the discontinuous density gradient might yield a more complete separation of discrete cell populations, continuous gradients are much easier to prepare and are more reproducible. The Percoll method with a continuous, spontaneously generated density gradient is relatively simple to perform and yields highly enriched preparations of single proximal tubular cells (over 97% homogeneity) and distal tubular cells (over 88% homogeneity) (Lash, 1992, 1993; Lash and Tokarz, 1989).

Uses and Limitations of Freshly Isolated Renal Cells

Freshly isolated renal cells, prepared by collagenase perfusion and then separated into discrete cell populations by Percoll density-gradient centrifugation, have proved to constitute an extremely useful in vitro system for biochemical toxicology studies (Lash, 1990, 1992, 1993; Lash and Tokarz, 1989, 1990; Lash and Woods, 1991; Lash and Zalups, 1992; Lash et al., 1993). Validation of these cells as in vitro models that reflect accurately the biochemical, physiologic, toxicologic and pathologic events that occur in the proximal tubular and distal tubular segments in vivo is by measurements of several biochemical properties, respiratory characteristics, and drug-metabolizing enzymes. Examples of biochemical properties that can be measured include facilitated diffusion and active transport processes for organic ions, glutathione (GSH) metabolism, adenine and pyridine nucleotide status, and intracellular calcium distribution. Measurement of cellular respiration assesses the ability of the cells to integrate metabolic pathways and to regulate energy metabolism.

Freshly isolated renal cells are excellent models for studying acute mechanisms of chemical injury. The cells are particularly useful for studying the enzymology of xenobiotic bioactivation. For example, cysteine S-conjugates of various halogenated alkanes and alkenes are potent and specific nephrotoxicants that target the proximal tubular region of the nephron. Recent work in freshly isolated renal cortical cells demonstrated the central role of the cysteine conjugate beta-lyase in the generation of reactive thiol-containing metabolites that are ultimately responsible for the nephrotoxicity observed in vivo (Jones et al., 1986; Lash and Anders, 1986; Lash et al., 1986a). That similar enzymatic activation pathways occur in human renal tissue (Lash et al., 1990) suggests that similar mechanisms of nephrotoxicity operate. Studies in isolated rat renal cells can thereby provide information on human risk of nephrotoxicity associated with halogenated hydrocarbons that are metabolized by this pathway.

A major advantage of the Percoll density-gradient centrifugation method is that it allows study of biochemical processes and responses in specific regions of the nephron. It allows investigation of the hypothesis that each nephron cell population exhibits its own characteristic susceptibility to chemical injury. To understand how the various differences in each cell populations contribute to the toxic response, two strategies can be used. One obvious strategy is to study the mechanism of action of a known nephrotoxic chemical that has a specific target cell in a purified preparation of that target cell. Such studies, however, do not address directly the issue of nephron heterogeneity and the associated toxicologic implications. A second strategy that does address that issue directly is to expose the different cell populations to cytotoxic chemicals or pathologic conditions that will produce injury by a specific chemical or biochemical mechanism but that do not themselves have specific renal target cells (Lash, 1990; Lash and Tokarz, 1990; Lash and Woods, 1991; Lash et al., 1993). The strategy allows analysis of the inherent susceptibility of isolated cells to such processes as oxidative stress, covalent-adduct formation, cellular-energy depletion, and oxygen deprivation. The contribution of such processes as GSH metabolism, mitochondrial function, drug metabolism, and active transport to nephrotoxicity can then be studied, and the approach can also help to increase our basic understanding of why chemicals are toxic and how cells respond to xenobiotics.

The period over which lactate dehydrogenase release can be a reliable measure of cellular viability in freshly isolated cells is generally limited to 4 hours. Beyond that period, isolated cells generally do not maintain adequate functional integrity to serve as valid controls for toxicologic studies. Similarly, many of the other measures that are used to assess cellular function can be valid measures of cellular integrity for only short periods. For most of the types of responses that occur as a consequence of acute exposure to chemical toxicants, the survival time is adequate to observe markers of exposure or to observe changes that can provide information about risk and biochemical mechanisms of action—information that is most useful if clinically observed toxicity is to be treated or prevented.

One potential limitation in the use of single cells in suspension that must be considered is the physical loss of membrane polarity that exists in these cells. Because the physical separation of the luminal and basal-lateral membranes by the epithelial tight junctions no longer exists, one cannot measure transepithelial fluxes or other types of processes that depend on an intact epithelium, as can be done with the isolated perfused kidney or, provided that lumens remain open, with isolated tubule fragments. However, for many toxicologic studies, the critical issue is not the physical separation of the membrane surfaces but the retention of enzymatic and transport activities on these membranes when the individual epithelial cells are placed into suspension. Ample evidence from studies with isolated proximal tubular cells from rats (Hagen et al., 1988; Jones et al., 1979; Lash and Anders, 1989; Lash and Jones, 1985; Lash and Tokarz, 1989, 1990) support the conclusion that these cells are metabolically competent and contain transport activities and other membrane-associated enzymatic activities at levels that are comparable with those in both in vivo and various in vitro model systems. One can often functionally distinguish processes that occur on the two membrane surfaces by the use of selective inhibitors. For example, transport of glucose across brush-border and basal-lateral plasma membranes can be readily distinguished by use of either phlorizin or phloretin to inhibit brush-border or basal-lateral transport, respectively. In the case of transport of glutathione across renal proximal tubular plasma membranes, the two transport processes can be readily distinguished from each other by virtue of their distinct kinetic and energetic properties and differential substrate specificity (Lash et al., 1988). That is not to minimize, however, the functional importance of plasma-membrane polarization in an intact epithelium. Many processes, as stated above, require retention of the physiologic cell-cell contacts to allow measurement to occur. Depending on the type of biochemical or physiologic process being examined, either single cells in suspension or tubule fragments might be the more appropriate in vitro model system.

Renal-Cell Cultures as Models of Chronic Injury and Carcinogenesis

Limitation of maintenance of viability of isolated tubules and isolated renal cells to, at most, 4 hours restricts investigations to processes that occur on a scale of minutes to hours and necessitates the development of other types of model systems to study processes that occur on a scale of hours to days. For example, mutagenesis, induction of renal tumors, and alterations in expression of some genes are common responses to many chemical toxicants, including carcinogens. Those processes change, for the most part, on a scale of hours to days; even longer-term models might be necessary to obtain a complete picture of the regulation of the processes.

A biologic model system needs to be developed, therefore, to meet two critical requirements. The first requirement is that the model consist of material whose renal-cell type of origin can be unambiguously defined; the second is that the model be maintained in a viable state for at least about 7-10 days. An excellent model system is primary cell culture that uses cellular material previously enriched in a particular renal-cell population. Although several renal-cell lines are available for such long-term in vitro studies (Table 5-3), their use raises many questions. Normal in vivo characteristics of the tissue must be adequately reflected and expressed in the culture models if they are to be useful indicators of exposure or useful tools for studying mechanisms of chronic toxicity; otherwise, no reliable conclusions concerning cell-type-specific responses or in vivo relevance can be drawn. A major problem with the existing renal-cell lines and with immortalized cells in general is that they dedifferentiate in culture. Therefore, cultured cells can express some properties of the original tissue but often do not express many other characteristics that are essential for the model to reflect in vivo cellular

TABLE 5-3 Some Immortalized Renal-Cell Culture Lines

Presumed Cell Population of Origin	Cell Line	Species
Glomerulus	SGE_1	Wistar rat
Proximal tubule	$LLC-PK_1$	Hampshire pig
	NRK52E	Rat
	OK	American opossum
	JTC-12	Cynomolgus monkey
	RK-L	Sprague-Dawley rat
Medullary thick ascending limb	GRB-MAL	Rabbit
	M-m TAL-lc	Mouse
Distal tubule and collecting duct	A6	African clawed toad
	MDCK	Cocker spaniel
Unclear	BSC-1	Monkey

function accurately. Another major deficiency is that cell lines from several nephron segments are not available from a single species but are derived from several species. Ambiguity can arise in studies that use these various cells. Interspecies differences in biochemical properties and susceptibilities to chemical injury, however minor, can make it difficult to obtain a clear understanding of cell-type-specific mechanisms of injury with established cell lines.

Primary cell culture is by no means simple or straightforward. Many pitfalls are inherent in its performance, so many investigators use cell-culture lines to mimic cellular function in a given cell type or tissue. Apart from the problems and limitations, primary culture of specific renal epithelial-cell populations is an ideal approach for biochemical toxicology studies involving chronic or long-term exposures or expression of genetic or developmental processes. The three principal challenges to successful primary culture of renal epithelial cells are to obtain starting material of a high-enough purity derived from a specific nephron-cell population, to maintain differentiated function during the period of culture, and to prevent fibroblast overgrowth.

To validate a primary epithelial cell-culture procedure, several criteria must be satisfied. First, the cells must express key markers identified with the cell population of origin, thereby indicating expression of differentiated function; and expression of differentiated function must be reasonably well maintained during the course of culture. Second, the primary cultures must maintain expression of epithelial markers and must not be overtaken in the culture by fibroblasts. Third, the cells must grow logarithmically long enough (at least about 5-7 days). Taub and colleagues (Aleo et al., 1989; Chung et al., 1982; Taub et al., 1989) have pioneered primary cell culture of rabbit proximal tubules, and several groups (Boogaard et al., 1990; Elliget and Trump, 1991; Lash, 1994; Smith et al., 1986; Tokarz and Lash, 1993) have recently developed procedures for primary culture of rat proximal-tubular cells for use in toxicity studies. In using cell cultures to study mechanisms of chemically induced toxicity, it is important to consider that placement of cells in culture can alter their inherent susceptibility to chemical toxicants. Hence, it is critical to validate in vitro cell-culture models so that their general susceptibility to chemicals or pathologic conditions, such as anoxia, are known. With that foundation, the models can be used as investigative tools.

One of the key markers of differentiated function in renal proximal tubular cells is the maintenance of high activity of mitochondrial oxidative phosphorylation and low glycolytic activity. Two studies (Aleo and Schnellmann, 1992b; Dickman and Mandel, 1989) reported that diminished oxygenation of stationary cell cultures produces the often-documented shift of cellular energy metabolism to glycolytic from mitochondrial ATP generation. That is a critical point for maintenance of a long-term cell-culture model. Although use of a shaking system for cell cultures is one means of improving oxygenation and hence maintenance of cellular energetics (Aleo and Schnellmann, 1992b; Dickman and Mandel, 1989), other investigators have used changes in glucose content of growth media (Blais et al., 1992) or a more complete addition of hormonal supplements to growth media to improve cellular energetics (Lash 1994; Tokarz and Lash, 1993).

Although most of the studies with primary renal-cell culture have used animal tissue as starting material, recent advances have been made in the use of human renal tissue as starting material for primary culture of proximal tubular cells (Chen et al., 1990b; Detrisac et al., 1984; Trifillis et al., 1985). Such methods will enable direct comparison between animal and human in vitro models.

Because primary cell cultures can be maintained for up to 10 days, two types of experimental designs can be used in toxicologic investigations: acute and chronic (Table 5-4). The cell-culture model could be particularly useful for assessing carcinogenic potency of chemical toxicants, and this would make it an excellent tool for risk assessment.

In Vitro Models of Nonrenal Urinary Epithelia

In vitro models for study of toxic and pathologic processes in regions of the urinary tract besides the kidneys have recently been developed. Because most of these studies have focused on growth, development, and neoplastic transformation, cell cultures derived from either

TABLE 5-4 Experimental Protocols Used with Primary Cell Cultures for Toxicologic Investigations

Characteristic	Acute Model	Chronic Model
Dose	Low-to-high	Single or multiple low
Exposure/Time	≤ 24 h	1-10 days
Examples of processes measured	Intermediary metabolism Enzyme activities Respiration Active transport Metabolite concentrations (e.g., ATP, glutathione) Lactate dehydrogenase release or trypan blue uptake	Gene expression Protein, DNA, and RNA synthesis Mutagenesis DNA repair Processes measured in acute model

bladder or ureter epithelium, rather than freshly isolated cells, have been used.

Reznikoff and colleagues (1983) cultured normal human uroepithelial cells from tissue explants consisting of transitional epithelial cells from ureter. In addition to fetal bovine serum (at 7%, by volume) hormone and growth factor supplements were included to optimize growth and differentiation. Culture of uroepithelial cells from experimental animals actually lagged behind culture of cells from humans. Johnson et al. (1985) developed a successful primary culture and later a serial cultivation process for bladder epithelial cells from normal rats. Fetal bovine serum and hormonal supplements were used for the primary culture, but completely defined medium (i.e., with no serum) was used for serial cultures derived from the primary culture. Johnson et al. (1985) noted that optimal culture conditions differ between rat and human uroepithelial cells although structure and biochemical and physiologic functions are very similar.

Acute cytotoxicity of bladder carcinogens—such as biphenyls, nitrofurans, and 3-methylcholanthrene—has been examined in cultured normal human uroepithelial cells (Reznikoff et al., 1986). Reduction in cell number during culture was used as the index of toxicity. The prevalence of urinary carcinogenesis has stimulated interest in use of the uroepithelial-cell culture models for study of growth regulation and neoplastic transformation. Most published studies in which these cell cultures are used have focused on processes related to tumorigenesis and growth regulation. The studies have included investigation of effects of polyamine-synthesis inhibition on growth (Messing et al., 1988), effects of bacterial endotoxicants on cell survival and growth (Wille et al., 1992), role of activation of the EJ/*ras* oncogene in neoplastic transformation (Pratt et al., 1992), characterization of chromosomal deletions induced by bladder carcinogens in SV40-immortalized human uroepithelial cells (Meisner et al., 1988; Reznikoff et al., 1988; Wu et al., 1991), and characterization of second-messenger mechanisms and the carcinogenic process in SV40-immortalized human uroepithelial cells (Jacob et al., 1991).

Chopra and colleagues (Chowdhury et al., 1989) have developed cell culture methods for prostatic epithelial cells from mouse. Cells were derived from the ventral prostate of normal adult mice. Primary cultures and serial propagation were performed in serum-free, hormonally defined media to optimize differentiation. The cells grew and exhibited tissue-specific markers, including prostatic acid phosphatase activity and prostate-specific antigen.

The issue of markers of exposure and susceptibility, which is a major focus of this report, has not been addressed directly in the in vitro systems. Tissue samples, such as those used as starting material for human uroepithelial-cell cultures (Jacob et al., 1991; Meisner et al., 1988; Messing et al., 1988; Pratt et al., 1992; Reznikoff et al., 1983, 1986, 1988; Wu et al., 1991), can be readily obtained and analyzed for transformation. Cytogenetic analysis should reveal the presence of chromosomal changes that indicate exposure to a mutagenic or carcinogenic agent. For prostatic epithelial cells, examination of tissue-specific markers as described by Chowdhury et al. (1989) might be a useful means of detecting exposure of the cells to toxic agents.

IMPROVED RISK-ASSESSMENT EXTRAPOLATION

Biologic markers are the key to improving risk-assessment extrapolation. Both qualitatively and quantitatively, biologic markers are crucial in determining the best available basis for and approach to cross-species, low-dose, and route-to-route extrapolation.

Relevance of Test Systems

The major qualitative question facing a risk assessment concerns the relevance of an experimental system or animal model to the human situation. Simply stated, one can address this concern by examining the sequence of markers that occurs in humans—the sequence that is associated with exposure to a toxicant and that indicates progression to a disease state. If that sequence or a portion of it is observed in the test system, that system can be regarded as relevant for human risk estimation. Or examination of the markers, particularly markers of susceptibility, associated with disease progression in an animal model might indicate substantial similarities or dissimilarities to humans and thus help to determine relevance. In the case of renal tumors mediated by alpha$_{2u}$-globulin (see Chapter 6), for example, the rat model might not be appropriate for assessing the risk of human renal tumors associated with some chemicals. Other examples of similarities and dissimilarities are discussed below. It is imperative that the relevance of proposed test systems be established to enhance the scientific basis of risk assessment.

Metabolism and Kinetics

The concept of metabolism as a means by which chemicals are biotransformed to reactive and toxic species is central to an understanding of chemically induced nephrotoxicity. Mammalian kidneys have numerous enzymatic activities that can activate a diverse array of chemicals. An important concept is that many activation and detoxification pathways are present simultaneously and therefore compete for the same substrates. The balance between competing pathways is often the primary determinant of the ultimate biologic response. Knowledge of the biochemical regulation of the pathways and of the interorgan and intrarenal distribution of the various enzymes involved is necessary for correlation of in vitro data with the in vivo situation and of data from experimental animals with human risk assessment and development of markers of human exposure.

The importance of bioactivation in determining toxicity is not a new idea, but it is critical to understand the prevalence of enzymatic activation reactions in chemical-induced toxicity (Miller and Miller, 1985). Many toxic or carcinogenic chemicals do not produce their effects directly but must be metabolized by cellular enzymes to generate reactive, electrophilic intermediates. The metabolites are responsible for the interactions with cellular components that lead to toxicity. Much of the target-organ specificity of many toxic chemicals is due to the tissue-specific distribution of activation pathways. Other important factors in determining risk are the tissue-specific patterns in types and concentrations of protective molecules, such as glutathione (GSH) and alpha-tocopherol; in activity of detoxification enzymes, such as catalase and GSH peroxidase; and in types of and activities of transport systems that provide access to intracellular sites.

When the relative roles of all those factors have been assessed in an experimental species, one must consider how differences in one or more of them will alter susceptibility to injury. This point is central to realistic assessments of human risk because differences between humans and laboratory animals and between several laboratory animal species in activation and detoxification enzymes in specific tissues have been documented. As a consequence of qualitative and quantitative species differences in drug-metabolizing enzymes and transport activities, patterns observed and conclusions reached in one species might not be applicable to another laboratory animal species or, more important, to humans.

Study of nephrotoxicants is complicated further by the pharmacokinetics and interorgan pathways involved in their disposition. Several chemicals are initially metabolized by the liver or other tissues and then, by enterohepatic and renal-hepatic pathways, reach the kidneys where they are processed further and cause nephrotoxicity. As described in Chapter 2, the kidneys are a frequent target of chemical toxicants because of high blood flow that reaches them; the large numbers, high activities, and overlapping substrate specificities of membrane transport systems; the high basal metabolic requirements; and the presence in renal epithelial cells of several activation enzymes, some of which are peculiar to kidneys (Anders, 1980, 1989; Commandeur and Verneulen, 1990; Jones et al., 1980; Lash et al., 1988; Rush and Hook, 1986; Rush et al., 1984).

The mammalian kidney is very active in numerous pathways of drug metabolism, and renal metabolism plays a quantitatively important role in overall metabolism of a large number and variety of chemicals (Table 3-1). The kidneys are sometimes overlooked as important sites of drug metabolism for several reasons: the liver has a quantitatively large role in metabolizing a huge number of xenobiotics, the kidneys make up only 1-2% of total body weight and so are thought not to contribute importantly to metabolism, and there is pronounced heterogeneity in the distribution of enzymes in the cell populations that constitute the nephron, so detection of enzyme activities in a tissue homogenate or in a mixed population of renal cells can be difficult if the enzymes are localized to a discrete cell population. For example, activities of some isozymic forms of cytochrome P-450 in proximal tubular cells from the rat kidney are comparable with those in the rat liver (Commandeur and Verneulen, 1990; Jones et al., 1980). Some phase II enzymes, particularly those in the GSH conjugation pathway, are found at very high activities in the renal proximal tubule but are nearly absent in other regions of the nephron.

Species and strain differences need to be evaluated for each chemical or class of chemicals. It needs to be determined whether the biochemical or physiologic responses obtained in test species will be the same as those in humans. Differences in isozymic forms or the presence or absence of metabolic pathways can have effects that vary from chemical to chemical. In some cases, the differences can produce entirely different pharmacokinetics and hence different biologic effects; in other cases, the species-dependent differences might produce only quantitative differences but not yield different overall biologic effects. Differences that are noted between responses in humans and in animal species do not necessarily invalidate an animal model, nor do they imply that an appropriate animal model is unavailable for a specific toxicant. Rather, the differences highlight that we do not have much mechanistic information about many urinary toxicants and that we have not yet developed an appropriate animal model. Species differences and resulting difficulties in extrapolating to humans also highlight the need for additional research.

The mercapturic acid pathway (see Chapter 3) and associated activation and detoxification pathways illustrate many of the above principles. Although the GSH S-transferases, which catalyze the initial reaction in this pathway, are common, the succeeding enzymatic reactions show considerable variability. The variability occurs within a given species, with respect to specificity and activity, and occurs between species for a given tissue. The tissue distribution of γ-glutamyltransferase, which is the first enzyme in the pathways of GSH S-conjugate metabolism, is the primary determinant of the pattern of interorgan metabolism (Lash et al., 1988). This initial, hydrolytic reaction determines which metabolite reaches the target organ. Whether the GSH S-conjugate is metabolized to the cysteine S-conjugate or to the N-acetylcysteine S-conjugate determines how actively the protoxicant is transported into renal cells for further metabolism to a reactive intermediate.

Because most biochemical studies on the metabolism and transport of GSH, GSH S-conjugates, and related metabolites have used rats, the view that has evolved from numerous investigations is based on the extremely high γ-glutamyltransferase activity of rat renal proximal-tubular brush-border membranes and the nearly complete absence of the enzyme in rat hepatic canalicular membranes (Lash et al., 1988). Liver and biliary epithelium of humans and other mammals, such as rabbits and guinea pigs, contain substantial γ-glutamyltransferase activity (Ballatori et al., 1988; Hinchman and Ballatori, 1990). In species that have higher activities of hepatic γ-glutamyltransferase, the liver will make a larger contribution and the kidneys a correspondingly smaller contribution to interorgan metabolism of GSH and GSH S-conjugates.

For reactive and therefore toxic electrophiles that are metabolized by GSH conjugation, it is the final step in the pathway that converts the metabolite to a highly polar N-acetylcysteine conjugate that is readily excreted in urine. γ-Glutamyltransferase also shows marked species differences in velocity and substrate specificity and is absent in guinea pigs. Data on metabolite distribution and interorgan metabolism of GSH S-conjugates cannot be directly extrapolated from laboratory animals to humans without consideration of those potential differences. Therefore, although pathways studied in rats or other laboratory animals will occur in all mammals, the relative importance of each pathway will differ considerably from species to species.

Mechanism of Action

Not as much is known about the mechanisms of toxicity as is known about metabolism and kinetics. Furthermore, the relevance of some findings from animal studies to prediction of human risk is controversial because it is not clear in some cases that the mechanism or mode of action by which toxicity is produced in the animals emulates what occurs in humans. Nevertheless, in recent times, more sophisticated scientific data have provided useful information for postulating mechanisms of toxicity in test animals for many chemicals that are potentially toxic to humans.

Understanding mechanisms of toxicity in animals can lead to new methods for estimation of human risk, although exactly how to interpret scientific advances is not always clear and so how to incorporate the information into risk assessment has not found general agreement. It is evident from the mechanistic data, however, that different risk-assessment methods might be used to model human risk associated with exposure to different chemicals. Progress in the elucidation of underlying mechanisms of toxicity indicates possible high-dose phenomena), whereas comparative biology suggests differences among species. In some cases, markers in animals might not be relevant to humans; in other cases, only indirect extrapolation from animal studies may be inappropriate. For example, there is considerable discussion among scientists regarding chemicals that induce cancer in rodent bioassays but do not exhibit classical genotoxicity and regarding the role of cell proliferation in the development of cancer, particularly for these "nongenotoxic" chemicals (Ames and Gold, 1990; Cohen and Ellwein, 1991). Whether events related to cell proliferation induced by high doses of a specific chemical are limiting in carcinogenesis and whether these events are likely to occur in humans is important in predicting the likelihood of carcinogenesis in humans exposed to low doses of the chemical. Most important, if cell proliferation, particularly in response to toxicity, is integrally involved in the induction of tumors, then only the dosages and mechanisms that cause toxicity will produce a proliferative response and result in tumor formation.

Quantitative Issues

The qualitative issues discussed above are related primarily to the hazard-identification component of risk assessment. Secondarily, consideration of those issues focuses attention on the experimental results suitable for quantitative extrapolation. Traditionally, quantitative risk assessment that has involved dose-response modeling (almost all of which has until recently been cancer risk assessment) has relied on extrapolation of the relationship between administered dose and observable disease outcome (e.g., cancer incidence). Lately, with the consideration of physiologically based pharmacokinetic models and biologically based dose-response models, that has begun to change. The role of biologic markers is fundamental to the progress that has been made and will be crucial to further progress.

Consider the simplified flowchart of classes of biologic markers shown in Figure 5-1. The figure shows the sequence of markers paralleling the processes leading from exposure to clinical disease and, more important, indicates the relationships between the markers. The pictorial representation can be translated into semiquantitative form as follows:

$$DI = f1(E, s1),$$
$$DE = f2(DI, s2),$$
$$BE = f3(DE, s3),$$
$$A = f4(BE, s4), \text{ and}$$
$$C = f5(A, s5),$$

where E, DI, DE, BE, A, and C are exposure magnitude, internal dose, biologi-

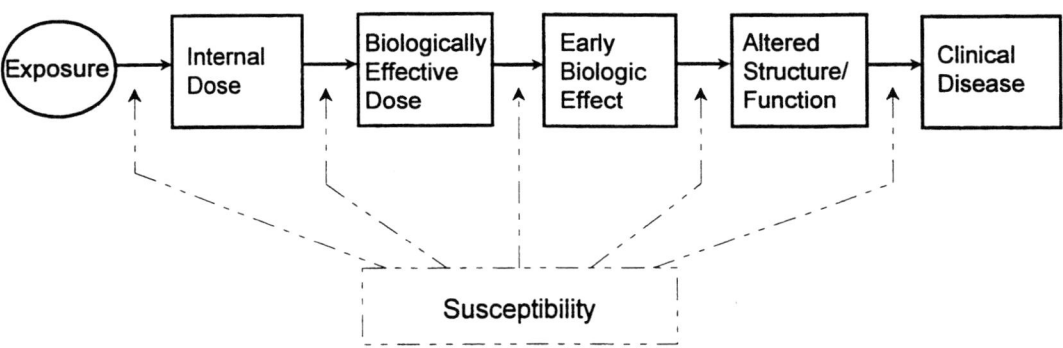

FIGURE 5-1 Simplified flow chart of classes of biologic markers (indicated by boxes). Solid lines indicate progression, if it occurs, to the next class of marker. Dashed lines indicate that individual susceptibility influences the rates of progression, as do other variables. Biologic markers represent a continuum of changes, and the classification of change might not always be distinct. Source: Adapted from Committee on Biological Markers of the National Research Council, 1987.

ically effective dose, early effective dose, early biologic effect, altered structure or function, and clinical disease, respectively. The functions relating the markers, f_i ($i = 1, \ldots, 5$) are represented simply but might be complex systems of equations (e.g., a pharmacokinetic model). In this context, the markers of susceptibility, s_i ($i = 1, \ldots, 5$), which might include biologic and nonbiologic components, can be considered indicators of all the factors that can modify the relationships among the markers of exposure and effect. In most cases, it is more appropriate to consider the probability of clinical disease (and perhaps of altered structure or function, of early biologic effect, or of the dose markers) to be functions of the other variables—e.g., $P(C) = f_5(A, s_5)$, where $P(C)$ is the probability of clinical disease—but the important features of this discussion are not altered in either case.

An example is provided by the development of bladder tumors as a result of exposure to MOCA. In that case, the markers can be defined as follows:

DI: absorbed MOCA,
DE: concentrations of MOCA in urine,
BE: DNA adducts in bladder epithelial cells,
A: increase in red blood cells in urine, and
C: bladder cancer.

The traditional estimation of the risk of bladder cancer in humans would have relied on extrapolating from animals a relationship of the form

$$P(C) = g(DI),$$

where $g(DI)$ is some function of administered dose expressed in terms like mg/kg/day or milligrams per kilogram per day or milligrams per square meter per day. It is assumed here that the conversion from exposure magnitude (e.g., atmospheric concentration to which the animals were exposed) to internal dose, DI, is appropriate, although in practice the estimation of an absorbed dose has typically been rather crude. The function g(DI) might be the multistage model routinely used in cancer risk assessment (Anderson and the Carcinogen Assessment Group of the U.S. Environmental Protection Agency, 1983; Crump, 1984) fitted to the observed cancer results in the animal test species.

If we refer to the relationships between the markers (the functions f1 to f5,) we see that the overall relationship P(C) = g(DI) ought to be representing the marker-to-marker relationships from an internal-dose marker to a marker of clinical disease. In fact, it will be noted that

$$\begin{aligned} P(C) &= g(DI) \\ &= f5(f4(f3(f2(DI, s2), s3), s4), s5). \end{aligned}$$

This convoluted form indicates why traditional risk-assessment extrapolations are problematic: the relatively simple functional relationship extrapolated from animals to humans should be representing a variety of biologically important processes. Moreover, given the variety of processes and differences among species with respect to markers of susceptibility, it is unlikely that the overall relationship is the same in humans as in animals.

The utility of an approach that considers biologic markers of effective dose, early biologic effect, and altered structure or function is related to our ability to model the relationships among those markers. The fundamental assumption of extrapolation—i.e., that relationships observed and modeled in one setting hold in another setting—can be applied at the more specific level of those marker-to-marker relationships. It might be possible to rely on extrapolation from experimental systems to humans for only a subset of those relationships.

In the case of MOCA, suppose that the observation of occupationally exposed people has provided sufficient information to derive a direct human model linking exposure and urinary MOCA concentrations. A physiologically based pharmacokinetic (PBPK) model derived from human data would serve that purpose; human PBPK models have been developed for other workplace contaminants, such as trichloroethylene (Allen and Fisher, 1993). The human PBPK model would provide the relationships represented by f1 and f2 in the above scheme.

Suppose also the availability of human data that relate the extent of hematuria with the probability of developing bladder cancer. Those data would not necessarily need to be specific to MOCA. This is another advantage of biologic markers in risk assessment: the relationship between two markers, especially markers of effect, need not be chemical-specific, so a wider base of information can be used to conduct an assessment.

If the links between exposure and

urinary concentration and between hematuria and bladder-cancer risk are established from human data, then the only relationships that need to be extrapolated from experimental systems are those between urinary concentration and adduct formation and between adduct formation and the degree of hematuria. Animal-based test systems—in vivo, ex vivo, or in vitro—can be used to elucidate those relationships. They can then be extrapolated to humans on the assumption that the relationships observed in the test systems are the same as in humans. That is the fundamental assumption of risk assessment, but in this context it is a much more particular and specific level than is typical in risk assessment. Because that is the case—i.e., because the extrapolations involve fewer variables and less "gap" between cause and effect—there is a greater chance of appropriately accounting for the modifying factors of susceptibility (e.g., species differences in DNA repair) and therefore a greater likelihood that the assumption is justified.

Even when human data are less direct than in the example just cited, the application of the extrapolation assumption to specific links in the chain leading from exposure to disease will improve the quality of risk-assessment estimates. The links in that chain can often be studied with experimental techniques or systems. Results of the studies can then be incorporated at the appropriate point, allowing even greater consideration of species differences or other differences that can be represented as differences in susceptibility. As mentioned above, the ability to separate the chemical-specific from the chemical-neutral relationships facilitates the incorporation of a larger base of data. The way to improve risk assessment is to give it a firmer scientific foundation. The use of biologic markers to break the disease process into manageable research pieces is a way to do that.

In risk-assessment practice, there will be a need to compromise between the ideal (all the markers one could want and a full understanding and representation of the sequence of events) and the real (less than complete information and less than perfect representation). The application of the basic assumption of extrapolation might not be at the level that one thinks is best; it might have to be at a level as close as possible to the best, given the data at hand. That introduces additional uncertainty, which can be reduced as additional research is conducted to provide more fundamental data, to understand more clearly how cause-effect relationships can be represented, or to elucidate the role of modifying factors.

For assessing risk, any model should account for individual variability in response and health or environmental status. Differences in those factors will alter susceptibility to potential chemical-induced injury.

SUMMARY

Extrapolation from animal models is a common and necessary component of risk assessment for humans. To improve the validity of such extrapolations, a better understanding of the relationship between these markers and disease is needed. In most cases, the scientific basis for assuming that animals are good surrogates for humans, and therefore a suitable basis for extrapolation to humans, is overwhelming. It is reasonable to use animal models for extrapolation to humans unless specific information on specific chemicals indicates otherwise. Identification of chemical hazards should include assimilation and evaluation of all relevant information, including appraisal of physical and chemical properties and structure-activity relationships, which can often provide important indications of potential toxicity. Difficulties in diagnosing renal injury and predicting its health consequences are considerable, primarily because the kidneys can undergo substantial chemically induced injury without any clinical manifestation, and subtle injury can be negligible because of the considerable functional reserve of the kidneys. Standard diagnostic criteria are needed that are sensitive enough to serve as markers of renal damage in the presence of renal functional reserve.

Only whole-animal studies or observations in humans can provide information on the operation of multiple cells, tissues, and organs under the influence of complicated feedback mechanisms. Animals are necessary in the study of chemically induced toxicity, because studies that involve modulation of cellular responses and tissue sampling cannot be performed in humans.

The first stage of any investigation of the nephrotoxicity of a xenobiotic should be in vivo studies. In the absence of any knowledge about potential toxicity and target-organ specificity, the first step should be to determine whether toxicity occurs and the tissue distribution of the toxic response. More detailed studies, both in vivo and in a variety of in vitro models, can then be pursued to elucidate modes of chemical action, specific mechanisms of toxicity, and potential protective or preventive strategies. Animal models developed as surrogates for humans in the study of renal and urinary function should conform to some general principles: the animal model should be reproducible within and among laboratories, the model should be specific to the part of the urinary tract under consideration, the model should be sensitive enough to differentiate normal from abnormal changes or functions, and the model should be able to measure alterations in renal function caused by exogenous agents.

A variety of experimental model systems are available for study of renal metabolism, renal function, and nephrotoxicity. They range from whole-animal studies to those in the isolated perfused kidney, kidney slices, isolated nephron segments, isolated tubule fragments, and isolated renal cells. Each model has advantages and limitations that must be taken into account when developing conclusions and extrapolating animal data to human risk assessment. In vitro models of nonrenal urinary tract epithelia have also been developed and applied primarily toward examination of carcinogenesis. Development of markers of exposure and susceptibility has not been addressed directly with such nonrenal models and should be pursued for better extrapolation of data for risk assessment.

The importance of enzymatic activation of toxic chemicals is central to an understanding of chemically induced renal injury. Species and strain differences in amounts and tissue distribution of various enzymes can be critical in determining the ultimate toxic response. Consequently, patterns observed and conclusions reached in one species might not apply to another species. We recommend that species and strain differences in disposition and metabolism be evaluated for each chemical or class of chemicals. For assessing risk, any experimental model should account for individual variability in response and in health and environmental status. Differences in those factors will alter susceptibility to potentially toxic chemicals.

6
NEW TECHNOLOGIES

The discussion that follows highlights several aspects of new technologies. It is not intended to be inclusive; rather, it illustrates some promising lines of investigation. To illustrate the potential difficulty of introducing a marker into clinical practice, we describe the use of prostate-specific antigen (PSA) in some detail. We also consider fully the importance of alpha$_{2u}$-globulin, a controversial topic involving the extrapolation of animal data to human conditions.

MARKERS OF CELL INJURY, REGENERATION, AND HYPERTROPHY

To a considerable extent, the identification of markers has been facilitated by an understanding of the molecular biology of the cell and the mechanisms that are responsible for both orderly growth and neoplastic transformation.

The Cell Cycle

Nongrowing, quiescent cells have low rates of DNA synthesis and are said to be in the G_0 phase of the cell cycle. These cells (Figure 6-1) can be stimulated to grow by the addition of one of several growth factors. After the binding of a growth factor to its receptor, growth-promoting signals are generated that commit the cell to a new phase of the cell cycle, G_1, where biosynthetic activities resume. During the transition to the G_1 phase, the cells express many genes that enable them to progress through this phase. Prominent among them are the so-called immediate early genes. The S phase starts when DNA synthesis starts and ends when the DNA content of the nucleus has doubled. Thereafter, the cells move into the G_2 phase of the cycle, where new protein and RNA synthesis occurs and the cells increase in size. The M phase begins with nuclear division and ends with cytoplasmic division (Safirstein, 1994). Various points in the cycle are influenced by environmental stress and can

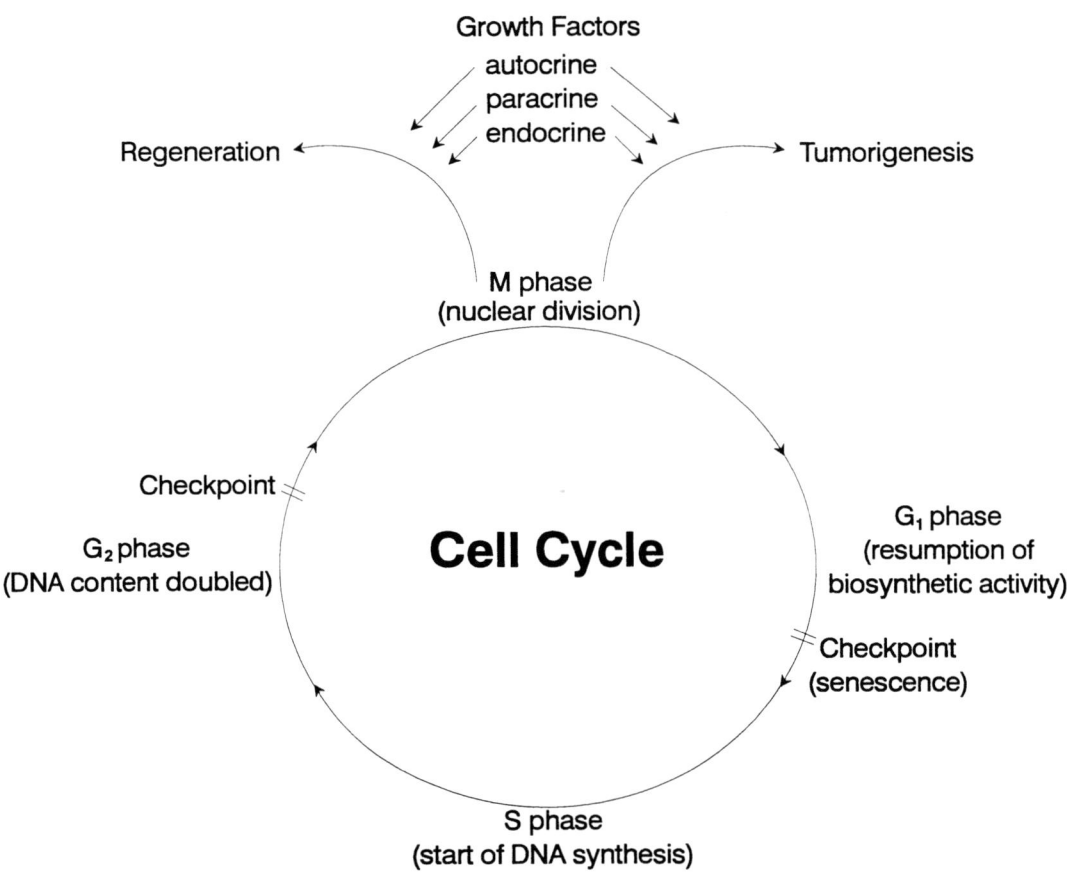

FIGURE 6-1 The cell cycle.

be targets for future biologic-marker research. The control of growth and differentiation involves sophisticated modes of cellular communication. Figure 6-2 illustrates actual and potential modes of growth control, as might be found in the prostate. *Endocrine* control is exemplified by the effects of androgen, which stimulates the production of growth factors in both epithelial and stromal cells. Fibroblasts secrete factors that bind to epithelial-cell receptors.

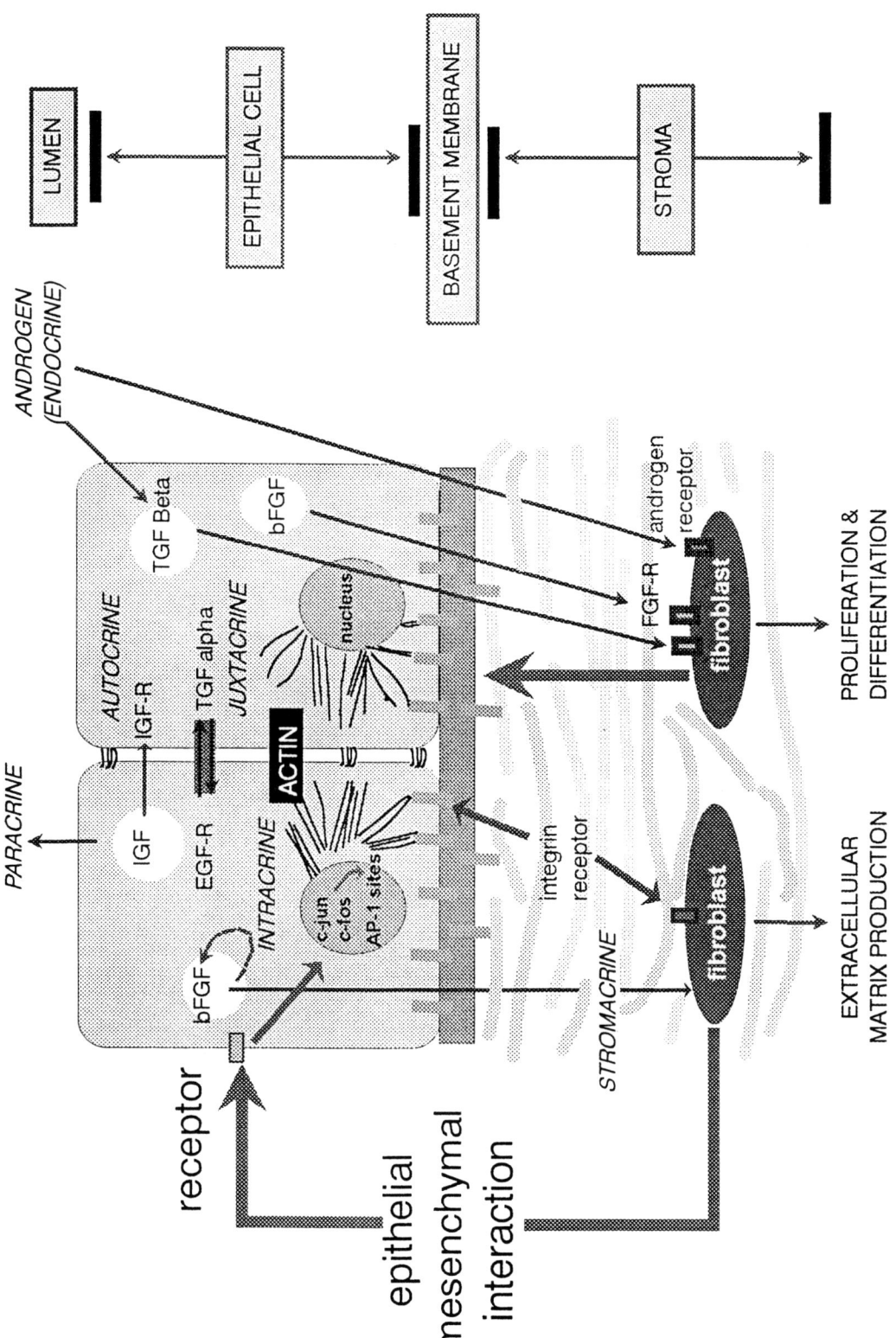

FIGURE 6-2 Illustration of actual and potential modes of growth control, as might be found in the prostate.

This represents a special case of *paracrine* growth control, "stromacrine" control. External signals, with specific growth peptide and cell-surface receptors producing tissue specificity, link with common intracellular signal pathways, which ultimately induce the growth genes c-*fos* and c-*jun* and transcription factors. Fibroblasts are responsible for controlling proliferation and differentiation, as well as for synthesizing extracellular matrix. Fibroblasts can also control epithelial-cell growth and differentiation through their effect on the basement membrane. The proteoglycan composition of basement membrane reflects whether it is growth-promoting or differentiation-promoting. If the membrane and matrix connect with the cell, differentiation generally results; thus, cell-adhesion molecules—such as integrins, fibronectin, and proteoglycans—that bind to cell receptors or matrix proteins are important inducers of differentiation. *Juxtacrine* growth in the epithelium occurs when a membrane-bound form of the growth peptide binds to a receptor on an adjacent cell. Juxtacrine growth control in the epithelium tends to favor differentiation. *Autocrine* control occurs when cells secrete growth peptides that induce their own growth by binding to their own receptors. *Intracrine* growth, which has so far been observed in model systems, occurs when the growth factors are prevented from being exported by mutations that inhibit cleavage of the pre-pro forms of the peptides.

Growth Factors

Renal growth is initiated by mechanisms that include changes in renal growth factors. Several of the growth factors have been found in kidney tissue itself or shown to act on renal cells in culture. The growth factors isolated from kidney tissue include epidermal growth factor (EGF), insulin-like growth factors (IGF), transforming growth factors (TGF), and platelet-derived growth factor (PDGF). Other substances, such as angiotensin II, have also been shown to have substantial growth-regulating properties.

Epidermal Growth Factor

EGF is a potent mitogen capable of stimulating the proliferation of cultured renal tubular cells (Goodyer et al., 1988; Norman et al., 1987). It is produced in the kidney (Olsen et al., 1984; Salido et al., 1986, 1989; Scott et al., 1985) with a precursor molecule, prepro-EGF (Bell et al., 1986; Fisher et al., 1989; Rall et al., 1985). Binding sites for EGF have been found in various portions of the renal tubules and the glomeruli. After injury to distal tubular cells, there might be an autocrine pathway which regulates synthesis, promotes receptor binding, and so regulates repair. Stimulation of proximal tubule replication most likely involves EGF released from other cells and the use of a paracrine pathway. EGF also stimulates the production of

IGF-I and acts directly to enhance IGF-I gene expression (Rogers et al., 1991).

In the case of renal ischemic injury, prepro-EGF mRNA synthesis or processing changes (Safirstein et al., 1989). There is a prolonged reduction of EGF excretion, a reduction in prepro-EGF mRNA, an increase in EGF-receptor density (Safirstein et al., 1990), and an increase in the binding of EGF in the postischemic kidneys. Indeed, exogenous EGF administered to rats after renal ischemic injury enhances renal tubular cell regeneration and accelerates the recovery of renal function (Humes et al., 1989; Tsau et al., 1989).

The synthesis or processing of prepro-EGF mRNA is also altered by exposure to nephrotoxic agents, such as cisplatin (Safirstein et al., 1989). In addition, an increase of EGF receptors has been seen during the renal hyperplasia that follows folic acid administration (Behrens et al., 1989) and ischemia (Saferstein et al., 1989).

Insulin-Like Growth Factors

The insulin gene family is made up of IGF-I, IGF-II, proinsulin, and relaxin. The IGF-I gene is on chromosome 12, and the insulin and IGF-II genes are on the short arm of chromosome 11. The insulin-like growth factors are bound to carrier proteins. IGF-I and IGF-II interact with a distinct cell-surface receptor. IGF-I is synthesized in most organs, where it is involved in the regulation of various metabolic and growth processes (Hammerman, 1989). IGF-I can be found in the cortical medullary collecting duct and the parts of the thin loop of Henle that are in the medulla (Andersson et al., 1988; D'Ercole et al., 1984). IGF-I mRNA has been isolated directly from the kidney; this implies that the IGF-I recovered from the kidney is synthesized there (Bortz et al., 1988). IGF-I receptors are present on glomerular mesangial cells and on proximal tubular cells (Hammerman and Rogers, 1987). Interaction with IGF-I can result in the stimulation of protein kinase activity.

In rats, IGF-I has been implicated in the compensatory hypertrophy of the kidney that follows unilateral nephrectomy, in the hypertrophy that accompanies the feeding of high-protein diets, and, more important, in regenerating tubular cells after ischemic injury. In compensatory hypertrophy, IGF-I gene expression is stimulated, and the renal concentration of IGF-I is substantially increased (El Nahas et al., 1989; Fagin and Melmed, 1987; Polychronakos et al., 1985; Stiles et al., 1985). When high-protein diets are fed, serum IGF-I increases (Hirschberg and Kopple, 1991; Isley et al., 1983; Maiter et al., 1988; Prewitt et al., 1982); this is circumstantial evidence of a causal role of IGF-I in hypertrophy. IGF-I immunoreactivity is transiently expressed by regenerating cells in the postischemic rat kidney. IGF-I immunoreactivity appears as early as 3 days after the injury. With increasing differentiation of the regenerating cells, IGF-I immunoreactivity vanishes; at 14 days, it is not detectable (Andersson and Jennische, 1988). The regenerative cells express IGF-I peptide and IGF-I mRNA in a transient manner that is thought to correlate with cell differentiation better than with cell division (Matejka and Jennische, 1992).

Transforming Growth Factor Beta

Transforming growth factor beta (TGF-B) constitutes a family of ubiquitous growth peptides that are intimately involved in extracellular matrix formation, cellular proliferation and differentiation, wound-healing, cartilage and bone formation, and possibly oncogenesis. Evidence is growing that TGF-B operates in an autocrine or paracrine fashion to bring about a multiplicity of cellular actions.

In the kidney, TGF-B has been implicated in the repair processes that follow renal ischemic injury (Humes and Daniel, 1990); it might be an important mediator in the genesis of renal fibrosis after the administration of anti-glomerular-basement-membrane IgG (Phan et al., 1990); it has been implicated in the pathogenesis of glomerulonephritis (Okuda et al., 1990) and it might contribute to the development of progressive kidney fibrosis (Yamamoto et al., 1994).

There is evidence that TGF-alpha expression is greater in prostatic epithelial cells of patients with benign prostatic hypertrophy (BPH) than in epithelial cells of normal persons (Harper et al., 1993). Extracts of human prostate with benign hypertrophy contain fibroblasts that have been shown to secrete TGF-B (Story et al., 1993). TGF-B has more effect on stromal growth than on epithelial growth (Sherwood et al., 1992), inasmuch as TGF-B concentrations are higher in stroma than in epithelium (Truong et al., 1993). The TGF family might have an important role in the pathology of BPH. For example, bFGF is known to stimulate fibroblasts and can increase expression of intracellular matrix proteins in other cell lines. aFGF is known to negate the inhibitory effects of TGFn1 in prostatic cell lines and is also androgen-responsive.

Another member of the FGF family, Int-2, has a sequence very similar to that of bFGF and in transgenic mice has been found to be associated with BPH. The BPH in those mice was dissimilar to human BPH in that the mice did not have accompanying stromal and fibromuscular hyperplasia; so it was difficult to study these model systems. The prostates from the transgenic mice did not grow when transplanted into nude mice; this indicates that alterations in hormonal and growth-factor concentrations might play a role in the pathogenesis of BPH. However, in transgenic mice, growth factors do not elicit histologic changes in the prostate.

Angiotensin II

Several lines of evidence indicate that

angiotensin II (AII) acts as a regulator of cell growth. For example, it has been shown that angiotensin-converting enzyme inhibitors attenuate renal hypertrophy (Anderson et al., 1986b). In cultures of rabbit proximal tubular cells, AII by itself has no direct effect on protein synthesis; but in the presence of EGF (Norman et al., 1987) or PDGF (Norman et al., 1987; Wolf and Neilson, 1991), it can enhance tubular epithelial proliferation. Although AII does not induce mitogenesis, it appears to be capable of promoting cellular hypertrophy and early oncogene expression (Rozengurt and Heppel, 1975). With EGF and PDGF, AII can induce the expression of various cellular oncogenes (Wolf and Neilson, 1990)—an effect that can be blocked with a competitive antagonist of AII. Indeed, AII has many of the features of growth factors and appears to be a growth regulator in the kidney. AII binds to specific cell-surface receptors on tubular cells and activates many of the intracellular signaling pathways associated with cell growth (Norman, 1991). For example, the AII-induced hypertrophic response might be related to its ability to stimulate Na^+/H^+ pump activity in proximal tubular cells.

Early-Response Genes

In cellular proliferation and hypertrophy, an orderly pattern of gene expression (Cowley et al., 1989) leads to the induction of mRNA and various protein products. The c-*fos* and c-*myc* genes are examples of so-called "immediate early" genes and encode DNA-binding proteins localized to the nucleus. The initial activity of the c-*fos* protein appears to be pivotal to the transcriptional activity of the cell and is necessary for the transition from the stable resting state of the cell cycle to the gap between the mitotic and S phases where DNA replication occurs. The protein product encoded by the c-*myc* gene serves a regulatory function in controlling DNA synthesis by binding DNA polymerase II and is necessary for the next transition, to the synthetic phase of the cell cycle. Later, other gene products appear with similar kinetic characteristics.

The pattern of expression of these gene products after stimuli that provoke either renal hypertrophy or hyperplasia has been examined. In studies of the compensatory renal hypertrophy induced by uninephrectomy (Beer et al., 1987; Norman et al., 1988), modest increases in c-*myc*, c-H-*ras*, and c-K-*ras* expression have been reported. Others have been able to detect both c-*fos* and c-*myc* transcripts within 15 minutes after uninephrectomy (Sawczuk et al., 1990) and *Egr-1* expression within 30 minutes (Ouellette et al., 1990). In rats with acute unilateral ureteral obstruction, within minutes there is a transient induction of c-*myc* transcripts and a somewhat more prolonged expression of c-*fos*. Although both decline by 24 hours, there is evidence of reinduction at 48 hours (Sawczuk et al., 1989). After nephrotoxic injury by a large parenteral dose of folic acid, expression of c-*fos*, c-*myc*, c-K1-*ras*, and c-Ha-*ras* is markedly increased (Asselin and Marcus, 1989; Cowley et al., 1989). After renal ischemic injury by bilateral renal arterial occlusion of 50 minutes (Safirstein, 1990), c-*fos* mRNA increases, reaches a peak at 1 hour, and declines rapidly to control values by 4 hours. A similar pattern has been observed for *Egr-1* mRNA. During the repair phase of ischemic acute renal failure, c-*myc* expression in the renal cortex increases by a factor of 10 (Humes and Daniel, 1990).

Nucleic Acid Synthesis and Hypertrophy

Although attention has been given to the cell cycle and DNA synthesis, other mechanisms are operative when cells hypertrophy. The hallmark of compensatory growth is a marked increase in cellular RNA content (Hayslett, 1979). For example, after unilateral nephrectomy in the rat, RNA and protein synthesis in the remaining kidney increases within 12-24 hours (Bucher and Malt, 1971) and persists for weeks as renal mass increases (Halliburton and Thomson, 1965). The increase in RNA synthesis is a useful index of the extent of hypertrophy. Less impressive alterations in DNA synthesis have been demonstrated within 6 hours after uninephrectomy (Toback and Lowenstein, 1974). This hyperplastic response reaches a peak at 2 days (Threfall et al., 1967) but is not of the same magnitude as the hypertrophic response. At two weeks, in contrast, total RNA content rises by 40% and total DNA content by only 25%; this shows that the predominant factor in compensatory growth is hypertrophy (Threfall et al., 1967).

During compensatory growth, the processing of nucleoplasmic RNA is accelerated (Willems et al., 1969), and more ribosomal RNA (rRNA) is produced (Ash and Cuppage, 1970; Halliburton and Thomson, 1965; Kurnick and Lindsay, 1968; Toback and Lowenstein, 1974). In fact, because rRNA constitutes more than 85% of total RNA in the kidney, the characteristic growth-induced increase in the renal RNA-to-DNA ratio results from an increase in ribosome number. The principal site of all these activities is the renal cortex. Indeed, there is direct evidence that the accretion of RNA occurs in the tubular cells of the cortex (Vancura et al., 1970).

Sodium-Pump Activity and Cell Growth

On the basis of studies performed in cell cultures, it has come to be recognized that a rapid increase in sodium entry into the cell stimulates Na^+/K^+ pump activity (Rozengurt and Heppel, 1975; Smith and Rozengurt, 1978) and increases intracellular K^+ (Tupper et al., 1977; Wolf and Neilson, 1990). The increases are associated with a commencement of DNA synthesis (Lopez-Rivas et al., 1982). Initially, an influx of Na^+ exceeds the rate at which Na^+ is extruded across the basolateral membrane by Na^+/K^+ pump activity with a reduction of intracellular pH (Schuldiner and Rozengurt, 1982). In proximal tubular cell cultures, the increase in Na^+/H^+ pump activity associated with the addition of hypertonic NaCl results in increases in cell volume and protein content without stimulation of DNA synthesis (Rozengurt, 1981); this effect suggests that cell hypertrophy is coupled with Na^+ influx (Fine et al., 1985). With a reduction in renal mass, growth stimulation and increase in Na^+/H^+ pump activity are closely related (Fine, 1986).

It also appears that Na^+ influx is necessary for the initiation of DNA synthesis stimulated by peptide growth factors. For example, mitogenic factors stimulate Na^+ influx (Rozengurt et al., 1981; Smith and Rozengurt, 1978) and H^+ efflux (Schuldiner and Rozengurt, 1982). The addition of NaCl to cultures of kidney epithelial cells results in an increase in the number of cells initiating DNA synthesis in response to EGF and insulin (Toback, 1980). Furthermore, a decrease in the rate of influx of Na^+ achieved by a reduction of Na^+ in the culture medium prevents the development of the mitogenic response to EGF and insulin (Burns and Rozengurt, 1984). In contrast, an increase in cellular Na^+ accumulation can be attenuated in the presence of a purified growth-inhibiting protein. The reversal by the addition of NaCl to the medium suggests that variation in the Na^+ flux during the onset of kidney epithelial cell growth can be related to the action of a specific cell protein that inhibits DNA synthesis (Toback, 1980).

Polyamines

The polyamines spermidine and spermine and their precursor diamine putrescine, omnipresent components of all living cells, are aliphatic polycations with three, four, and two positive charges, respectively, at physiologic pH. In the cell, they are bound to macromolecular anionic sites in nucleic acids, ribosomes, and membranes. Polyamines are believed to be essential for cellular growth, proliferation, and differentiation, although their physiologic function at the molecular level is still not well understood (Manteuffel-Cymborowska, 1993).

CANCER OF THE BLADDER

Cell Culture

In vitro models of toxic and pathologic processes of the urinary tract, besides the kidneys, have been recently developed. Because most of them have focused on growth, development, and neoplastic transformation, they have used cell cultures derived from either bladder or ureter epithelium, rather than freshly isolated cells, as an in vitro system. Reznikoff and colleagues (1983) cultured normal human uroepithelial cells from tissue explants that comprised ureteral transitional epithelial cells. Although fetal bovine serum (7% by volume) was included in the culture medium, hormone and growth factor supplements were also included to optimize growth and differentiation. Culture of uroepithelial cells from experimental animals actually lagged behind culture of cells from humans. Johnson et al. (1985) developed a successful primary culture and later a serial cultivation process for bladder epithelial cells from normal rats.

Acute cytotoxicity of bladder carcinogens—such as biphenyls, nitrofurans, and 3-methylcholanthrene—has been examined in cultured normal human uroepithelial cells (Reznikoff et al., 1986). Reduction in cell number during culture was used as the index of toxicity. The prevalence of urinary carcinogenesis has stimulated interest in uroepithelial-cell culture as a model for the study of growth regulation and neoplastic transformation, and most published studies in which these cell cultures have been used focused on processes related to tumorigenesis and growth regulation. The studies have included investigation of effects of polyamine synthesis inhibition on growth (Messing et al., 1988), of effects of bacterial endotoxicants on cell survival and growth (Wille et al., 1992), of the role of activation of the *EJ/ras* oncogene in neoplastic transformation (Pratt et al., 1992), of chromosomal deletions induced by bladder carcinogens in SV40-immortalized human uroepithelial cells (Meisner et al., 1988; Reznikoff et al., 1988; Wu et al., 1991), and of second-messenger mechanisms and the carcinogenic process in SV40-immortalized human uroepithelial cells (Jacob et al, 1991).

Quantitative Fluorescence Image Analysis

Quantitative fluorescence image analysis (QFIA) is an exciting recent technique for marker research (West, 1970; West et al., 1987). It is based on the recognition of a single cancer cell in a mass of normal cells through the quantitation of a fluorescent signal to obtain a measurement of the DNA, protein, or other macromolecular content of individual cells. Aneuploidy (the presence of more or fewer than the normal diploid number of chromosomes) is apparently the result of genetic instability and is a hallmark of carcinogenesis (McGowan et al., 1988). The use of QFIA permits more precise quantitation of DNA content in individual cells. Unlike flow cytometry, QFIA selects cells on the basis of composition or structure; unlike immunohistochemistry, it is capable of true quantitation (Parry and Hemstreet, 1988; Rao et al., 1993).

That cancer cells often have aneuploidy has been known for decades, and the technique of flow cytometry was developed originally to determine ploidy. The difficulty with using overall ploidy as a marker is that normal cells yield some measurement variability in the amount of DNA, and usually some cells are in the process of dividing and hence will have higher DNA content than normal. Consequently, it is very difficult to detect abnormal individual

New Technologies

cells in the vast sea of normal cells (Parry and Hemstreet, 1988).

Image analysis has been used to detect cancer cells for over a decade. The main advantage of the approach is that it provides objective, reproducible results. Image analysis is well suited to so-called rare-event analysis, the detection of objects that make up 1% or less of the cell sample.

In its initial application for bladder-cancer detection, QFIA urinary cytology successfully combined the attributes of classical cytopathology with those of DNA cytometry and thereby increased sensitivity to low-grade lesions (Bass et al., 1987; Bonner et al., 1993; Hemstreet et al., 1990; Hemstreet et al., 1983). The semiautomated QFIA uses both quantitative and qualitative analyses of urinary-tract specimens and is a new, highly sensitive alternative for early cancer detection. QFIA technology has an advantage over flow cytometry in requiring fewer cells for valuations and in being able to analyze individual cells for a one-cell diagnosis of cancer rather than requiring patterns of cells. Yet the two techniques can be complementary, and together they have greater sensitivity than classical cytopathologic methods in detecting cellular transformation.

QFIA technology has also been used in several large-scale occupational studies of people at risk for bladder cancer and has shown a specificity of 93-96% (Hemstreet et al., 1988). It is being increasingly recognized as a valuable tool for screening asymptomatic people to identify biologic risks associated with working in environments that contain toxic and possibly carcinogenic substances (Bi et al., 1993). QFIA has been applied extensively in both clinical detection of bladder cancer and several trials in worker cohorts exposed to bladder carcinogens (Bi et al., 1993; Hemstreet et al., 1988).

In addition, QFIA has been used successfully to monitor for the efficacy of treatment with chemopreventive retinoids in an animal model (Hemstreet et al., 1992; Hurst et al., 1991). The numbers and DNA content of bladder-wash cells obtained from rats treated with carcinogens and chemopreventive retinoids was found to be an accurate predictor of risk of developing bladder cancers (Hemstreet et al., 1992; Hurst et al., 1991).

There are newer directions of QFIA research to learn whether commonly seen atypical cells are the result of an inflammatory or irritative process or are truly progressing toward malignancy. Fluorescence probes for cytoskeletal oncogenes and growth factor receptors in addition to DNA and RNA are being incorporated (Rao et al., 1990, 1991). The ability to image multiple probes on a single cell greatly increases its clinical usefulness. The new capabilities are being added to the experimental armamentarium for determining and possibly reducing cancer risk or reversing malignant cell transformation.

The ability to study macromolecular changes in single living cells with fluorescent probes offers a tool that exceeds the capabilities of the electron microscope and the scanning electron microscope. The combination of QFIA technology with the use of monoclonal antibodies and DNA hybridization promises imaginative applications for early cancer screening and diagnosis.

Basic research with QFIA will continue to investigate the quantitation of cell-surface antigens, oncogenes, and cytoskeletal components (Jones et al., 1990; Rao et al., 1993). The findings can be correlated with nuclear DNA content and classical histopathology. The objectives will be to develop methods for understanding the fundamental oncogenic process at the single-cell level, to define biochemical markers, to identify preneoplastic conditions, and to differentiate tumors with metastatic potential (Rao et al., 1990, 1993).

Given the recognition that aneuploidy is a marker of genetic instability and is only indirectly related to changes in gene expression or deletion, QFIA was extended from DNA to protein markers, which might be more directly related to genetic changes than is aneuploidy. In addition, other studies have shown that using QFIA with G-actin (a precursor molecule of the cytoskeleton) yields an early marker of carcinogenesis that can guide chemoprevention studies and, in conjunction with tumor markers of aneuploidy and the presence of cells that react with antibodies against tumor-specific antigens, can be helpful in the diagnosis and management of bladder cancer in worker cohorts exposed to bladder carcinogens (Rao et al., 1990; Hemstreet al., 1992).

More important, those techniques might help to define individuals at risk for developing cancer and to identify and monitor for the effects of chemopreventive agents.

DIFFERENTIAL-DISPLAY POLYMERASE CHAIN REACTION

Differential-display polymerase chain reaction (DD-PCR), as first presented by Liang and Pardee (1992) and improved by Pardee (Liang et al., 1993) and Bauer and colleagues (1993) has facilitated the investigation of altered gene expression in a wide range of applications. It might well become the method of choice for the elucidation of the molecular mechanisms underlying the cellular response to xenobiotics. The responses that occur on xenobiotic challenge involve an increase ("up-regulation") or a decrease ("down-regulation") in the synthesis of structure- and differentiation-related proteins. Metabolism can be changed, various growth and signal peptides and their receptors can be altered, and the growth kinetics of the cell can be affected.

Individual sections of base pairs on DNA are necessary to encode a particular protein. Identification of the expression of these genes can use conventional techniques, such as Northern and Southern blots, but requires knowledge of the nature of the expressed genes and the availability of previously characterized or identified probes. Most mammalian cells express around 15,000 genes. Testing for each individually is a daunting task. In the cell nucleus, polymerase enzymes copy a sequence of

DNA by transcription to messenger RNA (mRNA). In the cell cytoplasm, proteins are ultimately synthesized on the basis of the information encoded in the mRNA. In theory, electrophoresis of the total RNA can separate the various mRNAs. In practice, some of the genes are expressed at very low levels, and their resolution is poor.

The general strategy of DD-PCR is to amplify partial DNA sequences copied from subsets of mRNA (cDNA) by reverse transcriptase enzymes and PCR (Bauer et al., 1993; Liang and Pardee, 1992; Liang et al. 1993). It is necessary somehow to divide all the cDNA sequences so that the ones of interest can be identified. To achieve this partitioning, a series of different primers is used. These primers initiate the reading of the DNA sequence. At the 3'end, which contains the poly(A) tail characteristic of mRNA, 12 different 3'primers are used; each one displays a different population of mRNAs. Further division can be obtained by the use of different upstream primers. A set of arbitrary primers has been shown to yield optimal results (Bauer et al., 1993; Liang et al., 1993). These delimit PCR at a specific length for each of the mRNAs. Each pair of 3' and upstream primers reproducibly anneals to the same sequences in each successive PCR cycle and selects a discrete set of mRNAs. A radioactive or fluorescent label is incorporated into the mixture to label the cDNAs and make them detectable. Differentially expressed genes are evident from differences in intensity of corresponding bands in the two gels. The sensitivity of the technique can be adjusted to any selected level of difference, thereby allowing those genes that are differentially expressed at the highest level to be detected first, then the genes at smaller levels. For example, genes with a difference in expression of a factor of 10 or higher could be identified first, then those with a difference of a factor of 5-10, and so forth. The cDNAs can be isolated, further amplified, directly cloned, or used as probes to isolate longer sequences.

DD-PCR has been widely applied to many fundamental problems in cancer research and other fields in which identification of changes in gene expression is important. The technique is likely to have a large impact on understanding of the underlying mechanisms of toxicity to the genitourinary tract and on the development of markers with which to investigate risk in human populations.

CANCER OF THE PROSTATE

Prostate-Specific Antigen

Carcinoma of the prostate is the second most common cause of cancer deaths in men. Prostate-specific antigen (PSA), a glycoprotein that is produced by the epithelial cells of the prostate and acts as a protease to liquefy the seminal coagulum (Wang et al., 1979), has been used in screening for prostatic cancer, for its staging, and as a marker of response to treatment. Although its utility has been demonstrated, controversy has developed about its proper use. It is appropriate to consider the questions about PSA in a discussion of new technologies and the development of other markers, which are likely to be associated with similar concerns when applied to clinical decision-making.

Major issues to be addressed are the effectiveness of current screening programs to detect occult prostatic cancer and the effectiveness of surgical intervention compared with conservative management. Only an estimated 13% of men will benefit from current prostatic-cancer detection programs. Studies have compared prostatic cancers found in radical cystoprostatectomy specimens (latent form) with prostatic-cancers detected in current prostatic-cancer screening programs. In the screening programs, prostate cancer was identified with PSA and digital rectal examination (DRE), and confirmed with ultrasonographically-guided needle biopsy of the prostate. The studies were based on the observation that prognosis is unequivocally related to the size of the tumor (McNeal et al., 1986), the stage of the disease, and the grade of the tumor. Tumors were organized into a hierarchy of risk for cancer death on the basis of division into 10 categories. Some 300 patients underwent whole-mount sections of the prostate with careful attention to the level of invasion and grade. At one end of the spectrum were patients with positive surgical margins, seminal-vesicle invasion, or lymph-node invasion (incurable). Only 4% of patients who came to radical prostatectomy were in the small-cancer or latent-cancer group, 35% were in the incurable group, and the rest were in the intermediate groups, whose cancers were potentially effectively controlled.

As longitudinal screening programs identify increased numbers of patients with organ-confined disease, patient survival will improve (Catalona et al., 1991, 1993, 1994). However, because of an increased number of multiple biopsies in these patients, a higher percentage of latent cancers will be detected. In some studies, only 5% of the patients with negative results of ultrasonography and DRE were in the group considered to have biologically inactive disease. That finding is supported by the observation that 1 g or 1 cm^3 of prostatic tissue raises serum PSA by 3.5 ng/mL. Furthermore, there is no relationship between reported tumor volume and PSA of less than 0.4 ng/mL. When tumor size, grade, and stage were compared with those in autopsy groups, 65% of the autopsy cases were in the very-low-risk group. These studies illustrate several important points. First, PSA is probably not detecting a large number of cases of latent prostatic cancer. Second, a high percentage of patients are being detected when they are in the high-risk group and it is too late for cancer control. Third, the identification of new biologic markers is necessary to decrease the length-time bias and to detect earlier cases that are likely to progress rapidly to advanced disease. With the current markers, it is likely that the introduction of screening programs will lead to a shift toward detecting cases in the treatable category, in that people will be screened more often. The identification of markers of prostatic cancer with greater specificity and sensitivity might be helpful in epidemiologic studies designed to determine the relationship between xenobiotic substances and the appearance or progression of prostatic cancer and to determine whether subsets of environmental and nutrient factors that predispose to the aggressive form of prostatic cancer can be identified.

It is worth asking whether prostatic cancer is being detected and treated in many patients who will ultimately die of other causes. Some studies indicate that most patients who are found to have prostatic cancer on the basis of increased PSA are those who should be effectively treated with radical surgical intervention. However, screening and treatment of this group must take several factors into consideration, including the cost of the evaluation, the morbidity associated with the tests themselves, and the morbidity of those who will not be cured, in contrast with those with progressive disease.

What would it mean for everyone with prostatic cancer to be treated? About 1-2% of men die of prostatic cancer, and 10-12% have clinically manifest disease. If all the men over the age of 50 with prostatic cancer had their cancers removed (30% of males), there would be an estimated 75,000 deaths due to surgery (assuming a complication rate of 1%), in contrast with the 30,000 who would have died from prostatic cancer. These screening programs do not take into consideration the morbidity of such an experiment, which would result in impotence in 250,000 men and 20,000 deaths in the first year alone (Hinman, 1991). Some argue that the surgical mortality is as low as 0.1%, but objective data from controlled trials are not available, and estimates consider mortality only in the immediate perioperative period (Catalona, 1994).

Debate continues with regard to the appropriateness of screening with PSA and the efficacy of surgical intervention, versus watchful waiting. A number of large clinical trials are being conducted to address these issues. The results could have an impact on individual risk assessment. Methods to assess an individual's natural life expectancy in relation to the biologic activity of the tumor are equally important.

Markers are useful to identify persons at risk, to detect preclinical disease, and to categorize clinical disease. Any reduction in mortality and morbidity achieved by the use of new markers must be superior to current diagnostic techniques. This includes cost assessments. Data are now available for comparing routine prostatic-cancer screening that uses DRE. In a study of 6,600 men comparing screening with DRE and PSA, Catalona et al. (1994) reported that DRE alone would miss 40% of prostatic cancers, and PSA alone would miss 23% of prostatic cancers. When PSA was added to DRE, there was a 68% increase in the detection of organ-confined prostatic cancers. The results confirm the complementary nature of the two tests.

In initial screening programs with PSA and DRE, 9% of the patients underwent biopsy evaluation (Catalona et al., 1993). Those who were negative were monitored every 6 months for 3 years. The longitudinal screening indicated that 40% of men will develop prostatic cancer and that about 80% of men with PSA over 10 ng/mL will have proven prostatic cancer. Screening doubles the identification of patients with organ-confined prostatic cancer. The identification of patients with organ-confined disease and a PSA between 4 and 10 ng/mL is controversial. Only 40% of patients with PSA between 4 and 10 ng/mL will have prostatic cancer (Benson et al., 1992a,b). Should we use PSA concentration or rate of its change as a guide to biopsy? Available data do not support a delay in biopsy except in patients with a small prostate, and the consensus is that the rate of change in PSA correlates poorly with prostatic cancer detection. The continued search for improved methods of interpreting PSA results attests to the limitation of this marker.

Marker Development

Several major advances in marker research would improve the outcome of prostatic cancer marker detection. The ideal marker would specifically detect potentially biologically active prostatic cancer in its premalignant stages. The identification of a marker with a low false-positive rate that would reduce the number of false-positive findings in patients with PSA concentrations between 4 and 10 ng/mL would reduce patient evaluation costs. The identification of a marker that would detect patients with biologically active tumors capable of rapid progression would identify the subgroup most in need of surgical intervention.

The frequent occurrence of multiple biopsies of the prostate in response to high false-positive PSA results provides ample tissue samples for marker evaluation. Prostatic intraepithelial neoplasia (PIN) and adenomatous hyperplasia are two premalignant lesions associated with prostatic tumorigenesis, and a grading scale has been developed for these lesions. The term PIN I refers to dysplastic changes, but aggressive clinical actions such as rebiopsy of the prostate, are not usually taken. In contrast, when PIN II or III is identified, the prostate is commonly rebiopsied; these lesions have been associated with increased PSA. Which of the premalignant lesions will develop into prostatic cancer is of current interest, and such markers as oncogenes, growth factors, and cytoskeletal markers are being evaluated (see Table 4-5). An understanding of the pathologic condition is important for proper interpretation of pathologic results in the development of new markers.

Scientists are searching for new markers that will improve the sensitivity and specificity of the current PSA test. PSA in blood predominantly forms a complex with alpha$_1$-antichymotrypsin. The PSA might be irreversibly inactivated in vitro by two protease inhibitors, alpha$_1$-antichymotrypsin and alpha$_{2u}$-globulin.

To improve the sensitivity of the PSA test, new assays have been developed to detect the different molecular forms of PSA in serum. One assay detects binding of PSA to proteinase inhibitors, a second assay detects PSA in complex with alpha$_1$-antichymotrypsin, and a third detects nonbound PSA. Figure 6-3 depicts the three assays. Continuing studies indicate that the specificity of the PSA test might be improved by quantitating the various forms of bound and nonbound PSA. The ratio of PSA bound to alpha$_1$-antichymotrypsin is higher in prostatic cancer patients; nonbound PSA is more commonly observed in patients with benign prostatic hyperplasia.

The development of monoclonal antibodies for other markers more specific for prostatic cancer is in progress. One major problem in prostatic cancer staging is seminal-vesical invasion. An assay that could specifically detect seminal-vesical invasion would be a useful prognostic indicator.

Neuroendocrine Cells in the Prostate

There are highly specialized neuroendocrine epithelial cells in the prostate. They have endocrine and paracrine functions and contain various peptides, including serotonin, calcitonin, and thyroid-stimulating hormone. Increasing evidence suggests that these cells are important during prostatic growth and differentiation. Cells with similar functions are found in the gastrointestinal tract and the respiratory system and are of local endodermal origin. The endocrine and paracrine cells in the prostate are slender apical processes that extend into the lumen and put out dendritic projections between adjacent epithelial cells.

The peptides are present in semen and are detectable with radioimmunoassay or high-performance liquid chromatography. The dendritic processes identified with electron microscopy might exert a paracrine function on adjacent epithelial cells. Of primary importance is that both afferent and efferent neurons have been observed in association with neuroendocrine cells. The potential of these molecules either as growth regulators or as markers of disease remains to be determined (see Table 4-5).

CANCER OF THE KIDNEY

This section discusses examples in urinary toxicology of the examination and use of biologic markers to determine the choice of relevant test species and aid in the interpretation of the results of toxicity studies.

Renal Tumors in Male Rats Mediated by Alpha$_{2u}$-globulin

A structurally diverse group of organic chemicals has been shown to cause a renal syndrome in male rats that is manifested acutely by the accumulation of marker protein in renal proximal-tubule cells visible under the light microscope as hyaline droplets (Hard et al., 1993; Health Effects Institute, 1985,1988; USEPA, 1991a,b). Mechanistic studies with a few select

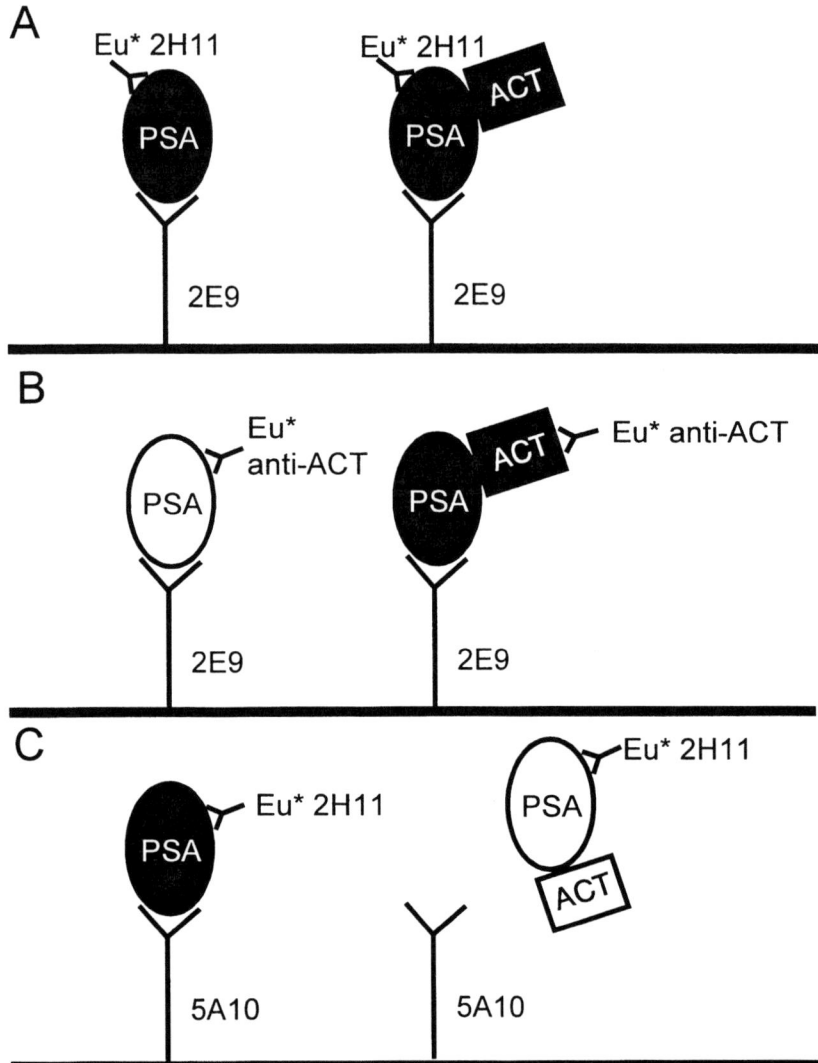

FIGURE 6-3 New assays developed to detect the different molecular forms of PSA in serum. Schematic representation of three sensitive two-site immunofluorometric procedures to measure PSA immunoreactivity in serum. (A) Assay T detects epitopes available on both the PSA molecule that did not form a complex and PSA that formed a complex with proteinase inhibitors such as alpha1-antichymotrypsin; (B) assay C, using the same catching antibody as assay T, measures PSA that forms a complex with alpha1-antichymotrypsin by a polyclonal tracing antibody against alpha1-antichymotrypsin; and (C) assay F detects PSA that has not formed a complex, but has difficulty recognizing PSA that has formed a complex with proteinase inibitors such as alpha1-antichymotrypsin. Source: Lilja, H., A.T.K. Cockett, and P-R. Abrahamsson. 1992. Reprinted with permission of *Cancer*.

chemicals have shown that the protein involved in the production of the hyaline droplets is alpha$_{2u}$-globulin (Alden et al., 1984; Dietrich and Swenberg, 1991; Lehman-McKeeman and Caudill, 1992a; Lock et al., 1987), the major urinary protein excreted by adult male rats. Compared with the female rat and some other species, including humans, the male rat is physiologically proteinuric (Neuhaus and Lerseth, 1979; Olson et al., 1990; Shapiro and Sachchidananda, 1982; Vandoren et al., 1983). A large amount of the alpha$_{2u}$-globulin is synthesized in the liver of the male rat under the control of several hormones, particularly androgen (Feigelson and Kurtz, 1977; Roy and Chatterjee, 1983; Roy and Neuhaus, 1967; Roy et al., 1983), but it has not been detected in the liver of normal female rats or in other species, including humans (MacInnes et al., 1986; Sippel et al., 1975). In addition to stimulating the synthesis of large amounts of alpha$_{2u}$-globulin in the male rat, testosterone suppresses the degradation of alpha$_{2u}$-globulin (Feigelson and Kurtz, 1977; MacInnes et al., 1986; Motwani et al., 1984; Roy and Chatterjee, 1983; Roy and Neuhaus, 1967; Roy et al., 1983; Sippel et al., 1975). The predominant hypothesis linking alpha$_{2u}$-globulin to renal carcinogenesis is that with continued exposure to such chemicals as d-limonene and trimethylpentane, which bind to alpha$_{2u}$-globulin and cause its accumulation, the acute protein overload progresses to renal-cell injury, which can be observed histologically as cell necrosis in the proximal tubule with the presence of granular casts in the outer medulla, chronic and progressive nephrosis, and papillary linear mineralization (Alden, 1986; Borghoff et al., 1990; Flamm and Lehmann-McKeeman, 1991; Hard et al. 1993; Swenberg et al., 1989; USEPA, 1991b).

That rather well-defined progression of toxicity in male rats is referred to as alpha$_{2u}$-globulin nephropathy (Alden, 1989; Borghoff et al., 1990; Swenberg et al., 1989). Ultimately, there is an increase in the incidence of tumors arising from the renal-tubule epithelial cells. It has been suggested that a chemical cannot induce alpha$_{2u}$-globulin nephropathy without having the potential for inducing tumors at some dose (Dominick et al., 1991). Because the chemicals that produce this type of toxicity and carcinogenicity are nongenotoxic, it is suggested that the renal-tubule tumors seen after chronic exposure to alpha$_{2u}$-globulin hyaline-droplet-inducing agents result from sustained target-organ toxicity that leads to increased renal-cell proliferation followed by promotion of spontaneously-initiated cells of increased opportunity for mutagenesis (Alden, 1989; Swenberg et al., 1989; Trump et al., 1984a); however, quantitative relationships between protein accumulation, renal disease, and sustained increases in cell proliferation are not understood (Melnick, 1992). An alternative hypothesis is that of a chemically mediated direct toxic effect on the kidney, where alpha$_{2u}$-globulin influences the dose of toxic chemical that reaches the target organ (Melnick, 1992).

The histopathologic sequence of this form of toxicity is distinctive, and the differences in potency and species and sex susceptibility suggest that these chemicals act by mechanisms different from classical renal carcinogens (Alden and Frith, 1991; Hard, 1987, 1990; Hard and Whysner, 1994; Lipsky and Trump, 1988). Nevertheless, renal-tubule tumors induced by $alpha_{2u}$-globulin accumulators are structurally indistinguishable from either spontaneous tumors or those induced by classical carcinogens (Hard, 1990; Lipsky and Trump, 1988; NTP, 1989, 1990). Although human renal-cell tumors are structurally similar to rodent tumors, the histopathology of the $alpha_{2u}$-globulin-induced nephropathy in affected kidneys in males differs remarkably from that in human renal carcinoma (Bennington and Beckwith, 1975; Tannenbaum, 1971; USEPA, 1991).

The mechanistic data developed to understand the formation of male rat renal tumors mediated by $alpha_{2u}$-globulin are reviewed below. As discussed elsewhere in greater detail (Melnick, 1992; USEPA, 1991b), gaps in data on specific mechanisms are considerable. For example, the crucial connections between cellular necrosis, increased cell proliferation, tubular hyperplasia, and renal-tubule tumors are not proved, and the possibility that similar proteins in humans lead to a common mechanism has not been completely ruled out. Chemical substances that induce renal-tubular tumors in male rats must be evaluated one by one. d-Limonene is used as an example of the accepted standard prototype $alpha_{2u}$-globulin accumulator (Hard and Whysner, 1994).

The Environmental Protection Agency has reviewed the data on a number of chemicals that cause increases in renal-tubule tumors in male rats and might increase $alpha_{2u}$-globulin hyaline-droplet nephropathy, and it has recommended criteria for the use of such information in risk assessment (USEPA, 1991b).

Acute Nephropathy

The renal effects of chemicals that induce $alpha_{2u}$-globulin accumulation constitute a continuum of changes that seem to be initiated by the appearance of eosinophilic hyaline droplets in the proximal-tubule epithelial cells and that with chronic exposure progress to severe hyperplasia and tumorigenesis. The features of acute nephropathy include accumulation of hyaline droplets in the proximal tubules; necrosis and regeneration, particularly in the P2 segment of the proximal tubule; linear mineralization in the papilla; and granular cast formation, primarily at the junction of the inner and outer stripes of the outer medulla (Alden et al., 1984, Halder et al., 1984; Hard and Whysner, 1994; HEI, 1988; Swenberg et al., 1989).

Studies that have involved the administration of chemicals capable of inducing $alpha_{2u}$-globulin accumulation in immature rats, aged rats, castrated male rats, male NCI Black Reiter (NBR) rats that do not synthesize $alpha_{2u}$-globulin in the liver, male rats given estrogen, or female rats given exogenous estrogen and $alpha_{2u}$-globulin show that development of the nephropathy occurs only in the presence of alpha$_{2u}$-globulin (Chatterjee et al., 1989; Dietrich and Swenberg, 1990; Garg et al., 1988, 1989b; Hobson et al., 1986; Logothetopoulos and Weinbren, 1955; Murty et al., 1988; Ridder et al., 1990; Roy and Neuhaus, 1967; Sippel et al., 1975).

A background of protein-droplet formation is seen in the male rat kidney (Goldsworthy et al., 1988a; Logothetopoulos and Weinbren, 1955; Maunsbach, 1966a). The droplets consist of protein that is being degraded within the phagolysosomal compartment of proximal-tubule epithelial cells. It is estimated that male rat kidneys filter 40-50 mg of alpha$_{2u}$-globulin each day and reabsorb about 25 mg of it into proximal-tubular cells (Lehman-McKeeman and Caudill, 1992b; Neuhauss et al., 1981). In contrast, there is no background incidence of hyaline droplets containing alpha$_{2u}$-globulin in female rats, which synthesize less than one-hundredth as much alpha$_{2u}$-globulin as male rats (Vandoren et al., 1983); this small amount of protein probably does not accumulate as histologically evident hyaline droplets. Thus, the physiologic processes underlying the spontaneous formation of hyaline droplets probably constitute a threshold phenomenon and result from the large quantities of alpha$_{2u}$-globulin that the male rat reabsorbs each day.

Agents capable of inducing hyaline-droplet formation increase both the size and the number of droplets in the male rat kidney—a process that requires the interaction of either the parent chemical or a metabolite with alpha$_{2u}$-globulin (Garg et al., 1989a; Short et al., 1987). The interaction is probably a rate-limiting step in the development of nephropathy. It has been shown that the binding of hyaline-droplet-inducing agents to alpha$_{2u}$-globulin is reversible but highly specific, inasmuch as the dissociation constant for the ligand-alpha$_{2u}$-globulin complexes is around 10^{-7}M (Borghoff et al., 1991; Lehman-McKeeman and Caudill, 1992b). It is the binding of a specific chemical to alpha$_{2u}$-globulin that disrupts the normal handling of the protein in the kidney, particularly its lysosomal degradation, and the protein-chemical complex accumulates within phagolysosomes (Lehman-McKeeman et al., 1990a). For example, the dose-response curve for d-limonene-induced hyaline-droplet nephropathy includes doses below which no exacerbation can be demonstrated (USEPA, 1991b).

Chronic Renal-Cell Injury and Proliferation

In addition to the manifestations just described of acute toxicity associated with alpha$_{2u}$-globulin nephropathy, chronic exposure to chemicals that cause the syndrome results in the development of more severe glomerular lesions, and prominent linear mineralization often occurs in the kidney (Alden, 1989; Alden and Frith, 1991; Bruner, 1984; Dominick et al., 1991; Trump et al., 1984b). The structural characteristics of the renal injury associated with alpha$_{2u}$-globulin nephropathy have been well described, but little has been learned about the mechanisms underlying the associated renal cell death (Melnick, 1992; USEPA, 1991). Nor has the intracellular concentration of alpha$_{2u}$-globulin required to trigger cell death been determined. Whether a role exists for the nonbound chemical or metabolite and whether the accumulation of alpha$_{2u}$-globulin is the actual cause of the renal toxicity have been questioned (Melnick, 1992). Renal injury can be initially identified as an impairment of proximal-tubule function with an increase in marker cellular casts or debris in the urine (Alden et al., 1984; Kanerva et al., 1987a; Short et al., 1987). In addition, a well-defined dose-response relationship indicates that a loss of renal function is observed only at dosages that cause alpha$_{2u}$-globulin nephropathy. Renal toxicity in both acute and subchronic types of d-limonene-induced alpha$_{2u}$-globulin nephropathy has been determined; it seems, in concordance with the hyaline-droplet response, that dosages of less than 5 mg/kg per day (for 91 days) do not alter renal-cortical function, as assessed by transport of organic anions and cations (Webb et al., 1989). In direct contrast, dosages known to induce renal-tubule tumor formation—from 75 to 150 mg/kg per day—cause marked loss of renal-cortical function (Lehman-McKeeman et al., 1989).

The cytotoxicity and loss of renal function associated with alpha$_{2u}$-globulin nephropathy cause a compensatory increase in renal-cell proliferation. For chemicals evaluated by the National Toxicology Program (NTP) for subchronic and chronic toxicity, the histologic appearance of renal-cell hyperplasia is commonly reported. Additional studies that used more-specific methods for measuring cell proliferation showed that unleaded gasoline or 2,4,4-trimethylpentane (Short et al., 1987, 1989), d-limonene (Dietrich and Swenberg, 1991), and 1,4-dichlorobenzene (Charbonneau et al., 1989) increase cell proliferation, particularly in the P2 segment of the proximal tubule. The shape of the dose-response curve for renal-cell proliferation induced by alpha$_{2u}$-globulin nephropathy indicates a nonlinear relationship. A quantitative dose-response relationship between protein accumulation, sustained cell proliferation, and tumorigenesis needs to be explored further.

Renal Neoplasia

Renal neoplasia occurs spontaneously in rats at a very low rate; tubule-cell adenoma and adenocarcinoma are not apparent until the age of 2 years and even then are observed in only 0.2 and 0.1%, respectively, of often studied strains of male rats and less than 0.1 and 0.1% of the female rats. If followed over the entire life span of the rat, those incidences increase to 0.2 and 0.4% of male rats and 0.4 and 0.6% of female rats (Solleveld et al., 1984). (See also National Toxicology Program (NTP) historical control data, e.g., USEPA, 1991c; NTP, 1990) Known genotoxic renal carcinogens, such as substituted nitrosamines, do not exhibit acute hyaline-droplet nephropathy or sex or species specificity (Alden and Frith, 1991; Hard, 1987). Moreover, renal tubule-cell tumor incidence after exposure to a genotoxic renal carcinogen is considerably higher, often approaching 100%. In direct contrast, the increased incidence of renal tumors associated with an alpha$_{2u}$-globulin hyaline-droplet inducer are generally low (the highest is 30%) even at maximum tolerated doses.

Genotoxicity

The mutagenicity of hyaline-droplet-inducing agents (both parent chemical and some of the metabolites that bind to alpha$_{2u}$-globulin) has been tested in a variety of assays, and these chemicals as a class are nongenotoxic (Flamm and Lehman-McKeeman, 1991; Hard et al., 1993; Hard and Whysner, 1994). For d-limonene, the major metabolite that binds to alpha$_{2u}$-globulin is d-limonene 1,2-epoxide. The identification of a potentially reactive epoxide intermediate that binds to alpha$_{2u}$-globulin might raise some concern that this epoxide could also react with DNA and be responsible for the renal carcinogenicity of d-limonene. However, that appears to be unlikely, inasmuch as the 1,2-epoxide is a very stable moiety that binds reversibly to alpha$_{2u}$-globulin, without forming covalent adducts (Lehman-McKeeman et al., 1989). In addition, this epoxide is not mutagenic in *Salmonella* species, does not induce sister-chromatid exchange in V79 cells, and causes no unscheduled DNA synthesis in rat hepatocytes (Flamm and Lehman-McKeeman, 1991; Vonder Hude et al., 1989; Watanabe et al., 1980). Collectively, the lack of genotoxicity of these chemicals, and their metabolites directly involved in the interaction with alpha$_{2u}$-globulin argues that "nongenotoxic" mechanisms must be involved in the chronic carcinogenicity associated with acute alpha$_{2u}$-globulin nephropathy.

Related Proteins and Their Role in Safety Evaluation

Some uncertainty remains about whether human subpopulations might be sensitive to similar proteins and, if so, the molecular identity, kinetics, and biology of these proteins in humans. The alpha$_{2u}$-globulin found in male rats is structurally related to a group of transport proteins, many of which are synthesized in humans. The proteins in this group of about 20 low-molecular-weight proteins, the alpha$_{2u}$-globulin superfamily of proteins called lipocalins, are similar in molecular weight, certain sequence homology, and tertiary structure (where it is known) to the alpha$_{2u}$-globulin (Lehman-McKeeman and Caudill, 1992b; Pervaiz and Brew, 1987). They include protein 1 (a minor urinary protein in humans, more abundant in male than female urine), retinol-binding protein (synthesized in the liver by all mammals), apolipoprotein D (isolated from human plasma), pregnancy-associated endometrial alpha$_2$-microglobulin (isolated from human pancreas), bovine beta-lactoglobulin, and mouse urinary protein (MUP), the protein most closely resembling alpha$_{2u}$-globulin (Åkerström and Lögdberg, 1990; Bernard et al., 1989; Brooks, 1987; Pervaiz and Brew, 1987; Pevsner et al., 1988; Snyder et al., 1988). The only member of this superfamily with a clearly defined physiologic function is retinol-binding protein, although all of them are thought to be carriers of lipophilic molecules. Although male rats have a much higher urinary protein concentration than humans (Flamm and Lehman-McKeeman, 1991), because concentrations of protein homologues in human urine are well below those of alpha$_{2u}$-globulin in male rats (Berggard, 1970; Ekstrom and Berggard, 1977; Peterson and Berggard, 1971), it is highly unlikely that enough protein could accumulate in the normal human kidney to result in hyaline-droplet formation (Hard and Whysner, 1994). Abnormal accumulation of hyaline droplets can be seen in disease processes, however. For example, large amounts of proteins are produced in multiple-myeloma patients (Pirani et al., 1983), lysozyme is produced in human mononuclear-cell leukemia (Muggia et al., 1969), and hyaline-droplet accumulation is observed in patients with epidemic hemorrhagic fever who are infused with large amounts of concentrated human serum albumin (Oliver and MacDowell, 1958).

Although mouse urinary protein is more abundant in males, females excrete it. In addition, mouse urinary protein accounts for 4% of the total hepatic mRNA in mice, whereas alpha$_{2u}$-globulin accounts for about 1% of the rat hepatic mRNA. Lehman-McKeeman and Caudill (1992a,b) reported that neither d-limonene 1,2-oxide nor the reactive metabolite of 2,4,4-trimethyl-2-pentanol binds to MUP or to protein 1. Binding of the reactive metabolite to the protein appears to be necessary for materials to induce a protein nephropathy related to protein accumulation. Differences in renal metabolite concentrations are not clearly due to binding, however, in that most of the metabolite in the kidney is not associated with alpha$_{2u}$-globulin. If the phenomenon involves a direct chemical action, the delivery of the active metabolite to the target is important, so characterization of the binding, turnover, and transport properties of low-molecular-weight, lipophilic ligand-binding proteins in humans might be more relevant than the quantities of protein.

Male Rat Renal Tumors in the Assessment of Risk to Humans

The preceding discussion has laid out the predominant view of the mode of action in the development of renal tumors caused by alpha$_{2u}$-globulin. An alternative mode of action has been proposed by Melnick (1992), who suggested that other factors besides alpha$_{2u}$-globulin accumulation might be involved in the proliferative response. A nongenotoxic chemical that causes renal tumors in male rats but not in female rats or in other species is probably acting through the alpha$_{2u}$-globulin mechanism if there is evidence of hyaline-droplet formation and of alpha$_{2u}$-globulin in the droplets, there is granular cast formation in the outer medulla and papillary linear mineralization, there is increased cell replication in the renal tubules, and there is no evidence of another plausible mechanism.

The renal carcinogens believed to work through the alpha$_{2u}$-globulin mechanism fulfill the first two criteria; several also fulfill the third criterion, and some fulfill the fourth. But d-limonene and 2,2,4-trimethylpentane are the only two of these chemicals that have been sufficiently investigated to establish binding of their metabolites to alpha$_{2u}$-globulin and, perhaps just as important, the absence of binding to other members of the superfamily of alpha$_{2u}$-globulin low-molecular-weight proteins (Lehman-McKeeman and Caudill, 1992a).

Tetrachloroethylene is an interesting example of a chemical that has been shown to cause alpha$_{2u}$-globulin accumulation and to induce renal tumors at a low incidence in male rats. The hyaline droplets appear only at very high doses—only at concentrations above the high dose of the carcinogenicity bioassay (Goldsworthy et al., 1988a; Green et al., 1990; Parker et al., 1992; USEPA, 1991c). In addition to the postulated alpha$_{2u}$-globulin mechanism, scientific evidence supports other alternative mechanisms. At least four different mechanisms of renal carcinogenesis have been postulated for tetrachloroethylene (Parker et al., 1992; USEPA, 1991c). Cytotoxic and potent mutagenic tetrachloroethylene metabolites that could contribute to the tumorigenesis of this chemical are formed in the kidney (Dekant et al., 1986; Green and Odum, 1985; Green et al., 1990; Vamvakas et al., 1989). Peroxisome proliferation occurs in the kidneys of rodents exposed to tetrachloroethylene and conceivably could contribute to the induction of renal tumors (Goldsworthy and Popp, 1989). In the case of tetrachloroethylene, the mode of renal tumorigenesis is not understood. Therefore, the renal tumors observed in male rats after tetrachloroethylene exposure would be considered as contributing to the overall weight of evidence in identifying potential cancer hazards to humans, although the tumors would not be used to derive a quantitative risk estimate because of the feasibility of their resulting in part from the alpha$_{2u}$-globulin nephropathy.

d-Limonene is the example of a tested chemical that meets all the criteria for the alpha$_{2u}$-globulin process of renal-tumor induction. The renal-tumor data on d-limonene do not indicate much likelihood of this chemical's producing human cancer, and it is appropriate to refrain from extrapolating the potential hazard and risk rate from the male rat renal tumors produced by d-limonene to humans (USEPA, 1991a,b; Hard and Whysner, 1994).

Studies have shown increased excretion of alpha$_{2u}$-globulin after low to moderate lead exposure (Mathias et al., 1992) and have identified renal lead-binding proteins (Fowler and DuVal, 1991) as the kidney-specific cleavage product of alpha$_{2u}$-globulin minus the first nine N-terminal residues. Recent studies have demonstrated the presence of similar lead-binding proteins in human kidney, liver, and brain cytosolic fractions (Kahng et al., 1992). Further research is needed to determine the possible roles of these proteins in mediating the observed increase (Steenland et al., 1992) in renal cancer in persons occupationally exposed to lead.

SUMMARY

New technologies using modern advances in molecular biology, genetic engineering, in vitro tissue-culture systems, and animal-model systems provide a wide variety of new methods that can be applied to biologic-marker research. The new technologies are rapidly being integrated into strategies for biologic-marker research pertaining to the genitourinary tract and, in a broader perspective, to other organ systems as well. The potential number of biologic markers is enormous. Each newly identified marker will require careful evaluation before its use in human population studies.

Several important lessons have been learned from the use of PSA as a clinical biologic marker for prostatic cancer. The uncertainties associated with its use have emphasized the need for the cooperation of various disciplines—including biostatistics and epidemiology, the basic sciences, and clinical specialties—to review the utility of specific biologic markers before their integration into clinical practice.

Extensive studies have attempted to extrapolate data concerning $alpha_{2u}$-globulin from animal studies to human conditions, such as nephrotoxicity and renal cancer. The results emphasize the need for caution in the extrapolation of in vitro and nonhuman in vivo studies to humans. At the same time, it is notable that potential biologic markers, such as $alpha_{2u}$-globulin, can be studied in relation to environmental stress. Some markers are quantifiable at the single-cell level, and the interactions of various cells can be investigated. New technologies—such as differential PCR, quantitative in situ PCR, and quantitative fluorescence image analysis (QFIA)—facilitate the use of biologic markers to describe genotypic and phenotypic changes associated with xenobiotic exposure.

New technologies can also promote the study of biologic markers at the cellular level. Particularly important are the changes that occur in nucleic acid synthesis, which regulate cell growth and death and which enhance our understanding of carcinogenesis, nephrotoxicity, and other diseases of the genitourinary tract.

7

CONCLUSIONS AND RECOMMENDATIONS

The recommendations of the Subcommittee on Biologic Markers in Urinary Toxicology follow in sequence the main sections of the report. This report brings into focus the limitations of current epidemiologic methods and laboratory tests for clearly defining the adverse effects of xenobiotics on the genitourinary tract. Current advances in epidemiology, biochemistry, molecular biology, and biologic-marker research have laid the groundwork for the development of programs of individual risk assessments and disease prevention. These programs will be most effective if implemented before the onset of irreversible nephrotoxicity or symptomatic cancer. To advance biologic-marker research, a multidisciplinary approach coupled with integration into a policy aimed at disease prevention could substantially reduce the adverse health effects associated with toxic exposures and the accompanying costs. The ability to assess an individual's risk, as opposed to a group risk, offers an additional approach to protecting the health of workers and the general public. In the conclusions and recommendations that follow, parallels have been drawn, when possible, between the effects of nephrotoxicants and those of other xenobiotics that are involved in carcinogenesis or have other adverse effects on the kidney, bladder, or prostate.

TOXIC EXPOSURE OF THE URINARY TRACT

The Kidney

Conclusion

In a substantial number of persons who have ESRD, the precise cause of kidney failure is unknown. Despite the data that have been collected on the nature of the patient population with ESRD, a definitive statement about the impact of occupational and environmental nephrotoxicants cannot be made. Specifically, the subcommittee noted the disproportionate incidences of ESRD among minority populations in the United States, and it has become known that several factors—including race and economic status—can be important predictors of ESRD. The lack of specific and comprehensive information on the impact of occupational and environmental nephrotoxicants on diseases of the kidney and urinary tract is troublesome. The possible correlations between the development of clinically significant renal disease and race, economic status, and exposure to occupational and environmental nephrotoxicants suggest the need for epidemiologic studies.

Recommendation

For patients entering programs for treatment of ESRD, details of occupational history or other factors that would show the impact of patients' environments on their condition should be obtained. As a first step, available information in relevant databases should be examined. Studies should be undertaken to determine whether the higher incidences of ESRD among minority groups and the economically disadvantaged are related to occupational or environmental exposure to nephrotoxicants. Epidemiologic studies need to focus on the various populations at risk; this focus should include not only the identification of the populations but their continued monitoring.

Conclusion

Some susceptible populations in the United States have anatomic or physiologic differences in the kidneys at birth that might help to explain their propensity for renal damage. As a result, they are less tolerant of various stresses, including exposure to environmental or occupational nephrotoxicants. Although emphasis has been placed on the toxicity of environmental and occupational nephrotoxicants, as well as diagnostic and therapeutic agents, a less well-defined problem is the frequency and severity of renal injury that results from recreational-drug use. Understanding the spectrum of urologic and renal diseases that might be produced by occupational or environmental agents requires integration of studies of the anatomy, biochemistry, and physiology of the kidney in conjunction with knowledge of the mechanisms of injury and disease.

Recommendation

Studies should be performed to determine whether an association between anatomic or physiologic differences of the kidney at birth and the later response to environmental or occupational nephrotoxicants leads to susceptibility to disease or to progression once disease occurs.

Data should be collected on the incidence of renal abnormalities among recreational-drug users to determine the influence of those substances on the rate of progression of renal disease due to other causes.

Basic studies are needed to determine the effects of occupational and environmental toxicants on specific segments of the kidney. These effects should be correlated with biochemical and anatomic changes.

The Bladder

Conclusion

Bladder cancer has been strongly associated with xenobiotic exposure. The bladder is an accessible site and is therefore an excellent model system for investigations of xenobiotic carcinogenesis.

Recommendation

Human bladder cancer induced by xenobiotic exposure in worker cohorts should be investigated to develop strategies of individual risk assessment, to formulate programs for prevention, and to evaluate new forms of therapy. Strategies of individual risk assessment need to be developed as the cornerstone of prevention. Once cohorts of at-risk persons are identified, they should be enrolled in long-term monitoring studies to assess the efficacy of prevention and treatment strategies.

Conclusion

Xenobiotics are potentially of importance in diseases of the bladder other than cancer.

Recommendation

Further research on the direct effect of xenobiotics on the bladder and the interactions of xenobiotics with the protective mechanisms of the bladder is very likely to uncover additional evidence that the bladder is a target organ. Markers associated with susceptibility should be identified to define the higher relative risk of disease in an exposed subset of the population.

The Prostate

Conclusion

In the United States, widespread screening for prostatic cancer has become accepted for white men over the age of 50 and black men over the age of 40. This screening has been associated with a high false-positive rate. In addition, quiescent disease is not sufficiently differentiated from biologically active disease.

Recommendation

Options for improving the efficacy of screening procedures should be studied. Tests with lower false-positive rates should be developed, as should tests able to detect premalignant changes and to separate quiescent from biologically active disease.

BIOLOGIC MARKERS OF EXPOSURE AND SUSCEPTIBILITY

The most efficient program for determining the importance of occupational and environmental toxicants and carcinogens in diseases of the urinary tract would include the identification of various susceptible populations and the correlation of disease processes with exposure to the agents. Linking markers of susceptibility and effect with markers of exposure is potentially a very powerful strategy for individual risk assessment. Biologic markers substantially increase the ability to evaluate the importance of low-level exposures, whether single or multiple.

Conclusion

To use markers to their greatest advantage, high-risk populations must be targeted.

Recommendation

Populations at risk of renal insults and carcinogenesis should be defined. Markers of human exposure and susceptibility should be sensitive (i.e., detectable before injury occurs), noninvasive, and chemically stable.

Conclusion

Susceptibility of particular populations is important in determining the onset of many diseases. Various factors modify human susceptibility to the effects of occupational and environmental nephrotoxicants. For example, specific genes that determine whether people are predisposed to develop disease have been identified. Some people inherit one defective copy of a tumor-suppressor gene and are at much greater risk for cancer than people who have two intact copies. Likewise, individual variations in metabolic pathways play a large role in susceptibility to cancer and toxicity. The goal of identifying markers of susceptibility and exposure is not to separate one population from another but rather to limit exposure to magnitudes that are tolerated by all.

Recommendation

Genetic and nongenetic factors that modify susceptibility to occupational and environmental genitourinary toxicants and carcinogens should be considered in the evaluation of individual susceptibility. These include sex, race, nutrition, socioeconomic factors, age, coexisting chronic disease, and drug abuse.

Conclusion

Some forms of renal disease are closely associated with chronic exposure to nephrotoxic agents, such as the renal injury that accompanies heavy use of analgesic agents, including nonsteroidal anti-inflammatory drugs. In others, the data relating exposure to disease are circumstantial but highly suggestive, such as the renal injury that has been associated with exposure to various hydrocarbons.

Recommendation

Markers of exposure and susceptibility should be identified to determine the relationship between coincident exposure to nephrotoxicants and the development or progression of chronic renal disease. Particular attention should be given to the role of widespread and sometimes excessive use of analgesics, including nonsteroidal anti-inflammatory drugs, in diseases of the urinary tract. Clinicians should be aware of the danger associated with abuse of these agents and should query renal-disease patients about their use. In particular, patients with established renal disease should be wary of exposure to these agents and other potential nephrotoxicants.

BIOLOGIC MARKERS OF EFFECT

Biologic markers of effect are pivotal in defining preclinical genitourinary and premalignant disease and are key to the prevention of nephrotoxicity and cancer. Biologic markers of effect are key to relating biologic markers of exposure and susceptibility to disease.

Conclusion

The "ideal" biologic marker of effect has been described in this report. No such ideal marker has been developed for the kidney, but a variety of markers have found wide acceptance. Because they might reflect different aspects of renal function, become detectable at different stages of exposure, or have different sensitivities and analytic reliability, no single test should be relied on for the demonstration of renal effects. Understanding the functional role of markers of effect is important in defining pathogenesis.

Recommendation

A battery of relatively simple and noninvasive tests should be used as a first step in screening populations at risk. On the basis of available information and technology, adequate initial screening results should be obtained by testing for proteinuria with dipsticks and then measuring urinary concentrating ability and serum creatinine and, for more sensitive measurements of tubular integrity, monitoring for an increase in urinary enzyme or low-molecular-weight protein excretion. Application of those tests to a population exposed, for instance, to diagnostic procedures or treatment with nephrotoxicants might identify early renal damage with adequate sensitivity.

Conclusion

Among established procedures for screening populations at risk, urinalysis and clearance measurements will continue to yield important functional markers of renal injury. The usefulness of clearances for evaluation of glomerular filtration would be much increased by the development of nonisotopic techniques for measuring, e.g., iodothalamate or chromium ethylenediaminetetraacetic acid in plasma or urine. A number of testing procedures in addition to the well-established ones have been cited.

Recommendation

Several of the newer testing procedures hold promise of future usefulness and should be further investigated. Among them are tests of urinary excretion of various growth factors, such as epidermal growth factor (EGF), and other tubular enzymes, such as intestinal alkaline phosphatase (IAP); both reflect some specificity of localization along the nephron and of cellular origin.

Conclusion

Molecular techniques have identified a variety of potentially useful markers of renal-cell injury. These include the products of expression of some early genes and changes in the expression of renal cytokines, growth factors, and growth-factor receptors.

Recommendation

It is highly likely that studies of these and similar molecular events will yield better markers of effect, and continued research in this area should be encouraged.

Conclusion

Fundamental advances have been made in the last decade in understanding the genetic basis of cancer and how it results from the subversion of normal growth controls by genetic and epigenetic mechanisms. Important species differences have been identified, and it is important to study cancer in humans, in whom stromal-epithelial interactions and the specific genetics of growth control are manifested.

Recommendation

Understanding fundamental cellular and molecular mechanisms of growth control in human tissues undergoing carcinogenesis (e.g., high-risk occupationally exposed populations or patients with premalignant processes) should be emphasized, because it is highly likely that more specific markers of effect will be identified and allow early intervention.

Conclusion

Cancer and many other diseases can have an important preclinical phase. Markers that define progress along the course of disease would be of value.

Recommendation

Markers that define preclinical disease should be identified.

Conclusion

Carcinogenesis involves mutational events that are primary and events that occur as a result of genetic instability.

Recommendation

Markers that detect xenobiotic-induced mutational events should be identified.

USE OF BIOLOGIC MARKERS IN EXTRAPOLATION

Conclusion

The human being is the best system to study for the identification of biologic markers. However, practical concerns and the expanding number of potential biologic markers mandate selected studies of nonhuman systems to develop an understanding of the biology underlying marker expression. Such studies have proved to be of considerable value and are necessary for a full appreciation of the effects of xenobiotics. Information gained from the study of environmental and occupational nephrotoxicants has been of fundamental value in identifying susceptible populations, reducing exposure, and modifying effects. Animal models and in vitro systems have utility for studying basic mechanisms and for identifying potential biologic markers and the effects of xenobiotics on whole organisms. The importance of living organisms for studies of xenobiotic tissues is recognized. Despite the advent of alternative research methods, animal studies continue to be necessary to prevent or minimize the impact on human populations.

Recommendation

Models for the identification and validation of markers should continue to be developed. The models must have sufficient sensitivity to distinguish between normal and abnormal function and must correlate well with known human toxicities. The models also must distinguish between functional alterations and pathologic changes. To obtain those characteristics, it will be necessary to develop and apply new technologies. Issues related to cost effectiveness should be considered.

Whole-animal studies should be used to establish target-organ specificity and to assess renal function in relation to survival. Species, sex, and strain differences must be taken into account in selecting animal models for particular uses.

In vitro methods should be used for mechanistic studies; the choice of models should depend on compatibility and validation with whole-animal studies.

Metabolic studies should be conducted to ascertain whether xenobiotics (or other agents) are biotransformed to reactive and toxic species and to identify sites of transformation, including renal tissue and other tissue in the urinary tract.

NEW TECHNOLOGIES

Two branches of study are central to the development of new markers: research into the mechanisms of cell growth, regeneration, and proliferation; and further study of the metabolic capacities of the kidney. Two categories of dispute are acknowledged: the problems that can emerge from the too rapid and widespread use of a single marker for a specific disease, as typified by the introduction of the test for prostate-specific antigen (PSA) for the detection of prostatic cancer; and the problems associated with extrapolation from animal studies to human conditions, as illustrated by the interpretation of the importance of renal accumulation of alpha$_{2u}$-globulin.

Conclusion

Cell repair can occur in response to cell injury. Thus, markers of cell growth, regeneration, and proliferation can indicate injury and be particularly useful when an injury is difficult to detect. In this circumstance, markers of repair may be the only indication that injury has occurred. This class of markers is not fully developed and holds promise as a new generation of markers.

Recommendation

Research should continue toward better understanding of the mechanisms of cell injury, because they can underlie the development of new markers. Emphasis should also be placed on understanding the mechanisms of cell growth, regeneration, and proliferation. Insight into the factors that control the cell cycle, regulate various growth factors, influence gene expression, and modulate nucleic acid synthesis might be critical in the development of new classes of markers.

Conclusion

Changes in metabolic pathways can

also occur in response to cell injury. Thus, biochemical markers associated with these pathways can be of value in detecting nephrotoxicity and carcinogenesis. Like markers of cell growth, regeneration, and proliferation, this class of markers is not fully developed and holds substantial promise.

Recommendation

Research should continue toward better understanding of the metabolic pathways of the kidney in relation to the effects of xenobiotics and susceptibility to them.

Conclusion

Understanding of the mechanisms of both cell growth and metabolism will allow further definition of the steps in the initiation and progression of various urinary tract cancers. It is anticipated that parallels will emerge that will yield insight into the progression of parenchymal renal disease.

Recommendation

Attention should be directed toward a deeper understanding of the mechanisms by which proto-oncogenes, tumor-suppressor genes, and epigenetic factors regulate the cell cycle and how damage to these mechanisms is related to disease. Attention should also be directed toward the elucidation of metabolic pathways, particularly as they are related to the production of toxic metabolites. Additional markers should be identified that help to identify populations at risk and to study the mechanisms by which environmental and occupational toxicants promote cancer.

Conclusion

Observations of the effects of growth factors on the prostate are conflicting. There are several reasons for the differences. First, assay methods are not uniform. Second, cells under study might have both low- and high-affinity receptors for a given peptide; cell response could depend on the strength of the signal and the concentrations of peptide. Third, both normal and neoplastic tissues are heterogeneous, and the expression of a given protein can vary widely among cells in different regions. Fourth, tumors might be hormonally sensitive or insensitive.

Recommendation

Research on the relation of growth factors to the prostate requires rigorous experimental approaches and designs and must consider multiple variables. Studies of biochemical changes in the areas next to a prostatic tumor might be more informative than analysis of the cancer itself.

Conclusion

Several agents share a curious relationship between nephrotoxicity and renal carcinogenesis. One example involves the response to lead exposure, which can lead to either acute or chronic lead nephropathy and under some circumstances can be associated with renal tumors.

Recommendation

The general relationship between nephrotoxicity and renal carcinogenesis should be explored.

Conclusion

The technology that is likely to yield new markers is complex. Equally complex is the identification of susceptible populations with the appropriate clinical assessment of exposure and effect. The use of biologic markers is essential in the examination of xenobiotic-induced diseases and other diseases of the human kidney, bladder, and prostate. Comprehending the sequences of events is an iterative process that involves a complex data set derived from scientific advances in molecular biology, epidemiology, pathology, biochemistry, and clinical medicine. Assembly of those data into an organized framework will be a major step toward individual risk assessment and should be a long-term objective.

Recommendation

To achieve the desired goal of identifying more-useful markers, cooperation between laboratory scientists, epidemiologists, and clinical researchers should be encouraged. Assays, particularly those involving enzymes or molecular probes, must be replicable in different laboratories.

REFERENCES

Abrahamsson, P.A., L.B. Wadstrom, J. Aluments, S. Falkmer, and L. Grimelius. 1987. Peptide-hormone-and serotonin-immunoreactive tumor cells in carcinoma of the prostate. Pathol. Res. Pract. 182:298-307.

Abrahamsson, P.A., S. Falkmer, K. Falt, and L. Grimelius. 1989. The course of neuroendorcrine differentiation in prostatic carcinomas. An immunohistochemical study testing chromogranin A as an endocrine marker. Pathol. Res. Pract. 185:373-380.

Acara, M.A., R.J. Mazurchuk, P.A. Nickerson, and R.J. Fiel. 1991. Magnetic resonance imaging and histopathology of hydronephrosis in the rat. Magnetic Resonanace Imaging 9:89-92.

Akaza, H., S. Kameyama, and Y. Aso. 1987. Significance of tumor markers in the treatment of urological malignancies [in Japanese]. Gan to Kagaku Ryoho 14:3034-3040.

Åkerström, B., and L. Lögdberg. 1990. An intruguing member of the lipwcalin protein family: α_1-microglobin. Trends Biochem. Sci. 15:240-243.

Akimoto, S., K. Akakura, and J. Shimazaki. 1988. Prostatic acid phosphtase (PAP), gamma-seminoprotein (gamma-Sm) and prostate specific antigen (PA) in prostatic cancer [in Japanese]. Hinyokika Kiyo (Acta. Urol. Jpn.) 34:1389-1396.

al-Abadi, H., and R. Nagel. 1988. Prognostic relevance of ploidy and proliferative activity of renal cell carcinoma. Eur. Urol. 15:271-276.

Alden, C.L. 1986. A review of unique male rat hydrocarbon nephropathy. Toxicol. Pathol. 14:109-111.

Alden, C.L. 1989. Male rat specific $alpha_{2u}$globulin nephropathy and renal tumorigenesis. Pp. 535-541 in Nephrotoxicity. In Vitro to In Vivo. Animals to Man. P.H. Bach, and E.A. Lock, eds. New York: Plenum Press.

Alden, C.L., and C.H. Frith. 1991. Urinary system. Pp 315 in Handbook of Toxicologic Pathology, W.M. Haschek, and C.G. Rousseaux, eds. San Diego: Academic Press.

Alden, C.L., R.L. Kanerva, G. Ridder, and L.C. Stone. 1984. The pathogenesis of the nephrotoxocity of volatile hydrocarbons in the male rat. Pp. 107-120 in Advances in Modern Environmental Toxicology. Vol. III. Renal Effects of Petroleum

Hydrocarbons. M.A. Mehlman, G.P. Hemstreet, J.J. Thorpe, and N.K. Weaver, eds. Princeton: Princeton Scientific Publishers, Inc.

Aleo, M.D., and R.G. Schnellman, 1992a. The neurotoxicants strychnine and biculline protect renal proximal tubules from mitochondrial inhibitor-induced cell death. Life Sci. 51:1783-1787.

Aleo, M.D., and R.G. Schnellman. 1992b. Regulation of glycolytic metabolism during long-term primary culture of renal proximal tubule cells. Am. J. Physiol. 262:F77-F85.

Aleo, M.D., M.L. Taub, P.A. Nickerson, and P.J. Kostyniak. 1989. Primary cultures of rabbit renal proximal tubule cells: I. Growth and biochemical characteristics. In Vitro Cell. Dev. Biol. 25:776-783.

Alfrey, A.C., and R.C. Tomford. 1982. Phosphate and prevention of renal failure. Pp. 31-38 in Prevention of Kidney Disease and Long-Term Survival, M.M. Avram, ed. New York: Plenum.

Allen, B.C., and J.W. Fisher. 1993. Pharmacokinetic modeling of trichloroethylene and trichloroacetic acid in humans. Risk Analysis 13:71-86.

Allen, B.C., K.S. Crump, and A.M. Shipp. 1988. Correlation between carcinogenic potency of chemicals in animals and humans. Risk Anal. 8:531-544.

Allen, J.K., E.A. Krauss, and E.G. Deeter. 1991. Dipstick analysis of urinary protein. A comparison of Chemstrip-9 and Multistix-10SG. Arch. Path. Lab. Med. 115:34-37.

Allison, M.E., E.M. Lipham, W.E. Lassiter, and C.W. Gottschalk. 1973. The acutely reduced kidney. Kidney Int. 3:354-363.

Almeder, R. and J. Humber. 1987. Introduction to Quantitative Risk Assessment. *Biomedical Ethics Reviews, 1986.* J. Humbar, R. Almeder, eds. Clifton, N.J.: Humana Press.

Almeida, A.R.P., D. Bunnachak, M. Burnier, J.F.M. Wetzels, T.J. Burke, and R.W. Schrier. 1992. Time-dependent protective effects of calcium channel blockers on anoxia- and hypoxia-induced proximal tubule injury. J. Pharmacol. Exp. Ther. 260:526-532.

Ambrose, S.S. 1983. Ureterosigmoidostomy. Pp. 511-520 in Urologic Surgery, 3rd edition, J.F. Glenn, ed. Philadelphia: Lippincott.

Ambrum, J.L., C.M. Ambrus, P. Forgach, S. Stadler, J. Haplern, S. Sayyid, P. Niswander, and C. Toumbis. 1992. Studies on tumor induced angiogenesis. EXS 61:436-444.

Ames, B.N., and L.S. Gold. 1990. Too many rodent carcinogens: Mitogenesis increases mutagenesis. Science 249:970-971.

Amico, S., J.C. Liehn, B. Desoize, H. Larbre, G. Deltour, and J. Valeyre. 1991 Comparison of phophotase isoenzymes PAP and PSA with bone scan in patients with prostate carcinoma. Clin. Nucl. Med. 16:643-648.

Anders, M.W. 1980 Metabolism of drugs by kidney. Kidney Int. 18:636-647.

References

Anders, M.W., ed. 1985. Bioactivation of Foreign Compounds. New York: Academic Press.

Anders, M.W. 1989. Biotransformation and bioactivation of xenobiotics by the kidney. Pp. 81-97 in Intermediary Xenobiotic Metabolism in Animals: Methodology, Mechanisms and Significance. D.H. Hutson, J. Caldwell, and G.D. Paulson, eds. London: Taylor & Francis.

Anders, M.W., L.H. LAsh, W. Dekant, A.A. Elfarra, and D.R. Dohn. 1988 Biosynthesis and metabolism of glutathione conjugates to toxic forms. CRC Crit. Rev. Toxicol. 18:311-341.

Anderson, R.L. 1985. Some changes in gastro-intestinal metabolism and in the urine and bladders of rats in responce to sodium saccharin ingestion. Fd. Chem. Toxic 23:457-463.

Anderson, R.L. 1988. An hypothesis of the mechanism of urinary tumorigenesis in rats ingesting sodium sacchrin. Fd. Chem. Toxic 26:637-644.

Anderson, R.L. 1991. Early indicators of bladder carcinogenesis produced by non-genotoxic agents. Mut. Res. 248:261-270.

Anderson, R.L., and C.L. Alden. 1989. Risk assessment for nitrilotriacetic acid (NTA). Pp. 390-426 in The Risk Assessment of Environmental Hazards. A Textbook of Case Studies. D.J. Paustenbach, ed. New York: John Wiley & Sons.

Anderson, P.M., and M.O. Schultze. 1965. Cleavage of S-(1,2-dichlorovinyl)-L-cysteine by and enzyme of bovine origin. Arch. Biochem. Biophys. 111:593-602.

Anderson, E., and the Carcinogen Assessment Group of the U.S. Environmental Protection Agency. 1983. Quantitative approaches in use to assess cancer risk. Risk Analysis 3:277-295.

Anderson, R.L., W.E. Bishop, and R.L. Campbell. 1985. A review of the environmental and mammalian toxicology of nitrolotriacetic acid. CRC Crit. Rev. Toxicol. 15:1-102.

Anderson, R.L., R.L. Kanerva, F.R. Lefever, and W.R. Francis. 1986a. Effect of N-nitrose-*n*-butyl-(4-hydroxybutyl)amine exposure on the changes in mineral deposition caused by trisodium nitrilotriacetate. Fd. Chem. Toxic 24:229-235.

Anderson, S., H.G. Rennke, and B.M. Brenner. 1986b. Therapeutic advantage of converting enzyme inhibitors in arresting progressive renal disease associated with systemic hypertension in the rat. J. Clin. Invset. 7:1993-2000.

Anderson, R.L., F.R. Lefever, and J.K. Maurer. 1988. The effect of various sacchrin forms on gastro-intersinal tract, urine and bladder on male rats. Fd. Chem. Toxic. 26:665-669.

Andersson, G., and E. Jennische. 1988. IGF-I immunoreactivity is expressed by regenerating renal tubular cells after ischaemic injury in the rat. Acta. Physiol. Scand. 132:4653-457.

Andersson, G.L., A. Skottner, and E. Jennische. 1988. Immunocytochemical and biochemical localization of insulin-like growth factor I in the kidney of rats before and after uninephrectomy. Acta. Endocrinol. 119:555-560.

Anglard, P., K. Tory, H. Brauch, G.H. Weiss, F. Latif, M.J. Merino, M.I. Lerman, B. Zbar, and W.M. Linehan. 1991. Molecular analysis of genetic changes in the origin and development of renal cell carcinoma. Cancer Res. 51:1071-1077.

Annab, L.A., J.T. Dong, P.A. Futreal, H. Satoh, M. Oshimura, and J.C. Barrett. 1992. Growth and transformation suppressor genes for BHK Syriam hamster cells on human chromosomes 1 and 11. Mol. Carcinog. 6:280-288.

Anonymous. 1985. Acid phosphatases and other tumor markers in the management of prostatic cancer. Proceedings of the Workshop of the Scandinavian Prostatic Cancer Group, Scandinavian Committee on Enzymes. Copenhagen, March 1985. Scand. J. Clin. Lab. Invest. Suppl. 179:1-117.

Anonymous. 1990. Bladder cancer screening in high-risk groups. Proceedings of the International Conference. September 13-14, 1989. J. Occup. Med. 32:787-945.

Arndt, R. J. Durkopf, H. Huland, F. Donn, T. Loening, and H. Kalthoff. 1987. Monoclonal antibodies for characterization of the heterogeneity of normal and malignant transitional cells. J. Urol. 137:758-763.

Ash, S.R., and F.E. Cuppage. 1970. Shift toward anaerobic glycolysis in the regererating rat kidney. Am. J. Pathol. 60:385:402.

Askergren, A., L.G. Allgen, and J. Berstrom. 1981. Studies on kidney function in subjects exposed to organic solvents. 1. Excretion of albumin and beta-2-microglobulin in the urine. Acta. Med. Scand. 209:479-483.

Askergren, A. 1984. Urinary protein and cell excretion in constructions workers exposed to organic solvents.

Asselin, C., and K.B. Marcus. 1989. Mode of c-myc gene regulation in folic acid-induced kidney regeneration. Oncogene Res. 5;67-72.

Aten, J., C.B. Bosman, J. Rozing, T. Stynen, P. Hoedemacker, and J.J. Weening. 1988. Mercuric chloride-induced autoimmunity in the Brown Norway Rat. Am. J. Path. 133:127-138.

Aulizky, W.K., P.N. Schlogel, D. Wu, C.Y. Cheng, C.C. Chen, P.S. Li, M. Goldstein, M. Reidenberg, and C.W. Bardin. 1992. Measurement of urinary clusterin as an index of nephrotoxicity. Proc. Soc. Exp. Biol. Med. 199:93-96.

Azzopardi, J.G., and D.J. Evans. 1971. Argentaffin cells in prostatic carcinoma: Differentiation from lipofuscin and melanin in prostatic epithelium. J. Pathol. 104:247-251.

Baisch, H., U. Otto, and G. Kloppel. 1986. Malignancy index based on flow cytonetry and histology for renal cell carcinomas and its correlation to prognosis. Cytometry 7:200-204.

Baisch, H., G. Kloppel, and U. Otto. 1990. Pp. 249-260 in DNA analysis in relan neoplasia. New York: Churchill Livingstone.

Balitskaia, O.V., N.K. Berdinskikh, and N.G. Kononenko. 1992. The possibilities of using the free polyamines of the peripheral blood as biochemical tumor markers in nephroblastoma in children [in Russian]. Vopr. Onkol. 38:674-682.

Ballardi, F.W., ed. 1992. Autoimmunity in Nethritis. Chur, Switzerland: Harwood Academic Publishers.

Ballatori, N., R. Jacob, C. Barret, and J.L. Boyer. 1988. Biliary catabolism of glutathione and differential reabsorption of its amino acids constituents. Am. J. Physiol. 254:G1-G7.

Bander, N.H., C.L. Finstad, C. Cordon-Cardo, R.D. Ramsawak, E.D. Vaughan, W.F. Whitmore, H.F.Oettgen, M.R. Melames, and L.J. Old. 1989. Analysis of a mouse monoclonal antibody that reacts with a specific region of the human proximal tubule and subsets renal cell carcinomas. Cancer Res. 49:6774-6780.

Bandyk, M., I. Sawczuk, C.A. Olsson, A.E. Katz, and R.Buttyan. 1990. Characterization of the products of a gene expressed during androgen-programmed cell death and their potential use as a marker of urogenital injury.

Banki, K., and M.W. Anders. 1989. Inhibition of rat kidney mitochondrial DNA, RNA and protein synthesis by halogenated cystein S-conjugates. Carcinogenesis 10:767-772.

Baretton, G., B. Kuhlmann, R. Krech, U. Lohrs. 1991. Intratumoural heterogeneity of nuclear DNA-content and proliferation in clear cell type carcinomas of the kidney. Virchows Arch B: Cell Pathol. Mol. Pathol. 61:57-63.

Baricordi, O.R., A. Sensi, C. DeVinci, L. Melchorri, G. Fabris, E. Marchette, F. Corrado, P.L. Mattiuz, and G. Pizza. 1985. A monoclonal antibody to human transitional cell carcinoma of the bladder cross reaction with a differentiation antigen of neutrophilic lineage. Int. J. Cancer 35:781-786.

Barret, J.C., and J.E. Huff. 1991. Cellular and molecular mechanisms of chemically induced renal cancinogenesis in nephretoxicity. Pp. 287-306 in *Mechanisms, Early Diagmosis, and Therapeutic Management.* P.H. Bach, N.J. Gregg, M.F.Wiks, and L. Delactuz, eds. New York: Marcel Dekker, Inc.

Balslov, J.T., and H.E. Jorgensen. 1963. A survey of 499 patients with acute anuric renal insufficience. Causes, treatment, complications, and mortality. Am. J. Med. 34:75.

Bass, R.A., G. Hemstreet, N.A. Honker, R.E. Hurst, R.S. Doggett. 1987. DNA cytometry and cytology by quantitative fluorescence inage analysis in sympltomatic bladder clncer patients. Int. J. Cancer 40(5):698-705.

Basta, M.T., A.M. Attallah, M.N. Seddek, H. el-Mohamady, E.S. Al-Hilaly, N. Atwaan, and M. Ghoneim. 1988. Cytokeratin shedding in urine: A biological marker for bladder cancer? Br. J. Unol. 61:116-121.

Bauer, H.W. 1988. Acid phosphatase, alkaline phosphatase and prostate-specific antigen--usefulness in the diagnosis of netastatic disease and follow-up. Prog. Clin. Biol. Res. 269:33-42.

Bauer, D., Müller, J. Reich, H. Riedel, V. Ahrenkiel, P. Warthoe, and M. Strauss.

1993. Indentification of differentially expressed mRNA species by an improver display technique (DDRT-PCR). Nucleic Acids Res. 21:4272-4280.

Beckett, M.L., G.B. Lipford, C.L. Haley, P.F. Schellhammer, and G.L. Wright. 1991. Monoclonal antibody PD41 recognizes and antigen restricted to prostate adencarcinomas. Cancer Res. 51:1326-1333.

Beer, D.G., K.A. Zweifel, D.P. Simpson, and H.C. Pitot. 1987. Specific gene expression during compensatory renal hypertrophy in the rat. J. Cell Physiol. 131:29-35.

Beggard, I., and A.G. Bearn. 1968. Isolation and properties of a low molecular weight β2-globulin occuring in human biological fluids. J. Biol. Chem. 243:4095-4103.

Bethrens, M.T., A.L. Corbin, and M.K. Hise. 1989. Epidermal growth factor receptor regulation in rat kidney: Two models of renal growth. Am. J. Physiol. 257:4095-4103.

Beirne, G.J., J.T. Brennan. 1972. Glomerulonephritis associated with hydrocarbon solvents. Arch. Environ. Health 24:365-369.

Bell, G.M., D. Doig, D. Thompson, J.L. Anderton, and J.S. Robson. 1985. End-stage renal disease associated with occupational exposure to organic solvent. Proc EDTA-ERA 22:725-729.

Bell, G.I., N.M. Fong, M.M. Stempien, M.A. Wormsted, D. Caput, L. Ka, M.S. Urdea, L.B. Rall, and R. Sanchez-Pescador. 1986. Human epidermal growth factor precursor: cDNA sequence, expression in vitro and gene organization. Nucleic Acids Res. 14:8427-8446.

Ben-Aissa, H., S. Paulie, and H. Koho. 1985. Specificities and binding properties of two monoclonal antibodies against carcinoma cells of the human urinary bladder. Br. J. Cancer 52:65-72.

Bennett, W.M., L.W. Elzinga, and G.A. Porter. 1991 Tubulointerstitial disease and toxic nephropathy. Pp. 1430-1496 in the Kidney. B.M. Brenner, F.C. Rector, and W.B. Saunders, eds. Philadelphia, London, Toronto, Montreal, Sydney, Tokyo.

Bennington, J.K., and J.B. Beckwith. 1975. Tumors of the Kidney, Renal Pelvis, and Ureter. Pp. 93-199 in Atlas of Tumor Pathology. Second Series, Fascicle 12. Washington, D.C.: Armed Forces Institute of Pathology.

Benson, M.C., I.S. Whang, C.A. Olsson, D.J. McMahon, And W.H. Cooner. 1992a. The use of prostate specific antigen density to enhance the predictive value of intermediate levels of serum prostate specific antigen. J. Urol. 147:813-821.

Benson, M.C., I.S. Whang, A. Pantuckk, K. Ring, S.A. Kaplan, C.A. Olsson, and W.H. Cooner. 1992b. Prostate specific antigen density: A means of distinuishing benign prostatic hypetrophy and prostate cancer. J. Urol. 147:815-816.

Bergerheim, U., M. Nordenskjold, and V.P. Collins. 1989. Deletion mapping in human renal cell carcinoma. Cancer Res. 49:1390-1396.

Bergerheim, U.S., K. Kunimi, V.P. Collins, and P. Ekman. 1991. Deletion mapping of chromosomes 8, 10, and 16 in human prostatic cancinoma. Genes Chromosom. Cancer 3(3):215-220.

Berggard, I. 1970. Plasma proteins in normal human urine. Pp. 7-19 in Proteins in Normal and Pathological Urine. Y. Manuel, J.P. Revillard, and H.Betuel. eds. S. Karger, Basel.

Berlyne, G.M. 1984. Toxic nephropathies and current methods for early dection of the toxicity of the kidney. Pp. 173-184 in Renal Effects of Petroleum Hydrocarbons. Advances in Moderm Environmental Toxicology, Vol. VII. M.A. Mehlman, C.P. Hemstreet, J.J. Thorpe, and N.K. Weaver, eds. Princeton: Princeton Scientific Publishers.

Berman, J., J. Seidman, R. Yetter, adn G. Moore. 1991. Clear cell dysplasia of the bladded. Report of a case with flow cytometric analysis. Anal. Quant. Cytol. Histol. 13(6):391-394.

Bernard, A., A. Vyskocyl, P. Mahieu, and R. Lauwerys. 1988. Effect of renal insufficiency on the concentration of free retinol-binding protein in urine and serum. Clin. Chim. Acta. 171:85-93.

Bernard, A.M., R.R. Lauwerys, A. Nöel, B. Vandeleene, and A. Lambert. 1989. Urine protein 1: A sex dependent marker of tubular or glomerular dysfunction. Clin. Chem. 35:2141-2142.

Berndt, W.O. 1981. Use of renal function tests in the evaluation of nephrotoxic effects. Pp. 1-29 in Toxicology of the Kidney, J.B. Hook, ed. New York: Raven.

Berrozpe, G., R. Miro, M.R. Caballin, J. Salvador, and J. Egozcue. 1990. Trisomy 7 may be a primary change in noninvasive transitional cell cancinoma of the bladded. Cancer Genet. Cytogenet. 50:9-14.

Bertani, T., M. Livio, D. Maconi, M. Morigi, G. Bisogno, C. Patrono, and G. Remuzzi. 1987. Platelet activating factor (PAF) as the mediator of injury in nephrotoxic nephritis. Kidney Int. 31:1248-1256.

Bi, W., J.Y. Rao, G.P. Hemstreet P. Fang. N.R. Asal, M. Zang, K.W. Min, Z. Ma, P. Fang, E. Lee, G. Li, R.E. Hurst, W. Wu, R.B. Bonner, Y. Wang, Y. Fradet, and S. Yin. 1993. Field molecular epidemiology: Feasability of monitering for the malignant bladder cell phenotype in a benzidine exposed occupational cohort. J. Occup. Med. 35(1):20-27.

Birk, H.W., S. Piberhofer, G. Schutterle, W. Haase, J. Kotting, and H. Koepsell. 1991. Analysis of Na^+-D-glucose cotransporter and other renal brush border proteins in human urine. Kidney Int. 40:823-837.

Blais, A., F. Jalal, P. Crine, J. Paiement, and A. Bertloot. 1992. Increased functional differentiation of rabbit proximal tubule cells cultured in glucose-free media. Am. J. Physiol. 263:F152-F162.

Blay, J.Y., S. Negrier. V. Combaret, S. Attali, E. Goillot, Y. Merrouche, A. Mercatello, A. Ravault, J.M. Tourani, J.F. Moskovtchenko, T. Phillip, and M. Favrot. 1992.

Serum level of interleukin 6 as a prognosis factor in metastatic renal cell cercinoma. Cancer Res. 52:3317-3322.

Blouin, P., M.C. Guiot, and S. Jothy. 1989. Definition of the human renal cell carcinoma phenotype using monoclonal and polyclonal antibodies: A tumor marker study. Exp. Pathol. 36:147-163.

Boag, A.H., and I.D. Young. 1992. Type IV collagenase expression in prostatic adenocarcinoma [abstract]. Mad. Pathol. 5:51A.

Bodenstab, W., J. Kaufman, and C.L. Parsons. 1983. Inactivation of antoadherence effect of bladder surface glycusaminoglycan by a complete urinary carcinogen (N-methyl-N-nitrosourea). J. Urol. 129.

Boehm, T., I. Lavenir, A. Forster, R.B. Wadey, J.K. Cowell, J. Harbott, F. Lampert, J. Waters, P. Sherrington, and P. Couillin. 1988. The T-ALL specific t(11;14)(p13;q11) translocation breakpoint cluster region is locatid near tho the Wilms' tumour prodisposition locus. Oncogene 3:691-695.

Boileau, M., D. Swartz, K. Schmidt, and W. Schmidt. 1987. Bladder cancer dection and surceillance: Carcinoembryonic antigen as a monitor of newplastic transformation. J. Surg. Oncol. 35:120-123.

Boldog, F., K. Arheden, S. Imreh, B. Strombeck, L. Szekely, R. Erlandsson, Z., Marcsek, J. Sumegi, F. Mitelman, and G. Klein. 1991. Involvement of 3p deletions in sporatic and hereditary forms of renal cell cancinoma. Genes Chromosomes Cancer 3:403-406.

Bombassei, G.J., and A.E. Kaplan. 1992. The association between hydrocarbon exposure and antiglomerular basement membrane antibody-mediated disease (Goodpasture's syndrome). Am. J. Indust. Med. 21:141-153.

Bonner, R.B., G.P. Hemstreet, Y. Fradet, J.Y. Rao, K.W. Min, and R.E. Hurst. 1993. Bladder cancer risk assessment with quantitative fluoroscence image analysis of tumor markers in exfoliated bladder cells. Cancer 72:2461-2469.

Boogaard, P.J., J.N.M. Commandeur, G.J. Mulder, N.P.E. Vermeulen, and J.F. Nagelkerke. 1989. Toxicity of cystein S-conjugates and mercapturic acids of four structurally related difluoroethylenes in isolated proximal tubular cells from rat kidney: Uptake of the conjugates and activation to toxic metabolites. Biochem. Pharmacol. 38:3731-3741.

Boogaard, P.J., J.P. Zoeteweij, T.J.C. van Berkel, J.M. van't Noorkende, G.J. Mulder, and J.F. Nagelkerke. 1990. Primary culture of proximal tubular cells from normal rat kidney as an in vitro model to study mechanisms of nephrotoxicity: Toxicity of nephrotoxicants at low concentrations during prolonged exposure. Biochem. Pharmacol. 39:1335-1345.

Bookstein, R., D. MacGrogan, S.G. Hilsenbeck, F. Sharkey and D.C. Allred. 1993. p53 is mutated in a subset of advanced-stage prostate cancers. Cancer Res. 53:3369-3373.

Borghoff, S.J., B.G. Short, and J.A. Swenberg. 1990. Biochemical mechanisms and pathobiology of α_{2u}-globulin nephropathy. Ann. Rev. Pharmacol. Toxicol. 30:349-367.

Borghoff, S.J., A.B. Miller, J.P. Bowen, and J.A. Swenberg. 1991 Characteristics of chemical binding to α_{2u}-globulin in vitro—evaluation structure-activity relationships. Toxicol. Appl. Pharmacol. 107:228-238.

Borland, R., C. Brendler, and W.B. Isaacs. 1992. Molecular biology of bladder cancer. Hematol. Oncol. Clin. NorthAm. 6(1):31-39.

Bortz, J.D., P. Rotwein, D. DeVol, et al. 1988. Focal expression of insulin-like growth factor I in rat kidney collection duct. J. Cell Biol. 107:811-819.

Bostwick, D.G. 1989. The pathology of early prostate cancer. Cancer J. Clin. 39:376-393.

Bostwick, D.G. 1989. Varients of prostatic carcinoma. Pp. 95-133 in Pathology of the Prostate, D.G. Bostwick, ed. New York: Churchill Livingstone.

Bostwick, D.G., Graham, P. Napalkov, P.A. Abrahamsson, P.A. di Sant'agnese, F. Algaba, P.A. Hoisaeter, F. Lee, P. Littrup, and F.K. Mostofi. 1993. Staging of early prostate cancer: A proposed tumor volume-based prognostic index. Urology 41:403-411.

Boswell, J.M., M.A. Yui, D.W. Burt, and V.E. Kelley. 1988. Increased tumor necrosis factor and IL-1β gene expression in the kidneys of mice with lupus nephritis. J. Immunol. 141:3050-3054.

Bot, F.J., J.C. Godschalk, K.K. Krishnadath, T.H. van der Kwast, and F.T. Bosman. 1994. Prognostic factors in renal-cell carcinoma: Immunohistochemical detection of p53 protein versus clinico-pathological paramenter. Int. J. Cancer 57:634-637.

Boucher, A., D. Droz, E. Adafer, and L-H. Noël. 1986. Characterization of mononuclear cellsubsets in renal cellular interstitial infiltrates. Kidney Int. 29:1043-1049/

Boffioux, C.R. 1980. Etiological and epidemiological considerations in prostatic cancer. Scand. J. Urol. Nephrol., Suppl. 55;9-16.

Bowers, M.A., L.D. Aicher, H.A. Davis, and J.S. Woods. 1992. Quantitative determination of porphyrins in rat and human urine and evaluation of urinary porphyrin profiles during mercury and lead exposures. J. Lab. Clin. Med. 120:272-281.

Boyce, N.W. P.G. Tipping, and S.R. Holdsworth. 1989. Glomerular macrophages produce reactive oxygen species in experimental glomerulonephritis. Kidney Int. 35:778-782.

Boyle, E. 1970. Biological patterns in hypertension by race, sex, body weight and skin color. JAMA 213:1637-1643

Brady, H.R. B.C. Kone, M.E. Stromski, M.L. Zeidel, G. Giebisch, and S.R. Gullans. 1990. Mitochondrial injury: An early event in cisplatin toxicity to renal proximal tubules. Am. J. Physiol. 258:F1181-F1187.

Brauch, H. K. Tory, W.M. Linehan, D.J. Weaver, M.A. Lovell, and B. Zbar. 1990.

Molecular analysis of the sort arm of chromosome 3 in five renal oncocytomas. J. Urol. 143:622-624.

Braver, D.J., M. Modan, A Chetrit, A. Lucky, and Z. Braf. 1987. Drinking, micturition habits, and urine concentration as potential risk factors in urinary bladder cancer. J. Natl. Cancer Inst. 78:437-440.

Brazy, P.C., L.J. Mandel, S.R. Gullans, and S.P. Soltoff. 1984. Interactions between phosphate and oxidative metabolism in proximal renal tubules. Am. J. Physiol. 247:F575-F581.

Brenner, B.M., and S. Anderson. 1989. Filtration surface area, salt intake, and hypertension. Pp. 45-59 in The Progressive Nature of Renal Disease: Myths and Facts. Contributions in Nephrology, Vol. 75, L. Oldrizzi, G. Maschio, and C. Rugiu, eds. Basel: Karger.

Brenner, B.M., and F.C. Rector, eds. 1986. The Kidney, 3rd edition. Philadelphia: W.B. Saunders Company.

Brenner, B.M., T.W. Meyer, and T.H. Hostetter. 1982. Dietary protein intake and the progressive nature of kidney disease. N. Engl. J. Med. 307:652-659.

Bretton, P.R., A. Myc, C. Cordon-Cardo, P. DeAngelis, W.R. Fair, and M.R. Melamed. 1989. Initial evaluation of a new epithelial antigen (T16) for bivariate flow cytometry of bladder irrigation specimens. Cytometry 10:339-344.

Breysse, P., W.G. Couser, C.E. Alpers, K. Nelson, L. Gaur, and R.J. Johnson. 1994. Membranous nephropathy and formaldehyde exposure.

Bringuier, P.P., R. Umbas, H.E. Schaafsma, H.F.M. Karthaus, F.M. J. DeBruyne, and J.A. Schalken. 199f3. Decreased e-cadherin immunoreactivity correlates with poor survival in patients with bladder tumors. Cancer Res. 53:3241-3245.

Bringuier, P.P., R. Bouvier, N. Berger, E. Piaton, J.P. Revillard, P. Perrin, and M. Devonec. 1993. DNA ploidy status and DNA content instability within single tumors in renal cell carcinoman. Cytometry 14:559-564.

Brochard, P., J. DePalmas, M. Martini, M. Bondet, and G. Largue. 1984. Etude de la prevalence des proteinuries dipistees chex des sujets exposes professionnellment aux solvants. XXI Int. Congr. Occup. Health, Dublin.

Broder, L.E., B.D. Weintraub, S.W. Rosen, M.H. Cohen, and F. Tejada. 1977. Placental proteins and their subunits as tumor markers in prostatic carcinoma. Cancer 40:211-216.

Brooks, D.E. 1987. The major androgen-regulated secretory proteins of the rat epidiymis bear sequence homology with members of the α_{2u}-globulin superfamily. Biochem. Int. 14:235-240.

Brooks, J.D., G.S. Bova, F.F. Marshall, and W.B. Isaacs. 1993. Tumor suppressor gene allelic loss in human renal cancers. J. Ruol. 150:1278-1283.

Bruner, R.H. 1984. Pathologic findings in laboratory animals exposed to hydrocarbon fuels of military interest. Pp. 133-140 in Advances in Modern Environmental Toxicology. Vol. Vii. Renal Effects of Petroleum Hydrocarbons. M.A. Mehlman,

References

G.P. Hemstreet, J.J. Thorpe, and N.K. Weaver, eds. Princeton: Princeton Scientific Publishers, Inc.

Bucher, N.L.R., and R.A. Malt. 1971. Regeneration of liver and kidney. Boston: Little, Brown.

Buchet, J.P., R. Lauwerys, H. Roels, A. Bernard, P. Bruaux, F. Claeys, G. Ducofere, P. De Plaen, J. Staessen, A. Amery, P. Lijnen, L. Thijs, D. Rondia, F. Sartor, A. Saint Remy, and L. Nick. 1990. Renal effects of cadmium body burden of the general populatoin. The Lancet 335:699-762.

Bullock, A.D., J.J. Becich, C.G. Klutke, and T.L. Ratliff. 1992. Experimental autoimmune cystitis: A potential murine model for ulcerative interstitial cystitis. J. Urol. 148:1951-1956.

Burg, M. b., and M.A. Knepper. 1986. Single tuble perfusion techniques. Kidney Int. 30:166-170.

Burns, C.P., and E. Rozengurt. 1984. Extracellular Na+ and initiation of DNA synthesis: Role of intracellular pH and K+. J. Cell Biol. 98:1082-1089.

Burton, B.T. and G.H. Hirschman. 1979. Demographic analysis: End-stage renal disease and its treatment in the United States. Clin. Nephrol. 11:47-51.

Bussemakers, M.J.G., W.B. Isaacs, B.S. Carter, W. Van de Ven, F.M.J. DeBruyne, and J.A. Schalken. 1991. Ecahedrin is a candidate trumor suppressor gene implicated in prostate cancer. J. Urol. 145:294A.

Buttyan, R., C.A. Olsson, J. Pintar, C. Chang, M. Bandyk, P. Ng., and I.S. Sawczuk. 1989. Induction of the TRPM-2 gene in cells undergoing programmed death. Molec. and Cell Biol. 9:3473-3481.

Byrne, C., J. Nedelman, and R.G. Luke. 1994. Race, socioeconomic status, and the development of end-stage renal disease. Amer. J. Kidney Dis. 23:16-22.

Cagnoli, L., S. Casanova, S. Pasquali, U. Donini, and P. Zucchelli. 1980. Relation between hydrocarbon exposure and the nephrotic syndrome. BMJ 280:1068-1069.

Cairns, P., A.J. Proctor, and M.A. Knowles. 1991. Loss of heterozygosity at the RB locus is frequent and correlates with muscle invasion in bladder carcinoma. Oncogene 6:2305-2309.

Cantrell, B.B., D.P. DeKlerk, J.C. Eggleston, J.K. Boitnott, and P.C. Walsh. 1981. Pathological factors that influence prognosis in state A prostatic cancer: The influence of extent versus grade. J. Urol. 125:516-520.

Capella, C., L. Usellini, R. Buffa, B. Frigerio, and e. Solcia. 1981. The endocrine component of prostatic carcinomas, mixed adenocarcinoid tumours and non-tumour prostate: Histochemical and ultrastructural identification of the endocrine cells. Histopathology 5:175-192.

Carbin, B.E., P. Ekman, P. Eneroth, and B. Nilsson. 1989. Urine-TPA (tissue polypeptide antigen), flow cytometry and cytology as markers for fumor invasiveness in urinary bladder carcinoma. Urol. Res. 17:269-272.

Carlisle, E.J., S.M. Donnelly, S. Vasuvattakul, K.S. Kamel, S. Tobe, and M.L. Halperin.

1991. Glue-sniffing and distal renal tubular acidosis: Sticking to the facts [review]. J. Amer Soc. Nephrol. 1:1019-1027.

Carter, H., S. Piantadosi, and J. Isaacs. 1990 Clinical evidence for an implications of the multistep development of prostate cancer. J. Ruol. 143(4):742-746.

Carter, B.S., T.H. Beaty, G.D. Steinberg, B. Childs, and P.C. Walsh. 1992. Mendelian inheritance of familial prostate cancer. Proc. Natl. Acad. Sci. U.S.A. 89(8):3367-3371.

Cartwright, R.A. 1986. Screening workers exposed to suspect bladder carcinogens. J. Occup. Med. 28(10):1017-1019.

Casalone, R., P. Granata Casalone, E. Minelli, P. Portentoso, R. Righi, E. Meroni, A. Guidici, D. Donati, C. Riva, and S. Salvatore. 1992. Siginificance of the clonal and sporadic chromosome abnormalities in non-neoplastic renal tissue. Hum. Genet. 90:71-78.

Catalona, W.J. 1992. Urothelial Tumors of the Urinary Tract. Pp. 1094-1158 in Campbell's Urology, Vol 2., P.C. Walsh, A.B. Retik, T.A. Stamye, and E. Darracott Vaughan, Jr., eds. W. B. Saunders Company.

Catalona, W.J. 1994. Screening for prostate cancer. Lancet 343:1436-1437.

Catalona, W.J., D.S. Smith, T.L. Ratliff, K.M. Dodds, D.E. Coplen, J.J. Yan, J.A. Petros, and G.L. Andriole. 1991. Measurement of prostate-specific antigen in serum as a screening test for rpostate cancer [published erratum appears in N.Engl. J. Med. 1991 Oct 31. 325(18):1324 [see comments]. N. Engl. J. Med. 324:1156-1161.

Catalona, W.J., J.P. Richie, F.R. Ahmann, M.A. Hudson, P.T. Scardino, R.C. Flanigan, J.B. deKernion, T.L. Ratliff, L.R. Kavoussi, B.L. Dalkin, W.B. Waters, M.T. MacFarlane, and P.C. Southwick. 1994. Comparison of digital rectal examination and serum prostate specific antigen in the early detection of prostate cancer: Results of a multicenter clinical trial of 6,630 men. J. Urol. 151:1283-1290.

Catalona, W.J., D.S. Smith, T.L. Ratliff, and J.W. Basler. 1993. Detection of organ-confined prostate cancer is increased through prostate-specific santigen-gased screening. J. Am. Med. Assoc. 270(8):948-954.

Cattell, V., T. Cook, and S. Moncada. 1990. Glomeruli synthesize nitrate in experimental nephrotoxic nephritis. Kisney Int. 38:1056-1060.

Chamberlin, M.E., A. LeFurgey, and L.J. Mandel. 1984. Suspension of medullary thick ascending limb tubules from the rabbit kdney. Am. J. Physiol. 247:F955-F964.

Charbonneau, M., J. Strasser, E.A. Lock, M.J. Turner, and J.A. Swenberg. 1989. Involvement of reversible binding to α_{2u}-globulin in 1,4 dichlorobenzene-induced nephrotoxicity. Toxicol. Appl. Pharmacol. 99:122-132.

Chasseaud, L.F. 1979. The role of glutathione and glutathione S-transferases in the metabolism of chemical carcinogens and other electrophilic agents. Adv. Cancer Res. 29:175-274.

Chatterjee, B., W.F. Demyan, C.S. Song, B.D. Garg, and A.K. Roy. 1989. Loss of androgenic induction of α_{2u}-globulin gene family in the liver of NIH black rats. Endocrinology 125:1385-1388.

Chavers, B.M., R.W. Bilous, E.E. Ellis, M.W. Steffes, and S.M. Mauer. 1989. Glomerular lesions and urinary albumin exretion in Type I diabetes without overt proteinuria.

Chen, Q., T.W. Jones, P.C. Brown, and J.L. Stevens. 1990a. The mechanisms of cysteine conjugate cytotoxicity in renal epithelial cells: Covalent binding leads to thiol depletion and lipid peroxidation. J. Biol. Chem. 265:21603-21611.

Chester, A.C., T.A. Rakowski, W.P. Argy, A. Giacalone, and G.E. Schreiner. 1979. Hemodialysis in the eight and ninth decade of life. Arch. Intern. Med. 139:1001-1005.

Cheung, C.K., and R. Swaminathan. 1989. Automated immunoturbidometric methods for the determination of retinol binding protein, prealbumin and transferrin in urine. Clin. Biochem. 22:425-427.

Chi, S.G., R.W. DeVere White, F.J. Meyers, D.B. Siders, F. Lee, and P.H. Gumerlock.1994. Prostate cancer: Frequent expressed transition mutations. J. Nat. Cancer Inst. 86:926-933.

Chin, T.Y., R.W. Tyl, J.A. Popp, and H. d'A. Heck. 1981. Clinical urolithiasis. 1. Characteristics of bladder stone induction by terephthalic acid and dimethyl terephthate in weanling Fischer-344 rats. Toxicol. Appl. Pharmacol. 58:307-321.

Chopin, D.K., J.B. deKernion, D.L. Rosenthal, and J.L. Fahey. 1985. Monoclonal antibodies against transitional cell carcinoma for detection of malignant urothelial cells in bladder washing. J. Urol. 134:260-265.

Chow. N.H., T.S. Tzai, S.N. Lin, S.H. Chan, and M.J. Tang. 1993. Reappraisal of the biological role of epidermal growth factor receptor in transitional cell carcinoma. Eur. Urol. 24:140-143.

Chowdhury, A., G.J. Harber, and D.P. Chopra. 1989. Characterization and serial propagation of mouse prostate epithelial cells in serum-free medium. Biol. Cell 67:281-287.

Chung, S.D., N. Alavi, D. Livingston, S. Hiller, and M. Taub. 1982. Characterization of primary rabbit kidney cultures that express proximal tubule functions in a hormonally defined medium. J. Cell Biol. 95:118-126.

Cisternino, A., F. Argona, A. Garbeglio, A. Ranieri, P. Bassi, and V. Pergoraro. 1986. CEA and ABO antigens as early signs of malignancy in urotheliomas of the upper urinary tract [in Spanish]. Arch. Esp. Urol. 39:529-533.

Clar-Blanch, F., L. Morell-Quadreny, B. Fenollosa-Entrena, P. Chuan-Nuez, M. Gil-Salom, P. Carretero-Gonzalez, and A. Llombart-Bosch. 1992. Correlation between serum values of prostatic acid phosphatase and morphometric analysis in the cytologic diagnosis of prostatic carcinoma. Eur. Urol. 21 Suppl. 1:75-78.

Clarke, A.R., C.A. Purdie, and D.J. Harrison. 1993. Thymocyte apoptosis induced by p53-dependent and independent pathways. Nature 362:849-852.

Clevenger, C., A. Epstein, and K. Bauer. 1987a. Quantitative analysis of a nuclear antigen in interphase and mitotic cells. Cytometry 8:280-286.

Clevenger, C., A. Epstein, and K. Bauer. 1987b Modulation of the nuclear antigen P105 as a function of cell-cycle progression. J. Cell. Physiol. 130:336-343.

Cockroft, D.W., and M.H. Gault. 1976. Prediction of creatinine clearance from serum creatinine. Nephron. 16:31-41.

Cohen, C., P.A. McCue, and P.B. Derose. 1988. Histogenesis of renal cell carcinoma and renal oncocytoma. An immunohistochemical study. Cancer 62:1946-1951.

Cohen, G.M. 1986. Basic principles of target organ toxicity. Pp. 1-6 in Target Organ Toxicity, Vol. 1, G.M. Cohen, ed. Boca Raton, Fla: CRC Press.

Cohen, R., G.S. Bedi, and M.E. Neider. 1990. Tissue distribution of an inducible cystatin in isoproterenal-treated rats. Lab. Invest. 62:452-457.

Cohen, S.M., and L.B. Ellwein. 1991. Genetic erros, cell proliferation, and carcinogenesis. Cancer Res. 51:6493-6505.

Cohen, P., D. Peehl, G. Lamson, and R. Rosenfeld. 1991a. Insulin-like growth factors (IGSs), IGF receptors, and IGF-binding proteins in primary cultures of prostate epithelial cells. J. Clin. Endocrinol. Metab. 73:401-407.

Cohen, S.M., L.B. Ellwein, T. Okamura, T. Masui, S.L. Johansson, R.A. Smith, J.M. Wehner, M.Khachob, C.I. Chappel, G.P. Schoenig, J.L. Emerson, and E.M. Garland. 1991b. Comparative bladder tumor promoting activity of sodium saccharin, sodium ascorbate, related acids, and calcium salts in rats. Cancer Res. 51:1766-1777.

Cohen, M.B., F.M. Waldman, P.R. Carroll, R. Kerschmann, K. Chew, and B.H. Mayall. 1993. Comparison of five histopathologic methods to assess cellular proliferation in transitional cell carcinoma of the urinary bladder. Hum. Pathol. 24:772-778.

Cojocel, C., K. Maita, D.A. Pasino, C.-H. Kuo, and J.B. Hook. 1983. Metabolic heterogeneity of the proximal and distal kidney tubules. Life Sci. 33:855-861.

Cole, P., and R. Hoover. 1971. Comments on turmor of the urinary bladder: An analysis of the occupation sof 1030 patients in Leeds, England. J. Nat. Cancer Inst. 46:1111-1113.

Cole., P., R.R. Monson, H. Haning, and G.H. Friedell. 1971. Smoking and cancer of the lower urinary tract. N. Eng. J. Med. 284:129-134.

Colle, A., C. Tonnelle, T. Jarry, C. Coirre, and Y. Manual. 1984. Isolation and characterization of post gamma globulin in mouse. Biochem. Biophys. Res. Comm. 122:111-115.

Commandeur, J.N.M., and N.P.E. Verneulen. 1990. Molecular and biochemical mechanisms of chemically induced nephrotoxicity: A review. Chem. Res. Toxicol. 3:171-194.

Committee on Biological Markers of the National Research Council. 1987. Biological Markers in environmental health research. Environ. Health Perspect. 74:3-9.

Connolly, J., and D. Rose. 1990. Production of epidermal growth factor and

References

transforming growth factor-α by the androgen-responsive LNCaP human prostate cancer cell line. Prostate 16:209-218.

Connolly, J., and D. Rose. 1992. Interactions between epidermal growth factor-mediated autocrine regulation and linoleic acid-stimulated growth of a human prostate cancer cell line. Prostate 20:151-158.

Cook, H.T., J. Smith, J.A. Salmon, and V. Cattell. 1987. Functional characteristics of macrophages in glomerulonephritis in the rat: O_2^- generation, MHC class II expressive and eicosanoid synthesis. Am. J. Pathol. 134:431-437.

Coombes, R.C., P.B. Greenberg, C. Hillyard, and I. MacIntyre. 1974. Plasma immunoreactive-calcitonin in patients with non-thyroid tumours. Lancet 1:1080-1083.

Coombs, L.M., D.A. Pigott, E. Sweeney, A.J. Proctor, M.E. Eydmann, C. parkinson, and M.A. Knowles. 1991. Amplification and over-expression of c-erbB-2 in transitional cell carcinoma of the urinary bladder. Br. J. Cancer 63:601-608.

Cordon-Cardo, C., Z. Fuks, M. Drobnjak, C. Moreno, L. Eisenbach, and M. Feldman. 1991. Expression of HLA-A,B,C antigens in primary and metastic tumor cell populations of human carcinomas. Can. Res. 51:6372-6380.

Cordon-Cardo, C., D.D. Wartinger, M.R. Melamed, W. Fair, and Y. Fradet. 1992. Immunopathologic analysis of human urinary bladder cancer. Characterization of two new antigens assocaited with low-grade superficial bladder tumors. Am. J. Pathol. 140:375-385.

Cornelius, C.E., A.S. Mia, and S. Rosenfield. 1965. Ruminant urolithiasis VII. Studies on the origin of Tamm-Horsfall urinary mucoprotein and its presence of ovine calculus matrix. Invest. Urol. 2:453-457.

Cowley, B.D., F.L. Smardo, J.J. Grantham, and J.P. Calvet. 1987. Elevated c-myc oncogene expression in autosomal recessive polycystic kidney disease. Proc. Natl. Acad. Sci. U.S.A. 84:8394-8398.

Cowley, B.D., L.J. Chadwick, J.J. Grantham, and J.P. Calvet. 1989. Sequential protooncogene expression in regenerating kidney following acute renal injury. J. Biol. Chem. 264:8389-8393.

Crump, K.S., 1984. An improved procedure for low-dose carcinogenic risk assessment from animal data. J. Environ. Pathol. Toxicol. 5:339-348.

Crump, K.S., B. Allen, and A. Shipp. 1989. Choice of dose measure for extrapolating carcinogenic risk from animals to humans: An empirical investigation of 23 chemicals. Health Physics 57(Suppl. 1): 387-393.

Cunningham, E.E., M.A. Zielezny, and R.C. Venuto. 1983. Heroin-associated nephropathy: A nationwide problem. J. Amer. Med. Assoc. 250:2935-2936.

Dal Cin, P., J. Gaeta, and R. Huben. 1989. Cytogenic characterization. Am. J. Clin. Pathol. 92:408-414.

Dal Cin, P., M.S. Aly, J. Delabie, J.J. Ceuppens, S. Van Gool, B. Van Damme, L. Baert, H. Van Poppel, and H. Van den Berghe. 1992. Trisomy 7 and trisomy 10 characterize subpopulations of tumor-infiltrating lymphocytes in kidney tumors and in the surrounding kidney tissue. Proc. Nat. Acad. Sci. U.S.A. 89:9744-9748.

Darmady, E.M., J. Offer, and M.A. Woodhouse. 1973. The parameters of the aging kidney. J. Pathol. 109:195-207.

Das, G. and R.W. Glashan. 1988. Correlation of urine cytology with ABO(H) antigenicity in transitional cell carcinoma of the bladder. J. Clin. Pathol. 41:538-539.

Das, G., N.J. Buxton, P.A. Stewart, and R.W. Glashan. 1986. Prognostic significance of ABH antigenicity of mucosal biopsies in superficial bladder cancer. J. Urol. 136:1194-1196.

Da Silva, J.L., C. Lancombe, P. Bruneval, N. Casadevall, M. Leporrier, J.P. Camilleri, J. Bariety, P. Tamourin, and B. Varet. 1990. Tumor cells are the site of erythropoietin synthesis in human renal cancers associated with polycythemia. Blood &5:577-582.

Daugaard, G., N.H. Holstein-Rathlou, and P.P. Leyssac. 1988. Effect of cisplatin on proximal convoluted and straight segments of the rat kidney. J. Pharmacol. Exp. Ther. 244:1081-1085.

Davies, P., C.L. Eaton, T.D. France, and M.E.A. Phillips. 1988. Growth factor receptors and oncogene expression in prostate cells. Am. J. Clin. Oncol. 11:S1-S7.

Davis, L.M., B. Zabel, G. Senger, H.J. Ludecke, B. Matzroth, K. Call, D. Housman, U. Claussen, B. Horsthemke, and T.B. Shows. 1991. A tumor chromosome rearrangement further defines the 11p13 Wilms tumor locus. Genomics 10:588:592.

DeCaprio, J.A., J.W. Ludlow, D. Lynch, and Y. Furukawa. 1989. The product of the retinoblastoma susceptibility gene has properties of a cell cycle regulatory element. Cell 58:1085-1095.

Decken, K., B. Schmitz-Drager, D. Rohde, S. Nakamura, T. Ebert, and R. Ackermann. 1992. Monoclonal antibody Due ABC 3 directed against transitional cell carcinoma. I. Production, specificity analysis, and preliminary characterization of the antigen. J. Urol. 147(1):235-241.

Deen, W.M. D.A. Maddox, C.R. Robertson, and B.M. Brenner. 1974. Dynamics of glomerular ultrafiltration in the rat. VII. Rersponse to reduced renal mass. Am. J. Physiol. 227:556-562.

Dekant, W., A. Schulz, M. Metzler, and D. Henschler. 1986a. Absorption, elimination and metabolism of trichloroethylene: A quantitative comparison between rats and mice. Xenobiotica 16:143-152.

Dekant, W. M. Metzler, and D. Henschler. 1986b. Identification of S-(1,1,2-tricholorovinyl)-N-acetylcysteine as a urinary metabolite of tetrachloroethylene: Bioactivation through glutathione conjugation as a possible explanation for its nephrocarcinogenecity. J. Biochem. Toxicol. 1:57-72.

Dekant, W., and D. Henschler. 1986c. Identification of S-(1,2-dichloroviny)-N-acetylcysteine as a urinary metabolite of trichloroethylene: A possible explanation for its nephrocarcinogenicity in male rats. Biochem. Pharmacol. 35:2455-2458.

Dekant, W., G. Martens, S. Vamvakas, M. Matzler, and D. Henschler. 1987a. Bioactivation of tetrachloroethylene: Role of glutathione S-transferase-catalyzed conjugation versus cytochrome P-450 dependent phospholipid alkylation. Drug Metab. Dispos. 15:702-709.

Dekant, W., L.H. Lash, and M.W. Anders. 1987b. Bioacativation of the cytotoxic and mephrotoxic S-conjugate S-(2-chloro-1,1,2-trifluoroethyl)-L-cysteine. Proc. Natl. Acad. Sci. U.S.A. 84:7443-7447.

Dekant, W., L.H. Lash, and M.W. Anders. 1988a. Fate of glutathione conjugates and bioactivation of cysteine S-conjugates by cysteine conjugate by-lyase. Pp. 415-557 in Glutathione Connugation: Its Mechanism and Biological Significance. H. Sies, and B. Ketterer, eds. London: Academic Press, Orlando.

Dekant, W., D. Schrenk, S. Vamvakas, and D. Henschler. 1988b. Metabolism of hexachloro-1,3-butadiene in mice: In vivo and in vitro evidence for activation by blutahtione conjugation. Xenobiotica 18:803-816.

Dekant, W., K. Berthold, S. Vamvakas, and D. Henschler. 1988c. Thioacylating agents as ultimate intermediates in the b-lyase catalyzed metabolism of S-(pentachlorobutadienyl)-L-cysteine. Chem. Biol. Interact. 67:139-148.

Dekant, W., K. Berthold, S. Vamvakas, D. Henschler, and M.W. Anders. 1988d. Thioacylating intermediates as metabolites of S-(1,2-dichlorovinyl)-L-cysteine and S-(1,2,2-trichlorovinyl)-L-cysteine formed by cysteine conjugate b-lyase. Chem. Res. Toxicol. 1:175-178.

Delaere, K., M. van Dieijen-Visser, A. Gijzen, and P. Brombacher. 1988. Is prostate-specific antigen the most useful marker for screening in prostate cancer? Am. J. Clin. Oncol. 11(Suppl 2): S65-S67.

Delahunt, B., P.B. Bhetwaite, J.N. Nacey, and J.L. Ribas. 1993. Proliferating cell nuclear antigen (PCNA) expression as a prognostic indicator for renal cell carcinoma: Comparison with tumor grade, mitotic index, and silver-staining nucleolar organizer region numbers. J. Pathol. 179:471-477.

D'Ercole, A.J., D. Stiles, and L.E. Underwood. 1984. Tissue concentrations of somatomedin C: Further evidence for multiple sites of synthesis and paracrine or autocrine mechanisms of action. Proc. Natl. Acad. Sci. U.S.A. 81:935-939.

De Riese, W.T., W.N. Crabtree, E.P. Allhoff, M. Werner, S. Liedke, G. Lenis, J. Atzpodien, and H. Kirchner. 1993. Prognostic significance of Ki-67 immunostaining in nonmetastatic renal cell carcinoma. J. Clin. Oncol. 11:1804-1808.

Detrisac, C.J., M.A. Sens, A.J. Garvin, S.S. Spicer, and D.A. Sens. 1984. Tissue culture of human kidney epithelial cells of proximal tubule origin. Kidney Int. 25:383-390.

Diamond, J.R., and I. Pesek. 1991. Glomerular tumor necrosis factor and interleukin-

1, during acute amino-nucleoside nephrosis: An immuno-histochemical study. Lab. Invest 64:21-28.

Dickman, K.G., and L.J. Mandel. 1989. Glycolytic and oxidative metabolism in primary renal proximal tubule cultures. Am. J. Physiol. 258:C333-C340.

Dickman, K.G., and L.J. Mandel. 1990. Differential effects of respiratory inhibitors on glycolysis in proximal tubules. Am. J. Physiol. 258:F1608-F1615.

Dieperink, H., H. Starkling, and P.P. Leyssac. 1983. Nephrotoxicity of cyclosporin A—An animal model. A study of the nephrotoxic effect of cyclosporin on overall renal and tubular function in conscious rats. Transplant Proc. 15(Suppl. 1):2736-2741.

Dierick, A.M., M. Praet, H. Roels, P. Verbeeck, C. Robyns, and W. Oosterlinck. 1991. Vimentin expression of renal cell carcinoma in relation to DNA content and histological grading: A combined light microscopic, immunocytochemical and cytophotometrical analysis. Histopathology 18:315-322.

Dietrich, P.Y., and J.P. Droz. 1992. Renal cell cancer: Oncogenes and tumor suppressor genes [in French]. Rev. Prat. 42:1236-1240.

Dietrich, D.R., and J.A. Swenberg. 1990. Lindane induces nephropathy and renal accumulation of α_{2u}-globulin in male but not in female Fischer 344 rats or male NBR rats. Toxicol. Lett. 53:179-181.

Dietrich, D.R., and J.A. Swenberg. 1991. The presence of α2u-globulin is necessary for d-limonene promotion of male rat kidney tumors. Cancer Res. 51:3512-3421.

Ding, Z., N. Sumrani, and J.H. Hong. 1991. Effect of timing of cyclosporine administration on recovery from renal ischemia in rats. J. Surg. Res. 51:341-343.

di Sant'Agnese, P.A. 1988. Neuroendocrine differentiation and prostatic carcinoma: The concept comes of age. Arch. Pathol. Lab. Med. 112:1097-1099.

di Sant'Agnese, P.A. 1992. Neuroendocrine differentiation in carcinoma of the prostate. Cancer 70:254-268.

di Sant'Agnese, P.A., and K.L. de Mesy Jensen. 1987. Neuroendocrine differentiation in prostatic carcinoma. Hum. Pathol. 18:849-856.

Doctor, V.M., A.R. Sheth, M.M. Shimha, N.J. Arbatti, and J.P. Aaveri. 1986. Studies on immunocytochemical localization of inhibin-like material in human prostatic tissue: Comparison of its distribution in normal, benign, and malignant prostates. Br. J. Cancer 53:547-554.

Dohn, D.R. and M.W. Anders. 1982. The enzymatic reaction of chlorotrifluoroethylene with glutathione. Biochem. Biophys. Res. Commun. 109:1339-1345.

Dohn, D.R., A.J. Quebbemann, R.F. Borch, and M.W. Anders. 1985a. Enzymatic reaction of chlorotrifluoroethylene with glutathione: 19F NMR evidence for stereochemical control of the reaction. Biochemistry 24:5137-5143.

Dohn, D.R., J.R. Leininger, L.H. Lash, A.J. Quebbemann, and M.W. Anders. 1985b. Nephrotoxicity of S-(2-chloro-1,1,2-trifluoroethyl) glutathione and S-(2-chloro-

References

1,1,2-trifluoroethyl)-L-cysteine, the glutathione and cysteine conjugates of chlorotrifluoroethene. J. Pharmacol. Exp. Ther. 235:851-857.

Dominick, M.A., D.G. Robertson, M.R. Bleading, R.E. Sigler, W.F. Bobrowski, and A.W. Gough. 1991. α-2u Globulin nephropathy without nephrocarcinogenesis in male Wistar rats administered 1-(amino methyl) cyclohexareactic acid. Toxicol. Appl. Pharmacol. 111:375.

Dosquet, C., A. Schaetz, C. Faucher, E. Lepage, J.L. Wautier, F. Richard, and J. Cabane. 1994. Tumour necrosis factor-alpha, interleukin-1 beta and interleukin-6 in patients with renal cell carcinoma. Eur. J. Cancer 30A:162-167.

Drabkin, H.A., C. Bradley, I. Hart, J. Bleskan, F.P. Li, and D. Patterson. 1985. Translocation of c-myc in the hereditary renal cell carcinoma associated with a t(3;8) p14.2;q24.13) chromosomal translocation. Proc. Nat. Acad. Sci. U.S.A. 82:6980-6984.

Droz, D., D. Zachar, L. Charbit, J. Gogusev. Y. Chretein, and L. Iris. 1990. Expression of the human nephron differentiation molecules in renal cell carcinomas. Am. J. Pathol. 137-895-905.

Dubach, U.C. and M. Le Hir. 1984. Critical evaluation of the diagnostic use of urinary enzymes. Contrib. Nephrol. 42: 74-80.

Dubach, U.C., M. Le Hir, and R. Gandhi. 1989. Use of urinary enzymes as markers of nephrotoxicity. Toxiol. Lett. 46:193-196.

Dubey, C., B. Bellon, F. Hirsch, J. Kuhn, M.C. Viol, M. Goldman, and P. Druet. 1991. Increased expression of class II major histocompatibility complex molecules on B cells in rats susceptible or resistant to $HgCl_2$-induced autoimmunity. Clin. Exp. Immunol. 86:118-123.

Duvall, E. and A.H. Wyllie. 1986. Death and the cell. Immunol. Today 7:115-119.

Easterling, R.E. 1977. Racial factors in the incidence and causation of end-stage renal disease (ESRD). Trans. Am. Soc. Artif. Intern. Organs 23:28.

Eaton, C.L., P. Davids, and M.E.A. Phillips. 1988. Growth factor involvement and oncogene expression in prostatic tumours. J. Steroid Biochem. 30:341-345.

Ebihara, I., P. Kellen, G. Laurie, T. Huang, Y. Yamoda, G. Martin, and K. Brown. 1981. Altered mRNA expression of basement membrane components in a murine model of polycystic kidney disease. Lab. Invest. 58:262-269.

Eble, J.N. and J.I. Epstein. 1990. Stage A carcinoma of the prostate. Pp. 61-82 in Pathology of the Prostate, D.G. Bostwick, ed. New York: Chruchill Livingstone.

Eguchi, J., K. Nomata, S. Kanda, T. Igawa, M. Taide, S. Koga, F. Matsuya, H. Kanetake, and Y. Saito. 1992. Gene expression and immunohistochemical localization of basic fibroblast growth factor in renal cell carcinoma. Biochem. Biophys. Res. Comm. 183:937-944.

Eklöv, S., K. Funa, H. Nordgren, A. Olofsson T. Kanzaki, K. Miyazono, and S. Nilsson. 1993. Lack of the latent transforming growth factor β binding protein in malignant, but not benign prostatic tissue. Cancer Res. 53:3193-3197.

Ekstrom, B. and I. Berggard. 1977. Human α_1-microglobulin: purification procedure, chemical and physicochemical properties. J. Biol. Chem. 252:8048-8057.

El-Aaser, A.A., M.M. el-Merzabani, and F.A. Abu-Bedair. 1985. Serum estrogen level in Egyptian breast cancer patients. Tumori 71:293-295.

Elfarra, A.A. and M.W. Anders. 1984. Renal processing of glutathione conjugates: Role in nephrotoxicity. Biochem. Pharmacol. 33:3729-3732.

Elfarra, A.A., I. Jakobson, and M.W. Anders. 1986a. Mechanism of S-(1,2-dichlorovinyl)glutathione-induced nephrotoxity. Biochem. Pharmacol. 35:283-288.

Elfarra, A.A., L.H. Lash, and M.W. Anders. 1986b. Metabolic activation and detoxication of nephrotoxic cysteine and homocysteine S-conjugates. Proc. Natl. Acad. Sci. U.S.A. 83:2667-2671.

Elfarra, A.A., L.H. Lash, and M.W. Anders. 1987. a-Ketoacids stimulate rat renal cysteine conjugate b-lyase activity and potentiate the cytotoxicity of S-(1,2-dichlorovinyl)-L- cysteine. Mol. Pharmacol. 31:208-212.

Elgebaly, S.A., M.E. Allam, M.P. Walzak, D. Oselinsky, C. Gillies, and H. Yamase. 1992. Urinary neutrophil chemotactic factors in interstitial cystitis patients and a rabbit model of bladder inflammation. J. Urol. 147:1382-1387.

Elinder, C.G., C. Edling, E. Lindberg, B. Kagedal, and O. Vesterberg. 1985. β_2-microglobulinuria among workers previously exposed to cadmium: Follow-up and dose-response analyses. Am. J. Ind. Med. 8:553-564.

Elliget, K.A. and B.F. Trump. 1991. Primary cultures of normal rat kidney proximal tubule epithelial cells for studies of renal cell injury. In Vitro Cell. Dev. Biol. 27A:739-748.

Ellis, W.J., K.D. Bauer, R. Oyasu, and K.T. McVary. 1992. Flow cytometric analysis of small renal tumors. J. Urol. 148:1774-1777.

el-Mohamady, H., M.T. Basta, M.N. Seddek, H. Helmy, E. al-Hilaly, A.M. Attallah, and M.A. Ghoneim. 1991. UNME/K1: An IgG2a monoclonal antibody specific to cytokeratin of human urinary bladder squamous cell carcinoma. Urol. Res. 19:145-150.

el-Naggar, A.K., H.D. van Dekken, L.G. Ensign, and S. Pathak. 1994. Interphase cytogenetics in paraffin-embedded sections from renal cortical neoplasms. Correlation with cytogenetic and flow cytometric DNA ploidy analyses. Cancer Genet. Cytogenet. 73:134-141.

El Nahas, A.M., J.E. Le Carpertier, A.H. Bassett, and D.J. Hill. 1989. Dietary protein and insulin-like growth factor-I content following unilateral nephrectomy. Kidney Int. 36 (Suppl. 27):S15-S19.

Emami, A., J.H. Schwartz, and S.C. Borkan. 1991. Transient ischemia or heat stress induces a cryoprotectant protein in rat kidney. Am. J. Physiol. 260:F479-F485.

Emanuel, A., S. Szucs, H.U. Weier, and G. Kovacs. 1992. Clonal aberrations of chromosomes X, Y, 7 and 10 in normal kidney tissue of patients with renal cell tumors. Genes Chromosomes Cancer 4:75-77.

References

Enomoto, T., J.M. Ward, and A.O. Perantoni. 1990. H-*ras* activation and *ras* p21 expression in bladder tumors induced in F3444/NCr rates by *N*-butyl-*N*-(4-hydroxybutyl)nitrosamine. Carcinogenesis 11:2233-2238.

EPA (U.S. Environmental Protection Agency). 1991a. Report of the EPA peer review workshop on alpha$_{2u}$-globulin: Association with renal toxicity and neoplasia in the male rat. EPA/625/3-91/021. August 1991. 78 pp.

EPA (U.S. Environmental Protection Agency). 1991b. Alpha$_{2u}$-globulin: Association with renal toxicity and neoplasia in the male rat. EPA/625-3-91/019F.

EPA (U.S. Environmental Protection Agency). 1991c. Response to issues and data submissions on the carcinogenicity of tetrachloroethylene (perchloroethylene). EPA/600/6-91/002F.

Epstein, M. 1979. Effects of aging on the kidney. Fed. Proc. 38:168-172.

Epstein, J.I., and J.M. Woodruff. 1986. Adenocarcinoma of the prostate with endometrioid features. Alight microscopic and immunohistochemical study of ten cases. Cancer 57:111-117.

Epstein, J.I., and P.H. Lieberman. 1985. Mucinous adenocarcinoma of the prostate gland. Am. J. Surg. Pathol. 9:299-306.

Ercole, C.J., Ph.H. Lange, M. Mathisen, R.K. Chious, P.K. Reddy, and R.L. Vessella. 1987. Prostatic specific antigen and prostatic acid phosphatase in the monitoring and staging of patients with prostatic cancer. J. Urol. 138: 1181-1184.

Erlandsson, R., U.S. Bergerheim, F. Boldog, Z. Marcsek, K. Kunimi, B.Y. Lin, S. Ingvarsson, J.S. Castresana, W.H. Lee, and E. Lee. 1990. A gene near the D3F15S2 site on 3p is expressed in normal human kidney but not or only at a severely reduced level in 11 of 15 primary renal cell carcinomas (RCC). Oncogene 5:1207-1211.

Erlandsson, R., F. Boldog, B. Persson, E.R. Zabarovsky, R.L. Allikmets, J. Sumegi, G. Klein, and H. Jornvall. 1991a. The gene from the short arm of chromosome 3, at D3F15S2, frequently deleted in renal cell carcinoma, encodes acylpeptide hydrolase. Oncogene 6:1293-1295.

Erlandsson, R., J. Szpirer, M.Q. Islam, F. Boldog, G. Klein, and S. Ingvarsson. 1991b. The most frequently lost allelic site in human renal cell carcinoma (D3F15S2) on the short arm of chromosome 3 has homologous sequences on rat chromosome 8. Cytogenet. Cell Genet. 57:149-150.

Eryigit, M., and Z. Kirkali. 1990. HLA antigens and transitional cell carcinoma of the bladder. Urol. Int. 45:75-77.

Escudero Barrilero, a., E. Fernandez, E. Garcia Cuerpo, F. Lovaco, and S. Navio. 1991. Infiltrating transitional cancer of the bladder (2). Prognostic value of the stage (Its influence on the therapeutic decision) [in Spanish]. Acta Urol. Esp. 15(3):213-230.

Eskelinen, M., P. Lipponen, R. Majapuro, K. Syrjanen, and S. Nordling. 1991. DNA ploidy, S phase fraction and G2 fraction as prognostic determinants in prostatic adenocarcinoma. Eur. Urol.20:62-66.

Esnard, A., F. Esnard, and F. Gauthier. 1988. Purification of the cystatin C-like inhibitors from urine of nephropathic rats. Bioch. Chem. Hoppe-Seyler 369:S219-222.

Evan, G.I., A.H. Wyllie, and G.S. Gilbert. 1992. Induction of apoptosis in fibroblasts by c-myc protein. Cell 69:119-128.

Evans, R.W., C.R. Blagg, and F.A. Bryan. 1981. Implications for health care policy: A social and demographic profile of hemodialysis patients in the United States. JAMA 345:487.

Eveloff, J., W. Haase, and R. Kinne. 1980. Separation of renal medullary cells: Isolation of cells from the thick ascending limb of Henle's loop. J. Cell Biol. 87:672-681.

Evrin, P.E., and L. Wibell. 1972. The serum levels and urinary excretion of β_2-microglobulin in apparently healthy subjects. Scand. J. Clin. Lab. Invest. 29:69-74.

Fagin, J.A., and S. Melmed. 1987. Relative increase in insulin-like growth factor I messenger ribonucleic acid levels in compensatory renal hypertrophy. Endocrinology 120:718-724.

Fanning, P., K. Bulovas, K.S. Saini, J.A. Libertino, A.D. Joyce, I.C. Summerhayes. 1992. Elevated expression of $pp60^{c-src}$ in low grade human bladder carcinomas. Cancer Res. 52:1457-1462.

Farrow, G.M. 1990. Urine cytology in the detection of bladder cancer: A critical approach. J. Occup. Med. 32:817-821.

Fauler, J., A. Wiemeyer, K.H. Marx, K. Kühn, K.M. Koch, and J.C. Frölich. 1989. LTB_4 in nephrotoxic serum nephritis in rats. Kidney Int. 36:46-50.

Fearon, E.R., A.P. Feinberg, S.H. Hamilton, and B. Vogelstein. 1985. Loss of genes on the short arm of chromosome 11 in bladder cancer. Nature 318:377-380.

Feigelson, P., and D.T. Kurtz. 1977. Hormonal modulation of specific messenger RNA species in normal and neoplastic rat liver. Adv. Enzymol. 47:275-312.

Fekete, M., T.W. Redding, A.M. Comaru-Schally, J.E. Pontes, R.W. Connelly, g. Srkalovic, and A.V. Schally. 1989. Receptors for luteinizing hormone-releasing hormone, somatostatin, prolactin, and epidermal growth factor in rat and human prostate cancers and in benign prostate hyperplasia. Prostate 14:191-208.

Feldman, H.K., M.J. Klag, A.P. Chiapelle, and P.K. Whelton. 1992. End-stage renal disease in U.S. minority groups. Amer. J. Kidney Dis. 19:397-410.

Fernandez-Repollet, E., S. Opava-Stitzer, and M. Martiez-Maldonado. 1992. Renal hemodynamics and urinary concentrating capacity in protein deprivation: Role of antidiuretic hormone. Am. J. Med. Sci.

Filmer, R.B., and J.R. Spencer. 1990. malignancies in bladder augmentations and intestinal conduits. J. Urol. 143:671-678.

Fine, L.G. 1986. The biology of renal hypertrophy. Kidney Int. 29:619-634.

Fine, L.G., B. Badie-Dezfooly, A.G. Lowe, A. Hamzeh, J. Wells, and S. Salehmoghaddam. 1985. Stimulation of Na+/H+ antiport is an early event in

hypertrophy of renal proximal tubular cells. Proc. Natl. Acad. Sci. U.S.A. 82:1736-1740.
Finn, W.F. 1982. Compensatory renal hypertrophy in Sprague-Dawley rats: Glomerular ultrafiltration dynamics. Renal Physiol. 5:222-234.
Finn, W.F. 1983. Compensatory hypertrophy of single nephrons following ischemic injury in the rat. Clin. Exp. Dial. Apheresis 7:101-114.
Finver, S.N., C. Martiniere, J. Kagan, W. Cavenee, and c.M. Croce. 1989. The chromosome 11 region flanking the t(11:14) breakpoint in human T-ALL is deleted in Wilms' tumor hybrids. Oncogene Res. 5:143-148.
Fiorelli, G., A. De Bellis, A. Longo, S. Giannini, A. Natali, A. Constantini, and G. Vannelli. 1991a. Insulin-like growth factor-I receptors in human hyperplastic prostate tissue: Characterization, tissue localization, and their modulation by chronic treatment with a gonadotropin-releasing hormone analog. J. Clin. Endocrinol. Metab. 72:740-746.
Fiorelli, G., A. De Bellis, A. Longo, P. Pioli, A. Constantini, S. Giannini, G. Forti, and M. Serio. 1991b. Growth factors in the human prostate. J. Steroid Biochem. Mol. Biol. 40:199-205.
Fisher, C.C., and B.M. Brenner, eds. 1989. Renal Pathology. Philadelphia: J.B. Lipincott.
Fisher, D.A., E.C. Salido, and L. Barajas. 1989. Epidermal growth factor and the kidney. Annu. Rev. Physio. 51:67-80.
Fjellestad-Paulsen, A., P.A. Abrahamsson, A. Bjartell, M. Grino, L. Grimelius, and H. Hedeland. 1988. Carcinoma of the prostate with Cushing's syndrome: a case report with histochemical and chemical demonstration of immunoreactive corticotropin-releasing hormone in plasma and tumoral tissue. Acta Endocrinol. (Copenh) 119:506-516.
Flamm, W.G. and L.D. Lehman-McKeeman. 1991. The human relevance of the renal tumor-inducing potential of *d*-limonene in male rats: Implications for risk assessment. Reg. Toxicol. Pharmacol. 13:70-86.
Fletcher, A.P., J.E. McLaughlin, W.A. Ratcliffe, and D.A. Woods. 1970. The chemical composition and electron microscope appearance of a protein derived from urinary casts. Biochim. Biophys. Acta 214:299-308.
Floege, J., R.J. Johnson, K. Gordon, H. Iida, P. Pritzl, A. Yoshimura, C. Campbell, C.E. Alpers, and W.G. Couser. 1991. Increased synthesis of extracellular matrix in mesangial proliferative nephritis. Kidney Int. 40:477-488.
Floege, J., M.W. Burns, C.E. Alpers, A. Yoshimura, P. Pritzl, K. Gordon, R. Seifert, D.F. Bowen-Pope, W.G. Couser, and R.J. Johnson. 1992a. Glomerular cell proliferation and PDGF expression precede glomerulosclerosis in the remnant kidney model. Kidney Int. 41:297-309.
Floege, J., C. Alpers, M. Burns, P. Pritzl, K. Gordon, W.G. Couser, and R.J. Johnson.

1992b. Glomerular cells, extracellular matrix accumulation, and the development of glomerulosclerosis in the remnant kidney model. Lab. Invest. 66:485-497.

Flyvbjerg, A., K.E. Bornfeldt, S.M. Marshall, H.J. Arnqvist, and H. Ørskov. 1990. Kidney IGF-1 mRNA in initial renal hypertrophy in experimental diabetes in rats. Diabetologia 33:334-338.

Ford, S.M., and J.B. Hook. 1984. Biochemical mechanisms of toxic nephropathies. Semin. Nephrol. 4:88-106.

Fournet, J.C., C. Beroud, E. Austruy, C. Leonard. 1992. Genetic aspects of renal tumors in adults [review] [in French]. Arch. Anat. Cytol. Pathol. 40:301-306.

Fowler, B.A. 1982. Ultrastructural and biochemical localization of organelle damage from nephrotoxic agents. Pp. 315-330 in Nephrotoxic Mechanisms of Drugs and Environmental Toxins, G.A. Porter, ed. New York: Plenum.

Fowler, B.A., and G.E. Duval. 1991. Effects of lead on the kidney: Roles of high-affinity lead-binding proteins. Environ. Hlth. Perspec. 91:77-80.

Fowler, J.J., W. Lynes, J. Lau, L. Ghosh, and A. Mounzer. 1988. Interstitial cystitis is associated with intraurothelial Tamm-Horsfall protein. J. Urol. 140:1385-1389.

Fowler, B.A., M.W. Kahng, and D.R. Smith. 1994. Role of lead-binding proteins in renal cancer. Environ. Health Perspect. 102(Suppl. 3):115-116.

Fox, S., P. Raj, J. Royds, R. Kore, P. Silcocks, and C. Collins. 1993. p53 and *c-myc* expression in stage A1 prostatic adenocarcinoma: useful prognostic determinants? J. Urol. 150:490-494.

Fradet, Y., C. Cordon-Cardo, T. Thomson, M.E. Daly, W.F. Whitmore, K.O. Lloyd, M.R. Melamed, and L.J. Old. 1984. Cell surface antigens of human bladder cancer defined by mouse monoclonal antibodies. Proc. Natl. Acad. Sci. U.S.A. 81:224-228.

Fradet, Y., C. Cordon-Cardo, W.F. Whitmore, M.R. Melamed, L.J. Old. 1986. Cell surface antigens of human bladder tumors: Definition of tumor subsets by monoclonal antibodies and correlation with growth characteristics. Cancer Res. 46:5183-5188.

Fradet, Y., N. Islam, L. Boucher, C. Parent-Vaugeois, and M. Tardif. 1987. Polymorphic expression of a human superficial bladder tumor antigen defined by mouse monoclonal antibodies. Proc. Natl. Acad. Sci. U.S.A. 84:7227-7231.

Fradet, Y., H. LaRue, C. Parent-Vaugeois, A. Bergeron, C. Dufour, L. Boucher, and L. Bernier. 1990a. Monoclonal antibody against a tumor-associated sialoglycoprotein of superficial papillary bladder tumors and cervical condylomas. Int. J. Cancer 46(6):990-997.

Fradet, Y., M. Tardif, L. Bourget, J. Robert, and the Laval University Urology Group. 1990b. Clinical cancer progression in urinary bladder tumors evaluated by multiparameter flow cytometry with monoclonal antibodies. Cancer Res. 50(2):432-437.

Franchini, I., A. Cavatorta, M. Falzoi, S. Lucertini, and A. Mutti. 1983. Early indicators of renal damage in workers exposed to organic solvents. Int. Arch. Occup. Environ. Health 52:1-9.

Frederick, C.B., K.L. Dooley, R.L. Kodell, W.G. Sheldon, and F.F. Kadlubar. 1989. The effect of lifetime sodium saccharin dosing in mice initiated with the carcinogin 2-acetlyaminofluorene. Fund. Appl. Toxicol. 12:346-357.

Friberg, L. 1948. Proteinuria and kidney injury among workmen exposed to cadmium and nickel dust. J. Indust. Hyg. Toxicol. 30:32-36.

Friedman, S.A., A.E. Raizner, H. Rosen, N.A. Solomon, and W. Sy. 1972. Functional defects in the aging kidney. Ann. Intern. Med. 76:41-45.

Fries, J.W.U., and T. Collins. 1992. Platelet-derived growth factor expression in a transgenic model. Kidney Int. 41:584-589.

Fuhrman, S.A., L.C. Lasky, and C. Limas. 1982. Prognostic significance of morphologic and parameters in renal cell carcinoma. Am. J. Surg. Path. 6:655-663.

Fujimoto, K., Y. Ichimori, T. Kakizoe, E. Okajima, H. Sakamoto, T. Sugimura, and M. Terada. 1991. Increased serum levels of basic fibroblast growth factor in patients with renal cell carcinoma. Biochem. Biophys. Res. Comm. 180:386-392.

Fujimoto, K., Y. Yamada, E. Okajima, T. Kakizoe, H. Sasaki, T. Sugimura, and M. Terada. 1992. Frequent association of p53 gene mutation in invasive bladder cancer. Cancer Res. 52:1393-1398.

Fukatsu, A., S. Matsuo, H. Tamai, N. Sakamoto, T. Matsuda, and T. Hirano. 1991. Distribution of interleukin 6 in normal and diseased human kidney. Lab. Invest. 65:61-66.

Fukutani, K., J.M. Libby, W.B. Panko, and P.T. Scardino. 1983. Human chorionic gonadotropin detected in urinary concentrates from patients with malignant tumors of the testis, prostate, bladder, ureter, and kidney. J. Urol. 129:74-77.

Fuse, H., K. Umeda, I. Mizuno, and T. Katayama. 1991. Multiple marker evaluation in prostatic cancer with prostatic acid phosphatase, gamma-seminoprotein and prostate-specific antigen. Int. Urol. Nephro. 23:455-463.

Fuse, H., I. Mizuno, M. Sakamoto, and T. Katayama. 1992. Epidermal growth factor in urine from the patients with urothelial tumors. Urol. Int. 48:261-264.

Fuzesi, L., M. Cober, and C. Mittermayer. 1992. Collecting duct carcinoma: Cytogenetic characterization. Histopathology 21:155-160.

Galand, P., and C. Degraef. 1989. Cyclin/PCNA immunostaining as an alternative to tritiated thymidine pulse labelling for marking S phase cells in paraffin sections form animal and human tissues. Cell Tissues Kinet. 22:383-392.

Gardiner, R.A., M.L. Samaratunga, M.D. Walsh, G.J. Seymour, and M.F. Lavin. 1992. An immunohistological demonstration of c-erbB-2 oncoprotein expression in primary urothelial bladder cancer. Urol. Res. 20:117-120.

Gardner, K.D., J.S. Burnside, L.W. Elzinga, and R.M Locksley. 1991. Cytokines in fluids from polycystic kidneys. Kidney Int. 39:718-724.

Garg, B.D., M.J. Olson, W.F. Demyan, and A.K. Roy. 1988. Rapid post exposure decay of α_{2u}-globulin and hyaline droplets in the kidneys of gasoline-treated male rats. J. Toxicol. Env. Health 24:145-160.

Garg, B.D., M.J. Olson, L.C. Li, and A.K. Roy. 1989a. Phagolysosomal alterations induced by unleaded gasoline in epithelial cells of the proximal convoluted tubules of male rats: Effect of dose and treatment duration. J. Toxicol. Env. Health 26:101-118.

Garg, B.D., M.J. Olson, L.C. Li, M.A. Mancini, and A.K. Roy. 1989b. Estradiol pretreatment of male rats inhibits gasoline-induced renal hyaline droplet and alph-2u-globulin accumulation. Research Publication, General Motors Research Laboratories. GMR-6557. March 1989.

Gattone, V.H., G.K. Andrews, N. Fu-Wen, L.J. Chadwick, R.M. Klein, J.P. Calvet. 1990. Defective epidermal growth factor gene expression in mice with polycystic kidney disease. Dev. Biol. 138:225-230.

Gemmill, R.M., J. Coyle-Morris, L. Ware-Uribe, N. Pearson, F. Hecht, R.S. Brown, F.P. Li, and H.A. Drabkin. 1989. A 1.5-megabase restriction map surrounding MYC does not include the translocation breakpoint in familial renal cell carcinoma. Genomics 4:28-35.

Gerber, W., P. Lenahen, A. Kendall, and W. Mercer. 1991. Computer-assisted microfluorometric detection of individual malignant bladder cells. Urology 38(5):466-472.

Gesek, F.A., D.W. Wolff, and J.W. Strandhoy. 1987. Improved separation method for rat proximal and distal renal tubules. Am. J. Physiol. 253:F358-F365.

Gesualdo, L., M. Pinzani, J.J. Floriano, M. Hassan, N.U. Nagy, F.P. Schena, S.N. Emancipator, and H.E. Abhoud. 1991. Platelet-derived growth factor expression in mesangial proliferative glomerulonephritis. Lab. Invest. 65:160-167.

Gesualdo, L., S.N. Emancipator, C. Kesselheim, and M.E. Lamm. 1992. Glomerular hemodynamics and eicosanoid synthesis in a rat model of IgA nephropathy. Kidney Int. 42:106-114.

Ghadirian, P., M. Cadotte, A. Lacroix, and C. Perret. 1991. Family aggregation of cancer of the prostate in Quebec: The tip of the iceberg. Prostate 19(1):43-52.

Ghiggeri, G.M., G. Candiano, G. Delfino, and C. Queirolo. 1985. Electrical charge of serum and urinary albumin in normal and diabetic humans. Kidney Int. 28:168-177.

Giampetro, O., and A. Clerico. 1990. Microalbuminuria in diabetes: Which method to employ, which sample to collect. J. Nucl. Med. Allied. Sci. 34:111-120.

Gietl, Y.S., and M.W. Anders. 1991. Biosynthesis and biliary excretion of S-conjugates of hexachlorobuta-1,3-diene in the perfused rat liver. Drug Metab. Dispos. 19:274-277.

Gittes, R.F. 1987. Prostate-specific antigen [editorial]. N. Engl. J. Med. 317:954-955.

Gleason, D.F. 1990. Histologic grading of prostatic carcinoma. Pp. 83-93 in Pathology of the Prostate, D.G. Bostwick, ed. New York: Churchill Livingstone.

References

Glenn, G.M., L.N. Daniel, P. Choyke, W.M. Linehan, E. Oldfield, M.B. Gorin, S. Hosoe, F. Latif, G. Weiss, and M. Walther. 1991. Von Hippel-Lindau (VHL) disease: distinct phenotypes suggest more that one mutant allele at the VHL locus. Hum. Genet. 87:207-210.

Gnarra, J.R., K. Tory, Y. Weng, L. Schmidt, M.H. Wei, H. Li, P. Latif, S. Liu, F. Chen, F.-M. Duh, I. Lubensky, R. Duan, C. Florence, R. Pozzatti, M.M. Walther, N.H. Bander, H.B. Grossman, H. Brauch, S. Pomer, J.D. Brooks, W.B. Isaacs, M.I. Lerman, B. Zbar, and W.L. Linehan. 1994. Mutation of the VHL tumor suppressor gene in renal carcinoma. Nature Genetics 7:85-90.

Goering, P.L., B.R. Fisher, P.P. Chaudhary, C.A. Dick. 1992. Relationship between stress protein induction in rat kidney by mercuric chloride and nephrotoxicity. Toxicol. and Appl. Pharmocol. 113:184-191.

Gohji, K., M. Ishii, H. Nagata, O. Matsumoto, and S. Kamidono. 1990. Serum basic fetoprotein in patients with renal cell carcinoma. Cancer 65:1405-1411.

Goldsworthy, T.L., and J.A. Popp. 1987. Chlorinated hydrocarbon-induced peroxisomal enzyme activity in relation to species and organ carcinogenicity. Toxicol. Appl. Pharmacol. 88:225-233.

Goldsworthy, T.L., O. Lyght, V.L. Burnett, and J.A. Popp. 1988a. Potential role of α-2u-globulin, protein droplet accumulation, and cell replication in the renal carcinogenicity of rats exposed to trichloroethylene, perchloroethylene, and pentachloroethane. Toxicol. Appl. Pharmacol. 96:367-379.

Goldsworthy, D.W., T. Smith-Oliver, D.J. Loury, J.A. Popp, and B.E. Buttherworth. 1988b. Assessment of chlorinated hydrocarbon-induced genotoxicity and cell replication in rat kidney cells. Env. Mol. Mutagen. 11(Suppl. 11):39.

Gomella, L.G., E.R. Sargent, T.P. Wade, P. Anglard, W.M. Linehan, and A. Kasid. 1989. Expression of transforming growth factor alpha on normal human adult kidney and enhanced expression of transforming growth factors alpha and beta 1 in renal cell carcinoma. Cancer Res. 49:6972-6975.

Goodrich, D.W., Y. Chen, P. Scully, and W.H. Lee. 1992. Expression of the retinoblastoma gene product in bladder carcinoma cells associates with a low frequency of tumor formation. Cancer Res. 52:1968-1973.

Goodyer, P.R., Z. Kachra, C. Bell, and R. Rozen. 1988. Renal tubular cells are potential targets for epidermal growth factor. Am. J. Physiol. 255:F1191-F1196.

Gospodarowicz, D. 1991. Biological activities of fibroblast growth factors. Ann. N.Y. Acad. Sci. 638:1-8.

GrafstrÜm, R., K. Ormstad, P. MoldÄus, and S. Orrenius. 1979. Paracetamol metabolism in the isolated perfused rat liver with further metabolism of a biliary paracetamol conjugate by the small intestine. Biochem. Pharmacol. 28:3573-3579.

Gram, T.E., L.K. Okine, and R.A. Gram. 1986. The metabolism of xenobiotics by certain extrahepatic organs and its relation to toxicity. Annu. Rev. Pharmacol. Toxicol. 26:259-291.

Grasso, R. 1952. Sobre las celulas argentafines de la uretra y de la glandula prostatica. (About the agrentaffin cells of the urethra and the prostrate gland.) Arch. Histol. Normal Pathol. 5:227-270.

Green, T., and J. Odum. 1985. Structure/activity studies of the nephrotoxic and mutagenic action of cysteine conjugates of chloro- and fluoroalkenes. Chem.-Biol. Interact. 54:15-31.

Green, T., J. Odum, and J.K. Foster. 1990. Perchloroethylene-induced rat kidney tumors: An investigation of the mechanisms involved and their relevance to humans. Toxicol. Apl. Pharmacol. 103:77-89.

Green, T., J. Odum, and J.R. Yates. 1994. Loss of heterozygosity on chromosone 16p13.3 in hamartomas from tuberous sclerosis patients. Nature Genetics 6:193-196.

Gregg, N.J., M.M. Elseviers, M.E. DeBroe, and P.H. Bach. 1989. Epidemiology and mechanistic basis of analgesic nephropathy. Toxicol. Lett. 46:141-151.

Grignon, D.J., M. Abdel-Malak, W.C. Mertens, W.A. Sakr, and R.R. Shepherd. 1994. Glutathione S-transferase expression in renal cell carcinoma: A new marker of differentiation. Mod. Pathol. 7:186-189.

Grignon, D.J., A. el-Naggar, L.K. Green, A.G. Ayala, J.Y. Ro, D.A. Swanson, P. Troncoso, D. McLemore, G.G. Giacco, V.F. Guinee. 1989. DNA flow cytometry as a predictor of outcome of stage I renal cell carcinoma. Cancer 63:1161-1165.

Grimmond, S.M., D. Raghavan, and P.J. Russell. 1992. Detection of a rare point mutation in Ki-ras of a human bladder cancer xenograft by polymerase chain reaction and direct sequencing. Urol. Res. 20:121-126.

Groeneveld, A.B.J., D.D. Tran, and J. van der Meulen. 1991. Acute renal failure in the medical intensive care unit: Predisposing, complicating factors and outcome. Nephron. 59:602.

Grollino, M.G., D. Cavallo, F. Di Silverio, M. Rocchi, and R. De Vita. 1993. Interphase cytogenetics and flow cytometry analyses of renal tumours. Anticancer Res. 13:2239-2244.

Gröne, H.J., K. Weber, E. Gröne, U. Helmchen, and M. Osborn. 1987. Coexpression of keratin and vimentin in damaged and regenerating tubular epithelia of the kidney. Am. J. Path. 129:1-8.

Grundy, P.E., P.E. Telezerow, N. Breslow, J. Moksness, V. Huff, and M.C. Paterson. 1994. Loss of heterozygosity for chromosones 16q and 1p in Wilms' tumors predicts an adverse outcome. Cancer Res. 54:2331-2333.

Gu, F.L., S.L. Cai, B.J. Cai, and C.P. Wu. 1991. Cellular origin of renal cell carcinoma--an immunohistological study on monoclonal antibodies. Scand. J. Urol. Nephrol. Suppl. 138:203-206.

Guarnieri, G., M. Ianche, and S. Lin. 1979. Renal enzyme and protein excretion after induction of diuresis. Br. Med. J. 2(6181):50-51.

References

Guder, W.G., and B.D. Ross. 1984. Enzyme distribution along the nephron. Kidney Int. 26:101-111.

Guder, W.G., and W. Hofmann. 1991. Future markers for the diagnosis of renal lesions. Pp. 575-580 in Nephrotoxicity. Mechanisms, Early Diagnosis and Therapeutic Management. P.H. Bach, N.J. Gregg, M.F. Wilks, and L. Delacruz, eds. Marcel Dekker.

Guinan, P., M. Shaw, P. Targonski, V. Ray, M. Rubenstein. 1989. Evaluation of cytokeratin markers to differentiation between benign and malignant prostatic tissue. J. Surg. Oncol. 42(3):175-180.

Gullans, S.R., P.C. Brazy, S.P. Soltoff, V.W. Dennis, and L.J. Mandel. 1982. Metabolic inhibitors: Effects on metabolism and transport in the proximal tubule. Am. J. Physiol. 243:F133-F140.

Gullans, S.R., S.I. Harris, and L.J. Madel. 1984a. Glucose-dependent respiration in suspensions of rabbit cortical tubules. J. Membr. Biol. 78:257-262.

Gullans, S.R., P.C. Brazy, V.W. Dennis, and L.J. Madel. 1984b. Interactions between gluconeogenesis and sodium transport in rabbit proximal tubule. Am. J. Physiol. 246:F859-F869.

Guo, Y. 1992. Production and characterization of a group of monoclonal antibodies to human transitional cell carcinoma [in Chinese]. Zhonghua Waike Zazhi (Chung Hua Wai Ko Tsa Chih) 29:777-780.

Gupta, R.K., J. van der Meulen, and K.V. Johny. 1991. Oliguric acute renal failure due to glue sniffing. Scand. J. Urol. Nephrol. 25:247-250.

Haber, D.A., and D.E. Housman. 1992. The genetics of Wilms' tumor. Adv. Cancer Res. 59:41-68.

Haber, D.A., and A.J. Buckler. 1992. WT1: A novel tumor supressor gene inactivated in Wilms' tumor [review]. New Biologist 4:97-106.

Hagen, T.M., T.Y. Aw, and D.P. Jones. 1988. Glutathione uptake and protection against oxidative injury in isolated kidney cells. Kidney Int. 34:74-81.

Hagood, P.G., F.E. Johnson, C.W. Bedrossian, and A.B. Silverberg. 1991. Small cell carcinoma of the prostate. Cancer 67:1046-1050.

Halder, C.A., T.M. Warne, and N.S. Hatoum. 1984. Renal toxicity of gasoline and related petroleum napthas in male rats. Pp. 73-88 in Advances in Modern Environmental Toxicology. Vol. VII. Renal Effects of Petroleum Hydrocarbons. M.A. Mehlman, G.P. Hemstreet, J.J. Thorpe, and N.K. Weaver, eds. Princeton, New Jersey: Princeton Scientific Publishers, Inc.

Halliburton, I.W., and R.Y. Thomson. 1965. Chemical aspects of compensatory renal hypertrophy. Cancer Res. 25:1882-1887.

Halloran, P., J. Urmson, V. Ramassae, C. Laskin, and P. Autenreid. 1988. Increased class I and class II MHC products and mRNA in kidneys of MRL/lpr mice during autoimmune nephritis and inhibition by cyclosporine. J. Immunol. 141:2303-2312.

Hammerman, M.R. 1989. The growth hormone-insulin-like growth factor axis in kidneys. Am. J. Physiol. 257(4 Pt. 2):F503-F514.

Hammerman, M.R., and S. Rogers. 1987. Distribution of IGF receptors in the plasma membrane of proximal tubular cells. Am. J. Physiol. 253:F841-F847.

Hankin, J.H., L. Zhao, L. Wilkens, and L. Kolonel. 1992. Attributable risk of breast, prostate, and lung cancer in Hawaii due to standard fat. Cancer Causes Control 3(1):17-23.

Hard, G.C. 1987. Chemically induced epithelial tumors and carcinogenesis of the renal parenchyma. Pp. 211-250 in Nephrotoxicity in the Experimental and the Clinical Situation, Part I. P.H. Bach and E.A. Lock, eds. Martinus Nijhoff Publishers, Lancaster.

Hard, G.C. 1990. Tumors of the kidney, renal pelvis and ureter. Pp. 301-344 in Pathology of Tumours in Laboratory Animals. Vol. 1 - Tumours of the Rat. Second Edition. V.S. Turusov and U. Mohr, eds. International Agency for Research on Cancer, Lyon. IARC Scientific Publications No. 99.

Hard, G.C., and J. Whysner. 1994. Risk assessment of d-Limonene. An example of male rat-specific renal tumorigens. CRC Crit. Rev. Toxicol. 24:231-254.

Hard, G.C., I.S. Rodgers, K.P. Baetcke, W.L. Richards, R.E. McGaughy, and L.R. Valcovic. 1993. Hazard evaluation of chemicals that cause accumulation of α2u-globulin, hyaline droplet nephropathy, and tubule neoplasia in the kidneys of male rats. Environ. Health Perspect. 99:313-.

Hargrove, J.L., and F.H. Schmidt. 1989. The role of mRNA and protein stability in gene expression. FASEB J. 3:2360-2370.

Hargus, S.J., and M.W. Anders. 1991. Immunochemical detection of covalently modified kidney proteins in S-(1,1,2,2-tetrafluoroethyl)-L-cysteine-treated rats. Biochem. Pharmacol. 42:R17-R20.

Harney, J.V., M. Liebert, S.P. Ethier, J.A. Stein, G.A. Wedemeyer, and R. Washington. 1991a. Down regulation of epidermal growth factor receptor in cultured human normal urothelial cells and in low and high grade human bladder cancer cell lines. J. Urol. 145:311A.

Harney, J.V., M. Liebert, G. Wedemeyer, R. Washington, J. Stein, D. Buchsbaum, Z. Steplewski, and H.B. Grossman. 1991b. The expression of epidermal growth factor receptor on human bladder cancer: Potential use in radioimmunoscintigraphy. J. Urol. 146:227-231.

Harper, M.E., L. Goddard, E. Glynne-Jones, D.W. Wilson, M. Price-Thomas, W.B. Peeling, and K. Griffiths. 1993. An immunocytochemical analysis of TGFα expression in benign and malignant prostatic tumors. Prostate 23:9-23.

Harrington, D.S., M. Fall, and S.L. Johansson. 1990. Interstitial cystitis: Bladder mucosa lymphocyte immunophenotying and peripheral blood flow cytometry analysis. J. Urol. 144:868-871.

References

Harris, C.C. 1993. p53. At the crossroads of molecular carcinogenesis and risk assessment. Science 262:1980-1981.

Harris, J.W. and M.W. Anders. 1991a. In vivo metabolism of the hydrochlorofluorcarbon 1,1-dichloro-1-fluoroethane (HCFC-141b). Biochem. Pharmacol. 41:R13-R16.

Harris, J.W. and M.W. Anders. 1991b. Metabolism of hydrochlorofluorocarbon 1,2-dichloro-1,1-difluoroethane. Chem. Res. Toxicol. 4:180-186.

Harris, C.C. and M. Hollstein. 1993. Clinical implications of the p53 tumor-supressor gene. N. Engl. J. of Med. 329:1318-1327.

Harris, C.C., A. Weston, J.C. Willey, G.E. Trivers, and D.L. Mann. 1987. Biochemical and molecular epidemiology of human cancer: Indicators of carcinogen exposure, DNA damage, and genetic predisposition. Environ. Health Perspect. 75:109-119.

Harris, J.W., W. DeKant, and M.W. Anders. 1992. In vivo detection and characterization of protein adducts resulting from bioactivation of haloethene cystein S-conjugates by 19F-NMR: Chlorotrifluoroethene and tetrafluoroethene. Chem. Res. Toxicol. 5:34-41.

Hasegawa, R., and S.M. Cohen. 1986. The effect of different salts of saccharin on the rat urinary bladder. Cancer Lett. 30:261-268.

Hashimura, T., R.R. Tubbs, R. Connelly, M.J. Caulfield, C.S. Trindade, J.T. McMahon, T.P. Galetti, M. Edinger, A.A. Sandberg, and P. Dal Cin. 1989. Characterization of two cell lines with distinct phenotypes and genotypes established from a patient with renal cell carcinoma. Cancer Res. 49:7064-7071.

Hasui, Y., K. Marutsuka, J. Suzumiya, S. Kitada, Y. Osada, and A. Sumiyoshi. 1992. The content of urokinase-type plasminogen activator antigen as a prognostic factor in urinary bladder cancer. Int. J. Cancer 50:871-873.

Hata, M. 1989. The study of plasminogen activator in renal cell carcinoma with special remarks on urokinase type plasminogen activator [in Japanese]. Nippon Hinyokika Gakkai Zasshi (Jpn. J. Urol.) 80:1558-1565.

Hattori, K., K. Uchida, T. Shimazui, R. Nemoto, K. Koiso, and M. Harada. 1988. Cell cycle kinetics of human bladder tumor in situ measured by bromodeoxyuridine (BrdU) [in Japanese]. Nippon Hinyokika Gakkai Zasshi (Jpn. J. Urol.) 79:1811-1817.

Haugen, A., D. Ryberg, I.L. Hansteen, and P. Amstad. 1990. Neoplastic transformation of a human kidney epithelial cell line transfected with v-Ha-ras oncogene. Int. J. Cancer 45:572-577.

Hayden, P.J. C.J. Welsh, Y.Yang, W.H. Schaefer, A.J.I. Ward, and J.L. Stevens. 1992. Formation of mitrochondrial phospholipid adducts by nephrotoxic cysteine conjugate metabolites. Chem. Res. Toxicol. 5:231-237.

Hayslett, J.P. 1979. Functional adaptation to reduction in renal mass. Physiol. Rev. 59:137-164.

HEI (Health Effects Institute). 1985. Gasoline Vapor Exposure and Human Cancer:

Evaluation of Existing Scientific Information on Recommendations for Further Research. Health Effects Institute, Report of the Institute's Health Review Committee, September 1985.

HEI (Health Effects Institute). 1988. An Update on Gasoline Vapor Exposure and Human Cancer: An Evaluation of Scientific Information Published between 1985 and 1987. Health Effects Institute, Report of the Institute's Health Review Committee, January 6, 1988.

Heim, M., B. Conte-Devolx, G. Pin, J.C. Manelli, G. Rougon-Rapuzzi, and A. Lagier. 1977. Syndrome de secretion in appropriee de vasopressine: A propos de trois observations. Sem. Hop. Paris 53:155-159; 1977.

Held, P.J., P.M. Hanno, A.J. Wein, M.V. Pauly, and M.A. Cann. 1990. Epidemiology of interstitial cystitis. Pp. 29-48 in Interstitial Cystitis. P. Hanno, D.R. Staskin, R.J. Krane, and A.J. Wein, eds. New York: Springer-Verlag.

Helmy, H., M.N. Seddak, M.T. Basta, A. Shaaban, M. el-Baz, S. el-Masry, E.S. Al-Hilaly, and M.A. Ghoneim. 1991. Cytokeratin shedding in urine as a biological marker for bladder cancer: Monoclonal antibody-based evaluation. Br. J. Urol. 68:248-253.

Helpap, B. 1980. The biological significance of atypical hyperplasia of the prostate. Virchows Arch. (A) 387:307-313.

Helpap, B. 1988. Observations on the number, size and localization of nucleoli in hyperplastic and neoplastic prostatic disease. Histopathology 13:203-211.

Hemminki, K. 1993. DNA adducts, mutations and cancer. (Commentary) Carcinogenisis 14:2007-2012.

Iesato, K., M. Wakastin, and Y. Wakastin. 1977. Renal tubular dysfunction in Minimata disease. Ann. Intern. Med. 86:731-733.

Iida, H., R. Seifert, C.E. Alpers, R. Gronwald, P.E. Phillips, P. Pritzl, K. Gordon, A.M. Gown, R. Ross, D. Bowel-Pope, and R.J. Johnson. 1991. Platelet-derived growth factor (PDGF) and PDGF receptor are induced in mesangial proliferative nephritis in the rat. Proc. Natl. Acad. Sci. U.S.A. 88:6560-6564.

Inoue, M., K. Okajima, and Y. Morino. 1981. Renal transtubular transport of mercapturic acid in vivo. Biochim. Biophys. Acta 641:122-128.

Inoue, M., K. Okajima, and Y. Morino. 1984. Hepato-renal cooperation in biotransformation, membrane transport, and elimination of cysteine S-conjugates of xenobiotics. J. Biochem. 95:247-254.

Inoue, S., D. Grant, and C.P. Leblond. 1989. Heparin sulfate proteoglycan is present in basement membrane as a double-tracked structure. J. Histochem. Cytochem. 37:597-602.

Inskeep, P., and F.P. Guengerich. 1984. Glutathione-mediated binding of dibromoalkanes to DNA: Specificity of rat glutathione S-transferases and dibromoalkane structure. Carcinogenesis 5:805-808.

Ishikawa, J., H.J. Xu, S.X. Hu, D.W. Yandell, S. Maeda, S. Kamidono, W.F. Benedict, and R. Takahashi. 1991. Inactivation of the retinoblastoma gene in human bladder and renal cell carcinomas. Cancer Res. 51:5736-5743.

Ishikura, K., M. Hasegawa, K. Nomura, T. Okamoto, S. Tanji, T. Abe, T. Fujioka, T. Ohhori, and T. Kubo. 1991. Epidermal growth factor in the urine of patients with renal cell carcinoma and bladder tumor [in Japanese]. Hinyokika Kiyo (Acta Urol. Jpn.) 37:1229-1234.

Isley, W.L., L.E. Underwood, and D.R. Clemmons. 1983. Dietary components that regulate serum somatomedin-C concentrations in humans. J. Clin. Invest. 71:175-182.

Ito, N., and S. Fukushima. 1989. Promotion of urinary bladder carcinogenesis in experimental animals. Exp. Pathol. 36:1-15.

Ito, H., Y. Nakagami, T.T. Lin, K. Ideda, and F. Oka. 1988. A comparative evaluation of PAP and gamma-Sm as the tumor markers of prostatic cancer. Nippon Ika Daigaku Zasshi 55:217-218.

Ito, N., S. Fukushima, and R. Hasegawa. 1989. Bladder cancers. Their process of development and its modification. Acta Pathol. Japan 39:1-14.

Jacob, A.K., D.J. Thomas, C.A. Reznikoff, T.V. Zenser, and B.B. Davis. 1991. Cyclic AMP response in SV40 immortalized human bladder cells. Carcinogenesis 12:1459-1463.

Jacobs, W.R., M. Sgambati, G. Gomez, P. Vilaro, M. Higdon, P.D. Bell, and L.J. Mandel. 1991. Role of cytosolic Ca in renal tubule damage induced by anoxia. Am. J. Physiol. 260:C545-C554.

Jakoby, W.B., and J.L. Stevens. 1984. Cysteine conjugate b-lyase and the thiomethyl shunt. Biochem. Soc. Trans. 12:33-35.

Jarrett, L., P.E. Lacy, and D.M. Kipnis. 1964. Characterization by immunofluorescence of an ACTH-like substance in nonpituitary tumors from patients with hyperadrenocorticism. J. Clin. Endocrinol. 24:542-549.

Javadpour, N., and R. Guirguis. 1992. Tumor collagenase-stimulating factor and tumor autocrine motility factor as tumor markers in bladder cancer--an update. Eur. Urol. 21 Suppl. 1:1-4.

Johnson, M.D., G.T. Bryan, and C.A. Reznikoff. 1985. Serial cultivation of normal rat bladder epithelial cells in vitro. J. Urol, 133:1076-1086.

Jones, D.P., G.B. Sundby, K. Ormstad, and S. Orrenius. 1979. Use of isolate kidney cells for study of drug metabolism. Biochem. Pharmacol. 28:929-935.

Jones, D.P., S. Orrenius, and S.W. Jakobson. 1980. Cytochrome P-450-linked monooxygenase systems in the kidney. Pp. 123-158 in Extrahepatic Metabolism of Drugs and Other Foreign Compounds, T.E. Gram, ed. New York: Spectrum.

Jones, T.W., A. Wallin, T. Thor, R.G. Gerdes, K. Ormstand, and S. Orrenius. 1986. The mechanism of pentachlorobutadienyl-glutathione nephrotoxicity studied with isolated rat renal epithelial cells. Arch. Biochem. Biophys. 251:504-513.

Jones, T.W., C. Qin, V.H. Schaeffer, and J.L. Stevens. 1988. Immunohistochemical localization of glutamine transaminase K, a rat kidney cystein conjugate b-lyase, and the relationship to the segment specificity of cysteine conjugate nephrotoxicity. Mol. Pharmacol. 34:621-627.

Jones, P.L., C.M. O'Hare, R.A. Bass, J.Y. Rao, G.P. Hemstreet, and R.E. Hurst. 1990. Quantitative immunofluorescence, anti-ras p21 antibody specificity, and cellular oncoprotein levels. Biochem. Biophys. Res. Comm. 167:464-470.

Jones, C.L., S. Buch, M. Post, L. McCulloch, E. Liu, and A.A. Eddy. 1991. Pathogenesis of interstitial fibrosis in chronic purine aminonucleoside nephrosis. Kidney Int. 40:1020-1031.

Jordan, A., J. Weingarten, and W. Murphy. 1987. Transitional cell neoplasms of the urinary bladder. Can biologic potential be predicted from histologic grading? [published erratum appears in Cancer 1988 Apr 1;61(7):1385]. Cancer 60(11):2766-2774.

Jordan, D.K., S.R. Patil, J.E. Divelbiss, S. Vemuganti, C. Headley, M.H. Waziri, and N.J. Gurll. 1989. Cytogenetic abnormalities in tumors of patients with von Hippel-Lindau disease. Cancer Genet. Cytogenet. 42:227-241.

Kadamani, S., N.R. Asal, and R.Y. Nelson, 1989. Occupational hydrocarbon exposure and risk of renal cell carcinoma. Am. J. Ind. Med. 15:131-141.

Kadlubar, F., M. Butler, K. Kaderlik, H. Chou, and N. Lang. 1992. Polymorphisms for aromatic amine metabolism in humans: Relevance for human carcinogenesis. Environ. Health Perspect. 98: 69-74.

Kageyama, Y., M. Katoh, K. Okada, K. Yoshida, and T.T suruo. 1991. Detection of P-glycoprotein in human urogenital carcinomas and its relationship to epidermal growth factor receptor expression. Eur.Urol. 20:58-61.

Kahng, M.W., E.A. Conner, and B.A. Fowler. 1992. Lead-binding proteins (PhBP) in human tissues. The Toxicologist 12:214.

Kaiser, U., M.L. Hansmann, and I. Papadopoulos. 1991. Does the immunophenotype of renal cell carcinoma correlate with its clinical stage? Urol. Int. 47:194-198.

Kakizoe, T., H. Komatsu, T. Niijima, T. Kawachi, and T. Sugimura. 1981. Maintenance by saccharin of membrane alterations of rate bladder cells induced by subcarcinogenic treatment with bladder carcinogens. Cancer Res. 41:4702-4705.

Kaloyanides, G.J. 1991. Metabolic interactions between drugs and renal tubulo-interstitial cells: Role in nephrotoxicity. Kidney Int. 39:531-540.

Kanerva, R.L., M.S. McCracken, C.L. Alden, and L.C. Stone. 1987a. Morphogenesis of decalin-induced renal alterations in the male rat. Food chem. Toxicol. 25:53-61.

Kanfer, A., D. deProst, C. Guettier, D. Nochy, V. LeFloch, N. Hinglais, and P. Druet. 1987. Enhanced glomerular procoagulant activity and fibrin deposition in rats with mercuric chloride-induced autoimmune nephritis. Lab. Invest. 57:38-143.

Kao, C., S. Wu, M. Bhattacharya, L.F. Meisner, and C.A. Reznikoff. 1991. Losses of 3p, 11p, and 13q in EJ/ras-transformable simian virus 40-immortalized human uroepithelial cells. Genes Chromosomes. Cancer 4(2):1-11, 1001.

References

Karthaus, H.F., M.J. Bussemakers, J.A. Schalken, K.H. Kurth, W.F. Feitz, F.M. Debruyne, H.P. Bloemers, and W.J. Van de Ven. 1987. Expression of proto-oncogenes in xenografts of human renal cell carcinomas. Urol. Res. 15:349-353.

Kawaguchi, A., M.H. Goldman, R. Shapiro, M. Foegh, P. Ramwell, and R. Lower. 1985. Increase in urinary thromboxane B_2 in rats caused by cyclosporine. Transplantation 40:214-216.

Kawamata, H., M. Azuma, S. Kameyama, L. Nan, and R. Oyasu. 1992. Effect of epidermal growth factor/transforming growth factor alpha and transforming growth factor beta 1 on growth in vitro of rat urinary bladder carcinoma cells. Cell Growth Differ. 3:819-825.

Kelley, V.E., S. Sneve, and S. Musinski. 1986. Increased renal thromboxane production in murine lupus nephritis. J. Clin. Ivest. 77:252-259.

Kerr, L., C. Huntoon, J. Donohue, P.J. Leibson, N.H. Bander, T. Ghose, S.J. Luner, R. Vassella, and D.J. McKean. 1990. Heteroconjugate antibody-directed killing of autologous human renal carcinoma cells by in vitro-activated lymphocytes. J. Immunol. 144:4060-4067.

Key, G., M.H.G. Becker, B. Baron, M. Duchrow, C. Schluter, and H.D. Flad. 1993. New Ki-67 murine monoclonal antibodies (MIB 1-3) generated against bacterially expressed parts of the Ki-67 cDNA containing three 62bp repetitive elements encoding for the Ki-67 epitope. Lab. Invest. 68:629-636.

Kiesewetter, F., O.P. HOrnstein, P.Hermanek, A. Herrlinger, S. Eberhard. 1987. Possibilities of DNA impulse cytophotometry in kidney cancer [in German]. Urologe A. 26:162-167.

Kiley, J.E., S.R. Powers, and R.T. Beebe. 1960. Acute renal failure: Eight cases of renal tubular necrosis. N. Engl. J. Med. 262:481.

Kimball, E.S., W.H. Bohn, K.D. Cockley, T.C. Warren, and S.A. Sherwin. 1984. Distinct high-performance liquid chromatography pattern of transforming growth factor activity in urine of cancer patients as compared with that of normal individuals. Cancer res. 44:3613-3619.

Kinouchi, T., S. Saidi, T. Naoe, A. Uenaka, T. Kotake, H. Shiku, and E. Nakayama. 1989. Correlation of c-myc expression with nuclear pleomorphism in human renal cell carcinoma. Cancer Res. 49:3627-3630.

Kleer, E., J.J. Larson-Keller, H. Zincke, and J.E. Oesterling. 1993. Ability of preoperative serum prostate-specific antigen value to predict pathologic stage and DNA ploidy. Influence of clinical stage and tumor grade. Urology 41:207-216.

Klingel, R., W. Dippold, S.Storkel, K.H. Meyer zum Buschenfelde, and H. Kohler. 1992. Expression of differentiation antigens and growth-related genes in normal kidney, autosomal dominant polycystic kidney disease, and renal cell carcinoma. Am. J. Kidney Dis. 19:22-30.

Klingelhutz, A.J., S.Q. Wu, J. Huang, and C.A. Reznikoff. 1992. Loss of 3p13-p21.2

in tumorigenic reversion of a hybrid between isogeneic nontumorigenic and tumorigenic human uroepithelial cells. Cancer Res. 52:1631-6134.

Kloppel, G., W.T. Knofel, H. Baisch, and U.Otto. 1986. Prognosis of renal cell carcinoma related to nuclear grade, DNA content and Robson stage. Eur. Urol. 12:426-431.

Klouda, P.T., J. Manos, and E.J. Acheson. 1979. Strong association between idiopathic membranous nephropathy and HLA-DRW3. Lancet 2:770-771.

Kluwe, W.M., and J.B. Hook. 1980. Effects of environmental chemicals on kidney metabolism and function. Kidney Int. 18:648-655.

Knudson, A.G. 1971. Mutation and cancer: Statistical study of retinoblastoma. Proc. Natl. Acad. Sci. U.S.A. 68:820-823.

Knudson, A.G., and L.C. Strong. 1972. Mutation and cancer: A model for Wilms' tumor of the kidney. J. Nat. Cancer Inst. 48:313-324.

Kochevar, G.J., J.A. Stanek, and E.B. Rucker. 1992. Truncated fibronectin. An autologous growth-promoting substance secreted by renal carcinoma cells. Cancer 69:2311-2315.

Kohler, H., E. Wandel, and B. Brunck. 1991. Acanthocyturia--A characteristic marker for glomerular bleeding. Kidney Int. 40:115-120.

Kohnon, T., T. Sekine, K. Tobisu, M. Oshimura, and J. Yokota. 1993. Chromosome 3p deletion in a renal cell carcinoma cell line established from a patient with von Hippel-Lindau disease. Jpn. J. Clin. Oncol. 23:226-231.

Koide, T. 1992. The clinical significance of tissue, blood and urine polyamine in renal cell carcinoma [in Japanese]. Nippon Hinyokika Gakkai Zasshi (Jpn. J. Urol.) 83:1228-1237.

Kolonel, L.N., A.M.Y. Nomura, M.W. Hinds, T. Hirohata, J.H. Hankin, and J. Lee. 1983. Role of diet in cancer incidence in Hawaii. Cancer Res. (Suppl.) 43:2397s-2402s.

Konchuba, A.M., M.C. Clements, P.F. Schellhammer, S. M. Schlossberg, and G.L. Wright. 1992. Failure of anticytokeratin 18 antibody to improve flow cytometric detection of bladder cancer. Cancer 70: 2879-2884.

Konen, J.C., L.G. Curtis, Z.K. Shihabi, and M.B. Dignan. 1990. Screening diabetic patients for microalbuminuria. J. Fam. Pract. 31:505-510.

Koo, C., B. Pauli, G. Friedell, and R.S. Weinstein. 1979. Induction of proliferative activity in urinary bladder epithelium by cystoscopy fluids. Lab. Invest. 40:265.

Koo, A.S., C. Armstrong, B. Bochner, T. Shimabukuro, C.L. Tso, J.B. deKernion, A Belldegrum. 1992. Interleukin-6 and renal cell cancer: Production, regulation, and growth effects. Cancer Immunol. Immunother. 35:97-105.

Koob, M., and W. Dekant. 1990. Metabolism of hexafluorpropene. Evidence for bioactivation by glutathione conjugate formation in the kidney Drug etab. Dispos. 18:911-922.

Koopman, M.G., R.T. Krediet, F.M.J. Zuijderhoudt, E.A.M. deMoor, and L. Arisz.

References

1987. Circadian rhythm or urinary beta2-microglobulin excretion in patients with a nephrotic syndrome. Nephron. 45:140-146.

Koss, L.G., and B. Czerniak. 1992. Image analysis and flow cytometry of tumors of prostate and bladder; with a comment on molecular biology of urothelial tumors. Monogr. Pathol. (34):112-128.

Kotake, T., and T. Kinouchi. 1994. Characterization of renal cell carcinoma: Current topics [review][in Japanese]. Gan to Kagaku Ryoho 21:5-11.

Kotake, T, S. Saiki, T. Kinouchi, H. Shiku, and E. Nakayama. 1990. Detection of the c-myc gene product in urinary bladder cancer. Jpn. J. Cancer Res. 81:1198-1201.

Kovacs, G. 1989. Papillary renal cell carcinoma. A morphologic and cytogenetic study of 11 cases. Am. J. Pathol. 134:27-34.

Kovacs, G., S. Szucs, W. deRiese, and H. Baumgartel. 1987. Specific chromosome aberration in human renal cell carcinoma. Int. J. Cancer 40:171-178.

Kovacs, G., P. Brusa, and W. deRiese. 1989. Tissue-specific expression of a constitutional 3;6 translocation: Development of multiple bilateral renal-cell carcinomas. Int. J. Cancer 43:422-427.

Kovacs, G., C. Welter, L. Wilkens, N. Blin, and W. deRiese. 1989. Renal oncocytoma. A phenotypic and genotypic entity of renal parenchymal tumors. Am. J. Pathol. 134:967-971.

Kovacs, G., L. Fuzesi, A. Emanuel, and H.F. Kung. 1991. Cytogenetics of papillary renal cell tumors. Genes Chromosom. Cancer 3:249-255.

Kovacs, G., A. Emanuel, H. Neumann, and H. King. 1991. Cytogenetics of renal cell carcinomas associated with von Hippel-Lindau disease. Genes Chromosomes Cancer 3:256-262.

Koziol, J.A., D.C. Clark, R.F. Gittes, and E.M. Tan. 1993. The natural history of interstitial cystitis: A survey of 374 patients. J. Urol. 149:465-469.

Kreisberg, J.I., G. Sachs, T.G. Pretlow, and R.A. McGuire. 1977a. Separation of proximal tubule cells from suspensions of rat kidney cells by free-flow electrophoresis. J. Cell. Physiol. 93:169-172.

Kreisberg, J.I., A.M. Pitts, and T.G. Pretlow. 1977b. Separation of proximal tubule cells from suspensions of rat kidney cells in density gradients of Ficoll in tissue culture medium. Am. J. Pathol. 86:591-602.

Kujubu, D.A., J.T. Norman, H.R. Herschman, and L.G. Fini. 1991. Primary gene expression in renal hypertrophy: Evidence for different growth initiation processes. Am. J. Physiol. 260:F823-F827.

Kumar, D., and S. Kumar. 1993. Adrenal cortical adenoma and adrenal metastasis of renal cell carcinoma: Immunohistochemical and DNA ploidy analysis. Mod. Pathol. 6:36-41.

Kumar, S., and A. Muchmore. 1990. Tamm-Horsfall protein -- uromodulin (1950-1990). Kidney Int. 37:1395-1401.

Kumar, S., S.M. Jakate, H.B. Marsden, P. Kumar, and B. Jasani. 1988. Tamm-

Horsfall protein is a marker of renal and extra-renal rhabdoid tumors. Int. J. Cancer 41:386-389.

Kumar, S., P.H. Hand, H.B. Marsden, P.Kumar, and A. Thor. 1991. Quantitation of enhanced expression of ras-oncogene product (p21) in childhood renal tumors. Anticancer Res. 11:1657-1662.

Kunes, J., K. Capek, J. Stejskal, and J. Jelinek. 1978. Age-dependent difference of kidney response to temporary ischemia in the rat. Clin. Sci. Mol. Med. 55:365-368.

Kurnick, N.B., and P.A. Lindsay. 1968. Nucleic acids in compensatory renal hypertrophy. Lab. Invest. 18:700-708.

Kurokawa, Y., A. Maekawa, M. Takahashi, and Y. Hayashi. 1990. Toxicity and carcinogenicity of potassium bromate--a new renal carcinogen [review]. Environ. Health Perspect. 87:309-335.

Kvist, N., and E. Nexo. 1989. Epidermal growth factor in urine after kidney transplantation in humans. Urol. Res. 17:225-228.

Lajara, R., P. Rotwein, J. Bortz, V. Hansen, J.L. Sadow, C.R. Betts, S.A. Rogers, and M.R. Hammerman. 1989. Dual regulation of insulin-like growth factor I expression during renal hypertrophy. Am. J. Physiol. 257:F252-F261.

Landberg, G., and G. Roos. 1991. Antibodies to proliferating cell nuclear antigent as S-Phase probes in flow cytometric cell cycle analysis. Cancer Res. 51:4570-4574.

Landrigan, P.J., R.A. Goyer, T.W. Clarkson, D.P. Sandler, J.H. Smith, M.J. Thun, and R.P. Wedeen. 1984. The work-relatedness of renal disease. Arch. Environ. Health 39:225-230.

Langkilde, N.C., H. Wolf, P. Meldgard, and T.F. Orntoft. 1991a. Frequency and mechanism of Lewis antigen expression in human urinary bladder and colon carcinoma patients. Br. J. Cancer 63:583-586.

Langkilde, N.C., H. Wolf, and T.F. Orntoft. 1991b. Lewis antigen expression in benign and malignant tissues from RBC Le(a-b-) cancer patients. Br. J. Haematol. 79:493-499.

Lanigan, D., P.A. McLean, B. Curran, and M. Leader. 1993. Comparison of flow and static image cytometry in the determination of ploidy. J. Clin. Pathol. 46:135-139.

Lanzafame, S., L. Puzzo, C. Di Naso, G. Di Marco, S. Ranno, and G. Minardi. 1986. Carcinoembryonic antigen (CEA) in carcinoma of the upper urinary tract and bladder. Immunohistochemical study [in Italian]. Pathologica 78:451-458.

Larsen, G.L. 1985. Distribution of cysteine conjugate b-lyase in gastrointestinal bacteria and in the environment. Xenobiotica 15:199-209.

Larsson, P., G. Roos, R. Stenling, and B. Ljungberg. 1993. Tumor-cell proliferation and prognosis in renal-cell carcinoma. Int. J. Cancer 55:566-570.

Lash, L.H. 1989. Isolated kidney cells in the study of chemical toxicity. Pp. 231-262 in In Vitro Toxicology: Model Systems and Methods, C.A. McQueen, ed. Caldwell, N.J.: Telford Press.

References

Lash, L.H. 1990. Susceptibility to toxic injury in different nephron cell populations. Toxicol. Lett. 53:97-104.

Lash, L.H. 1992. Nephrotoxicity studies with freshly isolated cells from rat kidney. Pp. 115-122 in In Vitro Methods of Toxicology, R.R. Watson, ed. Boca Raton, Fla: CRC Press.

Lash, L.H. 1993. Purification of renal cortical cell populations by Percoll density-gradient centrifugation. Pp. 397-410 in In Vitro Biological Systems: Preparation and Maintenance Methods in Toxicology, Vol. 1., Part A, C.A. Tyson, and J.M. Frazier, eds. San Diego: Academic.

Lash, L.H. 1994. Patterns of chemical-induced injury in primary cultures of rat renal proximal tubular (PT) and distal tubular (DT) cells. Toxicologist 14:168.

Lash, L.H., and M.W. Anders. 1986. Cytotoxicity of S-(1,2-dichlorovinyl)glutathione and S-(1,2-dichlorovinyl)-L-cysteine in isolated rat kidney cells. J. Biol. Chem. 261:13076-13081.

Lash, L.H., and M.W. Anders. 1989. Uptake of nephrotoxic S-conjugates by isolated rat renal proximal tubular cells. J. Pharmacol. Exp. Ther. 248:531-537.

Lash, L.H., and D.P. Jones. 1983. Transport of glutathione by renal basal-lateral membrane vesicles. Biochem. Biophys. Res. Commun. 112:55-60.

Lash, L.H., and D.P. Jones. 1984. Renal glutathione transport: Characteristics of the sodium-dependent system in the basal-lateral membrane. J. Biol. Chem. 259:14508-14514.

Lash, L.H., and D.P. Jones. 1985. Uptake of the glutathione conjugate S-(1,2-dichlorovinyl)glutathione by renal basal-lateral membrane vesicles and isolated kidney cells. Mol. Pharmacol. 28:278-282.

Lash, L.H., and J.J. Tokarz. 1989. Isolation of two distinct populations of cells from rat kidney cortex and their use in the study of chemical-induced toxicity. Anal. Biochem. 182:271-279.

Lash, L.H., and J.J. Tokarz. 1990. Oxidative stress in isolated rat renal proximal and distal tubular cells. Am. J. Physiol. 259:F338-F347.

Lash, L.H., and E.B. Woods. 1991. Cytotoxicity of alkylating agents in isolated rat kidney proximal tubular and distal tubular cells. Arch. Biochem. Biophys. 286:46-56.

Lash, L.H., and R.K. Zalups. 1992. Mercuric chloride-induced cytotoxicity and compensatory hypertrophy in rat kidney proximal tubular cells. J. Pharmacol. Exp. Ther. 261:819-829.

Lash, L.H., and R.K. Zalups. 1994. Activities of enzymes involved in renal cellular glutathione metabolism after urinephrectomy in the rat. Arch. Biochem. Biophys. 309:129-138.

Lash, L.H., A.A. Elfarra, and M.W. Anders. 1986a. S-(1,2-Dichlorovinyl-L-homocysteine-induced cytotoxicity in isolated rat kidney cells. Arch. Biochem. Biophys. 251:432-439.

Lash, L.H., T.M. Hagen, and D.P. Jones. 1986b. Exogenous glutathione protects intestinal epithelial cells from oxidative injury. Proc. Natl. Acad. Sci. U.S.A. 83:4641-4645.

Lash, L.H., A.A. Elfarra, and M.W. Anders. 1986c. Renal cysteine conjugate b-lyase: Bioactivation of nephrotoxic cysteine S-conjugates in mitochondrial outer membrane. J. Biol. Chem. 261:5930-5935.

Lash, L.H., D.P. Jones, and M.W. Anders. 1988. Glutathione homeostasis and glutathione S-conjugate toxicity in the kidney. Rev. Biochem. Toxicol. 9:29-67.

Lash, L.H., R.M. Nelson, R.A. Van Dyke, and M.W. Anders. 1990. Purification and characterization of human kidney cytosolic cysteine conjugate b-lyase activity. Drug Metab. Dispos. 18:50-54.

Lash, L.H., R.M. Nelson, R.A. Van Dyke, and M.W. Anders. 1990a. Purification and characterization of human kidney cytosolic cysteine conjugate b-lyase activity. Drug Metab. Dispos. 18:50-54.

Lash, L.H., A.A. Elfarra, D. Rakiewicz-Nemeth, and M.W. Anders. 1990b. Bioactivation mechanism of cytotoxic homocysteine S-conjugates. Arch. Biochem. Biophys. 276:322-330.

Lash, L.H., J.J. Tokarz, E.B. Woods, and B.M. Pedrosi. 1993. Hypoxia and oxygen dependence of cytotoxicity in renal proximal tubular and distal tubular cells. Biochem. Pharmacol. 45:191-200.

Lash, L.H., P.J. Sausen, R.J. Duescher, A.J. Cooley, and A.A. Elfarra. 1994. Roles of cysteine conjugate β-lyase and S-oxidase in nephrotoxicity: Studies with S-(1,2-dichlorovinyl)-L-cysteine and S-(1,2-dichlorovinyl)-L-cysteine sulfoxide. J. Pharmacol. Exp. Ther. 269:374-383.

Latif, F., K. Tory, J. Gnarra, M. Yao, F.-M. Duh, M.L. Orcutt, T. Stackhouse, I. Kuzmin, W. Modi, L. Geil, L. Schmidt, F. Zhou, H. Li, M.H. Wei, G. Glenn, F.M. Richards, P.A. Crossey, M.A. Ferguson-Smith, D. Le Paslier, I. Chumakov, D. Cohen, C.A. Chinault, E.R. Maher, W.M. Linehan, B. Zbar, and M.I. Lerman. 1993. Identification of the von Hippel-Lindau disease tumor suppressor gene. Science 260:1317-1320.

Lauwerys, R., A. Bernard, C. Viau, and J.P. Buchet. 1985. Kidney disorders and haematotoxicity from organic solvent exposure. Scand. J. Work Environ. Health 11(Suppl. 1):83-90.

le Coutre, P., S. Bock, G. Jakse, and P.E. Petrides. 1992. Immunoreactive low-molecular-weight epidermal growth factor in urine of patients with renal cell carcinoma. Urol. Res. 20:293-296.

Lefkowith, J.B., J. Pippin, T. Nagamatsu and V. Lee. 1992. Urinary eicosanoids and the assessment of glomerular inflammation. J. Am. Soc. Nephrol. 2:1560-1567.

Lehman-McKeeman, L.D., and D. Caudill. 1992. Biochemical basis for mouse resistance to hyaline droplet nephropathy: Lack of relevance of the α2u-globulin

References

protein superfamily in this male rat-specific syndrome. Toxicol. Appl. Pharmacol. 116:170.

Lehman-McKeeman, L.D., P.A. Rodriguez, R. Takigiku, D. Caudill, and M.L. Fey. 1989. d-Limonene-induced male rat-specific nephrotoxicity: Evaluation of the association between d-limonene and α_{2u}-globulin. Toxicol. Appl. Pharmacol. 99:250-259.

Lehman-McKeeman, L.D., D. Caudill, R. Takigiku, R.E. Schneider, and J.A. Young. 1990a. Comparative disposition of d-limonene in rats and mice: Relevance to male-rat-specific nephrotoxicity. Toxicol. Lett. 53:193-195.

Lehman-McKeeman, L.D., M.I. Rivera-Torres, and D. Caudill. 1990b. Lysosomal degradation of α_{2u}-globulin-xenobiotic conjugates. Toxicol. Appl. Pharmcol. 103:539-548.

Levin, I., T. Klein, J. Goldstein, O. Kuperman, J. Kanetti, and B. Klein. 1991. Expression of class I histocompatibility antigens in transitional cell carcinoma of the urinary bladder in relation to survival. Cancer 68:2591-2594.

Levin, I., B. Klein, S. Segal, A. Eyal, J. Gopas, et al. 1992. Expression of HLA class I-encoded cell surface antigens in transitional cell carcinoma of the urinary bladder. Tissue Antigens 39(1):19-22.

Lewis, R.A., R.H. Schwartz, and E.A. Schenk. 1972. Tamm-Horsfall mucoprotein. II. Ontogenetic development. Lab. Invest. 26(6):728-730.

Leyh, H., A. Lehmer, and R. Hartung. 1992. Patterns of ploidy in renal cell carcinoma. Prog. Clin. Biol. Res. 378:187-193.

Liang P., and A.B. Pardee. 1992. Differential display of eukaryotic messenger RNA by means of the polymerase chain reaction. Science 257:967-971.

Liang P., L. Averboukh, and A.B. Pardee. 1993. Distribution and cloning of eukaryotic mRNAs by means of differential display: Refinements and optimization. Nucleic Acids Res. 21:3269-3275.

Lianos E.A., and A. Zanglis. 1990. Glomerular platelet-activating factor levels and origin in experimental glomerulonephritis. Kidney Int. 37:736-740.

Lianos, E., G.A. Andres, and M.J. Dunn. 1983. Glomerular prostaglandin and thromboxane synthesis in rat nephrotoxic serum nephritis. J. Clin. Invest. 72:1439-1448.

Lianos, E., M.A. Rahman, and M.J. Dunn. 1985. Glomerular arachidonate lipoxygenation in rat nephrotoxic serum nephritis. J. Clin. Invest. 76:1355-1359.

Lianos, E.A., B. Bresnahan, and C. Pan. 1991. Mesangial cell immune injury: Synthesis, origin and role of eicosanoids. J. Clin. Invest. 88:623-631.

Libertino, J.A., L. Zinman, and E. Watkins. 1987. Long-term results of resection of renal cell cancer with extension into inferior vena cava. J. Urol. 137:21-24.

Liebert, M., G.A. Wedemeyer, J.A. Stein, R.W. Washington, A. Flint, L. Ren, and H.B. Grossman. 1989. Identification by monoclonal antibodies of an antigen shed by human bladder cancer cell. Cancer Res. 49:6720-6726.

Liebert, M., G. Wedemeyer, J.A. Stein, R. Washington, G. Faerber, A. Fling, and H.B. Grossman. 1993. Evidence for urothelial cell activation in interstitial cystitis. J. Urol. 149:470-475.

Lilja, H., A.T.K. Cockett, and P.-A. Abrahamsson. 1992. Prostate specific antigen predominantly forms a complex with α_1-antichymotrypsin in blood. Cancer (Suppl) 70:230-234.

Limas, C. 1990. A,B blood group antigens in tissues of AB heterozygotes. Emphasis on normal and neoplastic urothelium. Am. J. Pathol. 137:1157-1162.

Limas, C. 1991. Quantitative interrelations of Lewis antigens in normal mucosa and transitional cell bladder carcinomas. J. Clin. Pathol. 44:983-989.

Limon, J., K. Mrozek, S. Heim, P. Elfving, B. Nedoszytko, M. Babinska, N. Mandahl, R. Lundgren, and F. Mitelman. 1990. On the significance of trisomy 7 and sex chromosome loss in renal cell carcinoma. Cancer Genet. Cytogenet. 49:259-263.

Lin, C.-W., D.A. Young, S.D. Kirley, A.-H. Khaw, and G.A. Prout. 1988. Detection of tumor cells in bladder washings by a monoclonal antibody to human bladder tumor-associated antigen. J. Urol. 140:672-677.

Lipponen, P., M. Eskelinen, K. Hietala, K. Syrjanen, and R.A. Gambetta. 1994. Expression of proliferating cell nuclear antigen (PC10), p53 protein and c-erbB-2 in renal adenocarcinoma. Int. J. Cancer 57:275-280.

Lipponen, P., and M. Eskelinen. 1990. Nuclear morphometry in grading transitional cell bladder cancer compared with subjective histological grading. Anticancer Res. 10(6):1725-1730.

Lipponen, P., M. Eskelinen, and S. Nordling. 1991a. Relationship between DNA flow cytometric data, nuclear morphometric variables and volume-corrected mitotic index in transitional cell bladder tumors. Eur. Urol. 19(4):327-331.

Lipponen, P., M. Eskelinen, and S. Nordling. 1991b. Progression and survival in transitional cell bladder cancer: A comparison of established prognostic factors, S-phase fraction and DNA ploidy. Eur. J. Cancer 27(7):877-881.

Lipponen, P., M. Eskelinen, K. Jauhiainen, E. Harju, R. Terho, and H. Haapasalo. 1992. Prediction of superficial bladder cancer by nuclear image analysis. Eur. J. Cancer 29A(1):61-65.

Lizard, G., P. Roignot, L. Dusserre-Guion, F. Morlevat, D. Michiels-Marzais, N. Ferry, and J.C. Tremeaux. 1992. Characterization of seven kidney tumors by flow cytometry: Analysis of cell cycle, DNA content and P-glycoprotein expression. Eur. Urol. 21(Suppl. 1):39-42.

Ljungberg, B., R. Stenling, and G. Roos. 1956. DNA content in renal cell carcinoma with reference to tumor heterogeneity. Cancer 56:503-508.

Ljungberg, B., I. Nordenson, and G. Roos. 1991. Cytogenetic and flow cytometric DNA analysis in renal cell carcinoma. Eur. Urol. 19:59-64.

Lloyd, S.N., I.L. Brown, and R.E. Leake. 1992. Transforming growth factor-alpha expression in benign and malignang human prostatic disease. Int. J. Biol. Markers 7:27-34.

Lock, E.A., and J. Ishmael. 1985. Effect of the organic acid transport inhibitor probenecid on renal cortical uptake and proximal tubular toxicity of hexachloro-1,3-butadiene and its conjugates. Toxicol. Appl. Pharmacol. 81:32-42.

Lock, E.A., and R.G. Schnellman. 1990. The effect of haloalkene cysteine conjugates on rat renal glutathione reductase and lipoyl dehydrogenase activities. Toxicol. Appl. Pharmacol. 104:180-190.

Lock, E.A., J. Odum, and P. Ormond. 1986. Transport of N-acetyl-S-pentachloro-1, 3-butadienyl-cysteine by rat renal cortex. Arch. Toxicol. 59:12-15.

Longin, A., A. Huazi, N. Berger-Dutrieux, et al. 1989. A monoclonal antibody (BL2-10D1) reacting with a bladder cancer associated antigen. Int. J. Cancer 43:183-189.

Lopez-Rivas, A.E., A. Adelberg, and E. Rozengurt. 1982. Intracellular K+ and the mitogenic response of 3T3 cells to peptide factors in serum-free medium. Proc. Natl. Acad. Sci. U.S.A. 79:6275-6279.

Lose, G., and B. Frandsen. 1989. Eosinophil cationic protein in urine in patients with urinary bladder tumors. Urol. Res. 17:295-297.

Lovern, W.J., B.L. Fariss, J.N. Wettlaufer, and S. Hane. 1975. Ectopic ACTH production in disseminated prostatic adenocarcinoma. Urology 5:817-820.

Lower, G.M., T. Nilsson, C.E. Nelson, H. Wolf, T.E. Gamsky, and G.T. Bryan. 1979. N-acetyltransferase phenotype and risk in urinary bladder cancer: Approaches in molecular epidemiology. Preliminary results in Sweden and Denmark. Environ. Health Perspect. 29:71-79.

Lunec, J., C. Challen, C. Wright, K. Mellon, and D.E. Neal. 1992. c-erbB-2 amplification and identical p53 mutations in concomitant transitional carcinomas of renal pelvis and urinary bladder [letter]. Lancet 339:439-440.

Lynn, K.L., A. Shenkin, and R.D. Marshall. 1982. Factors affecting excretion of human urinary Tamm-Horsfall glycoprotein. Clin. Sci. 62:21-26.

Maack, T. 1986. Renal clearance and isolated kidney perfusion techniques. Kidney Int. 30:142-151.

Maack, T., C.Y. Park, M.S.F. Camargo. 1985. Renal filtration, transport and metabolism of proteins. Pp. 1773-1803 in The Kidney: Physiology and Pathophysiology. D.W. Seldin, and G. Giebisch, eds. New York: Raven Press.

MacDermott, J.P., C.H. Miller, N. Levy, and A.R. Stone. 1991. Cellular immunity in interstitial cystitis. J. Urol. 145:274-278.

MacFarlane, M., J.R. Foster, G.G. Gibson, L.J. King, and E.A. Lock. 1989. Cysteine conjugate b-lyase of rat kidney cytosol: Characterization, immunohistochemical localization, and correlation with hexachlorobutadiene nephrotoxicity. Toxicol. Appl. Pharmacol. 98:185-197.

MacInnes, J.I., E.S. Nozik, and D.T. Kurtz. 1986. Tissue-specific expression of the rat α_{2u}-globulin gene family. Molec. Cell. Biol. 6:3563-3567.

MacMahon, B., and T.F. Pugh. 1970. Pp. 261-262 in Epidemiology: Principals and Methods. Boston: Little, Brown.

Maddy, S.Q., G.D. Chisholm, A. Busuttil, and F.K. Habib. 1989. Epidermal growth factor receptors in human prostate cancer: Correlation with histological differentiation of the tumour. Br. J. Cancer 60:41-44.

Madrenas, J., N.A. Parfrey, and P.F. Halloran. 1991. Interferon gamma-mediated renal MHC expression in mercuric chloride-induced glomerulonephritis. Kidney Int. 39:273-281.

Mahadevia, P.S., A. Ramaswamy, E.S. Greenwald, D.I. Wollner, and D. Markham. 1983. Hypercalcimia in prostatic carcinoma: report of eight cases. Arch. Intern. Med. 143:1339-1342.

Maiter, D., M. Maes, L.E. Underwood, T. Fliesen, G. Gerard, and J.M. Ketelslegers. 1988. Early changes in serum concentrations of somatomedin-C induced by dietary protein deprivation in rats: Contributions of growth hormone receptor and post-receptor defects. J. Endocrinol. 118:113-120.

Malmstrom, P.U., J. Vasko, K. Wester, and C. Busch. 1991. Determination of DNA ploidy and ABH antigen reactivity in both frozen and formalin-fixed bladder tumor tissue. Urol. Res. 19:357-359.

Malmstrom, P.U., K. Wester, J. Vasko, and C. Busch. 1992. Expression of proliferative cell nuclear antigen (PCNA) in urinary bladder carcinoma. Evaluation of antigen retrieval methods. Acta Pathol. Microbiol. Immunol. Scand. **Sect. A,B or C** 100:988-992.

Maloney, K.E., R.W. Norman, C.L. Lee, O.H. Millard, and J.P. Welch. 1991. Cytogenetic abnormalities associated with renal cell carcinoma. J. Urol. 146:692-696.

Mandal, A.K. 1986. Transmission electron microscopy of urinary sediment in renal disease. Semin. Nephrol. 6:346-370.

Mandal, A.K., and W. Bennett. 1988. Transmission electron microscopy of urinary sediment in the assessment of aminoglycoside nephrotoxicity in the rat. Nephron 49:67-73.

Mandal, A.K., G.N. Mize, and D.B. Birnbaum. 1987. Transmission electron microscopy of urinary sediment in aminoglycoside nephrotoxicity. Renal Failure 10:63-81.

Mannens, M., R.M. Slater, C. Heyting, A. Geurts van Kessel, E. Goedde-Salz, R.R. Frants, G.J. Van Ommen, and P.L. Pearson. 1987. Regional localization of DNA probes on the short arm of chromosome 11 using aniridia-Wilms' tumor-associated deletions. Hum. Genet. 75:180-187.

Manteuffel-Cymborowska, M. 1993. Differential role of polyamines in hyperplasia and hypertrophy. Acta Biochim. Pol. 40:383-388.

References

Marcias-Nunez, J.E., and J.S. Cameron. 1987. Pp. 503 in Renal Function and Disease in the Elderly. Butterworth, London, Boston, Durban, Singapore, Sydney, Toronto, Wellington.

Martin, B.F. 1962. The effect of distension of the urinary bladder on the lining epithelium and on its histochemical reaction for alkaline phosphatase. Ann. Histochim. 1:51-.

Martin, E.W., W.E. Kibbey, L. DiVecchia, G. Anderson, P. Catalano, and J.P. Minton. 1976. Carcinoembryonic antigen: Clinical and historical aspects. Cancer 37:62-81.

Martinerie, C., and B. Perbal. 1991. Expression of a gene encoding a novel potential IGF binding protein in human tissues. Comptes Rendus de l Academie des Sciences - Serie Iii, Sciences de la Vie 313:345-351.

Martinerie, C., E. Viegas-Pequignot, I. Guenard, B. Dutrillaux, V.C. Nguyen, A. Bernheim, and B. Perbal. 1992. Physical mapping of human loci homologous to the chicken nov proto-oncogene. Oncogene 7:2529-2534.

Marx, J. 1991. How the retinoblastoma gene may inhibit cell growth. Science 252:1492.

Masters, J.R., R.S. Camplejohn, M.C. Parkinson, C.R. Woodhouse, and S.M. O'Reilly. 1992. Does DNA flow cytometry give useful prognostic information in renal parenchymal adenocarcinoma?. Br. J. Urol. 70:364-369.

Masuko, T., H. Gaigita, and Y. Hashimoto. 1984. Monoclonal antibodies against cell surface antigens present on human urinary bladder cancer cells. J. Natl. Cancer Inst. 72:523-530.

Masuko, T., K. Sugahara, T. Kamiya, and Y. Hashimoto. 1989. Increase in murine monoclonal-antibody-defined urinary antigens in patients with bladder cancer and benign urogenital disease. Int. J. Cancer 44:582-588.

Matejka, G.L., and E. Jennische. 1992. IGF-I binding and IGF-I mRNA expression in the post-ischemic regenerating rat kidney. Kidney Int. 42:1113-1123.

Mathiesen, E.R., B. Oxenbøll, K. Johansen, P.A. Svendsen, and T. Deckert. 1984. Incipient nephropathy in type I (insulin-dependent) diabetes. Diabetologia 26:406-410.

Mathiesen, E.R., E. Nexo, E. Hommel, and H. Parving. 1988. Reduced urinary excretion of epidermal growth factor in incipient and overt diabetic nephropathy. Diabetic Med. 6:121-126.

Matsumoto, K., and M. Hatano. 1989. Production of interleukin 1 in glomerular cell culture from rats with nephrotoxic serum nephritis. Clin. Exp. Immunol. 75:123-128.

Matsumoto, K., J. Dowling, and R.C. Atkins. 1988. Production of interleukin-1 in glomerular cell cultures from patients with rapidly progressive crescentic glomerulonephritis. Am. J. Nephrol. 8:463-470.

Matsusako, T., H. Muramatsu, T. Shirahama, T. Muramatsu, and Y. Ohi. 1991. Expression of a carbohydrate signal, sialyl dimeric Le(x) antigen, is associated with metastatic potential of transitional cell carcinoma of the human urinary bladder. Biochem. Biophys. Res. Commun. 181:1218-1222.

Mattila, A.L., I. Saario, L. Viinikka, O. Ylikorkala, and J. Perheentupa. 1988. Urinary epidermal growth factor concentrations in various human malignancies. Br. J. Cancer 57:139-141.

Maunsbach, A.B. Observations on the segmentation of the proximal tubule in the rat kidney. Comparison of results from phase contrast, fluorescence and electron microscopy. J. Ultrastruct. Res. 16:239-258.

Maw, M.A., P.E. Grundy, L.J. Millow, M.R. Eccles, R.S. Dunn, P.J. Smith, A.P. Feinberg, D.J. Law, M.C. Paterson, P.E. Telzerow, et al. 1992. A third Wilms' tumor locus on chromosome 16q. Cancer Res. 52:3094-3098.

McCormick, D., and P.A. Hall. 1992. The complexities of proliferating cell nuclear antigen. Histopathology 21:591-594.

McGowan, P., R.E. Hurst, R.A. Bass, G.P. Hemstreet, and R. Postier. 1988. Equilibrium binding of Hoechst 33258 and Hoechst 33342 fluorochromes with rat colorectal cells. J. Histochem. Cytochem. 36:757-762.

McKeehan, W.L., and P.S. Adams. 1988. Heparin-binding growth factor/prostatropin attenuates inhibition of rat prostate tumor epithelial cell growth by transforming growth factor type beta. In Vitro Cell Dev. Biol. 24:243-246.

McKinney, L.L., F.B. Weakley, A.C. Eldridge, R.G. Campbell, J.C. Cowan, J.C. Picken, and H.C. Beister. 1957. S-(dichlorovinyl)-L-cysteine: An agent causing fatal aplastic anemia in calves. J. Am. Chem. Soc. 79:3932-3933.

McKinney, L.L., J.C. Picken, F.B. Weakley, A.C. Eldridge, R.G. Campbell, J.C. Cowan, and H.E. Beister. 1959. Possible toxic factor of trichloroethylene-extracted soybean oil. J. Am. Chem. Soc. 81:909-915.

McLachlan, M.S.F., J.C. Guthrie, C.K. Anderson, and M.J. Fulker. 1977. Vascular and glomerular changes in the aging kidney. J. Pathol. 121:65-78.

McManus, L.M., M.A. Naughton, and A. Martinez-Hernandez. 1976. Human chorionic gonadotropin in human neoplastic cells. Cancer Res. 36:3476-3481.

McNeal, J.E., and D.G. Bostwick. 1986. Intraductal dysplasia: A premalignant lesion of the prostate. Hum. Pathol. 17:64-71.

McNeal, J.E., D.G. Bostwick, R.A. Kindrachuk, E.A. Redwin, F.S. Freiha, and T.A. Stamey. 1986. Patterns of progression in prostate cancer. Lancet 1:60-63.

McNeal, J.E., E.A. Redwine, F.S. Freiga, and T.A. Stamey. 1988a. Zonal distribution of prostatic adenocarcinoma. Am. J. Surg. Pathol. 12:897-.

McNeal, J.E., J. Alroy, I. Leav, E.A. Redwine, F.S. Freiha, and T.A. Stamey. 1988b. Immunohistochemical evidence for impaired cell differentiation in the premalignant phase of prostate carcinogenesis. Am. J. Clin. Path. 90:23-32.

McNichols, D.W., J.W. Segura, and J.H. DeWeerd. 1981. Renal cell carcinoma: Long-term survival and late recurrence. J. Urol. 126:17-23.

McQueen, E.G. 1962. The nature of urinary casts. J. Clin. Pathol. 15:367-373.

References

Medeiros, L.J., A.B. Gelb, and L.M. Weiss. 1988. Renal cell carcinoma. Prognostic significance of morphologic parameters in 121 cases. Cancer 61:1639-1651.

Medeiros, L.J., and L.M. Weiss. 1990. Renal adenocarcinoma. Pp. 35-70 in Tumors and Tumor-like Conditions of the Kidneys and Ureters, J.N. Eble, ed. New York: Churchill Livingstone, Inc.

Meisner, L.F., S.-Q. Wu, B.J. Christian, and C.A. Reznikoff. 1988. Cytogenetic instability with balanced chromosome changes in an SV40 transformed human uroepithelial cell line. Cancer Res. 48:3215-3220.

Mellon, K., S. Thompson, R.G. Charlton, C. Marsh, M. Robinson, D.P. Lane, A.L. Harris, C.H. Horne, and D.E. Neal. 1992. p53, c-erbB-2 and the epidermal growth factor receptor in the benign and malignant prostate. J. Urol. 147:496-499; 1992.

Melnick, R.L. 1992. An alternative hypothesis on the role of chemically induced protein droplet (α2u-globulin) nephropathy in renal carcinogenesis. Reg. Tox. Pharmacol. 16:111-125.

Merritt, S., P.D. Killen, S.H. Phan, and R.C. Wiggins. 1990. Analysis of α1 (I) procollagen, α1(IV) collagen and β-actin mRNA in glomerulus and cortex of rabbits with experimental antiglomerular basement membrane disease. Lab. Invest. 63:762-769.

Merz, V.W., G.J. Miller, T. Krebs, T.L. Timme, D. Kadmon, S. Park, S. Egawa, P. Scardino, and T. Thompson. 1991. Elevated transforming growth factor-β1 and β3 mRNA levels are associated with ras + *myc*-induced carcinomas in reconstituted mouse prostate: Evidence for a paracrine role during progression. Mol. Endocrinol. 5:503-513.

Messing, E.M. 1989. Interstitial Cystitis and Related Syndromes. Pp. 982-1005 in Campbell's Urology, 6th Edition. P.C. Walsh, A.B. Retik, T.A. Stamey, and E.D. Vaughan, eds. Philadelphia, PA: W.B. Saunders Company.

Messing, E.M., and A. Vaillancourt. 1990. Hematuria screening for bladder cancer. J. Occup. Med. 32:838-845.

Messing, E.M., J.E. Bubbers, K.E. Whitmore, J.B. de Kernion, M.S. Nestor, and J.L. Fahey. 1984. Murine hybridoma antibodies against human transitional carcinoma-associated antigens. J. Urol. 132:167-172.

Messing, E.M., P. Hanson, P. Ulrich, and E. Erturk. 1987. Epidermal growth factor--interactions with normal and malignant urothelium: In vivo and in situ studies. J. Urol. 138:1329-1335.

Messing, E.M., P. Hanson, and C.A. Reznikoff. 1988. Normal and malignant human urothelium: In vitro response to blockade of polyamine synthesis and interconversion. Cancer Res. 48:357-361.

Messing, E.M., T.B. Young, V.B. Hunt, E.B. Roecker, A.M. Vaillancourt, W.J. Hisgen, E.B. Greenberg, M.E. Kuglitsch, and J.D. Wegenke. 1992. Home screening for hematuria: Results of a multi-clinic study. J. Urol. 148:289-292.

Miao, T.J., Z. Wang, and N. Sang. 1991. Correlation between the expression of the

P21 ras oncogene product and the biological behavior of bladder tumors. Eur. Urol. 20:307-310.

Michael, A.F., E. Blau, and R.L. Vernier. 1970. Glomerular polyanion: Alteration in aminonucleoside nephrosis. Lab. Invest. 23:649-657.

Miki, T., M. Kuroda, H. Kiyohara, M. Usami, T. Nakamura, and T. Kotaka. 1980. Primary carcinoid tumor fo the prostate: Report of a case. Cancer 45:2580-2592.

Miki, S., M. Iwano, Y. Miki, M. Yamamoto, B. Tang, K. Yokokawa, T. Sonoda, T. Hirano, and T. Kishimoto. 1989. Interleukin-6 (IL-6) functions as an in vitro autocrine growth factor in renal cell carcinomas. FEBS Lett. 250:607-610.

Milani, M., R. Tacconi, G. Marsili, A. Trognoni, G. Centioni, T. Leoni, and A. Fumarola. 1986. Critical-laboratoristic correlation of the diagnostic importance of prostate specific antigen compared with prostatic acid phosphatase and carcinoembryonic antigen in the prostatic cancer. Quad. Sclavo Diagn. Clin. Lab. 22:428-432.

Miles, J., K. Michalski, M. Kouba, D.J. Weaver. 1988. Genomic defects in nonfamilial renal cell carcinoma. Possible specific chromosome change. Cancer Genet. Cytogenet. 34:135-142.

Miller, W.L., and B.J. McCarthy. 1979. Gene expression in normal and regenerating mouse kidney. J. Biol. Chem. 254:742-748.

Miller, E.C., and J.A. Miller. 1985. Some historical perspectives on the metabolism of xenobiotic chemicals to reactive electrophiles. Pp. 3-28 in Bioactivation of Foreign Compounds, M.W. Anders, ed. Orlando, Fla: Academic Press.

Miller, C.H., J.P. MacDermott, K.B. Quattrocchi, G.A. Broderick, and A.R. Stone. 1992. Lymphocyte function in patients with interstitial cystitis. J. Urol. 147:592-595.

Mogensen, C.E. 1987. Microalbuminuria as a predictor of clinical diabetic nephropathy. Kidney Int. 31:673-689.

Mogensen, C.E., A. Chachati, C.K. Christensen, et al. 1985-86. Microalbuminuria: An early marker of renal involvement in diabetes. Uremia Invest. 9:85-95.

Moll, R., J. Laufer, X.R. Wu, and T.T. Sun. 1993. Uroplakin III, a specific membrane protein of urothelial umbrella cells, as a histological markers for metastatic transitional cell carcinomas [in German]. Verh. Dtsch. Ges. Pathol. 77:260-265.

Molland, E.A. 1978. Prostatic adenocarcinoma with ectopic ACTH production. Br. J. Urol. 50:358.

Moller, E., J.F. McIntosh, and D.D. Van Slyke. 1929. Relationship between urine volume and the rate of urea excretion by normal adults. J. Clin. Invest. 6:427-465.

Monks, T.J., and S.S. Lau. 1987. Commentary: Renal transport processes and glutathione-mediated nephrotoxicity. Drug Metab. Dispos. 15:437-441.

Montironi, R., M. Scarpelli, S. Sisti, G. Ansuini, E. Pisani, and G. Mariuzzi. 1987. Prognostic value of computerized DNA analysis of noninvasive papillary carcinomas of the urinary bladder. Tumori 73:567-574.

References

Montironi, R., M. Scarpelli, S. Sisti, A. Braccischi, P. Gusella, E. Pisani, R. Alberti, and G.M. Mariuzzi. 1990. Quantitative analysis of prostatic intraepithilial neoplasia on tissue sections. Anal. Quant. Cytol. Histol. 12:366-372.

Moore, R.D., C.R. Smith, J.J. Lipsky, E.D. Mellitus, and P.S. Lietman. 1984. Risk factors for nephrotoxicity in patients treated with aminoglycosides. Ann. Inter. Med. 100:352-373.

Morita, R., S. Saito, J. Ishikawa, O. Ogawa, O. Yoshida, K. Yamakawa, Y. Nakamura. 1991. Common regions of deletion on chromosomes 5q, 6q, and 10q in renal cell carcinoma. Cancer Res. 51:5817-5820.

Moriyama, M., T. Akiyama, T. Yamamoto, T. Kawamoto, T. Kato, K. Sato, T. Watanuki, T. Hikage, N. Katsuta, and S. Mori. 1991a. Expression of c-erbB-2 gene product in urinary bladder cancer. J. Urol. 145:423-427.

Moriyama, M., I. Sugawara, H. Hamada, T. Tsuruo, T. Kato, K. Sato, T. Hikage, T. Watanuki, and S. Mori. 1991b. Elevated expression of P-glycoprotein in kidney and urinary bladder cancers. Tohoku J. Exp. Med. 164:191-201.

Moriyama-Gonda, N., H. Sumi, H. Shiina, Y. Himeno, and T. Ishibe. 1991. Unstable chromosome aberrations in bladder and renal cell carcinoma. Cancer Genet. Cytogenet. 56:65-72.

Morkve, O., and J. Hostmark. 1991. Influence of tissue preparation techniques on p53 expression in bronchial and bladder carcinomas, assessed by immunofluorescence staining and flow cytometry. Cytometry 12:622-627.

Morote Robles, J., and A. Ruibal Morell. 1987. Is the prostate-specific antigen useful in the detection of prostatic cancer? [letter] [in Spanish]. Med. Clin. 89(15):663-.

Morote, J., A. Ruibal, J. Palou, J.A. de Torres, and A. Soler-Rosello. 1986. Clinical utility of prostatic specific antigen and prostatic acid phosphatase serum levels in monitoring prostate cancer. Int. J. Biol. Markers 3(1):23-28.

Morote, J., M.A. Lopez-Pacios, G. Encabo, J.A. de Torres. 1990. Urine determination of CEA, CA50 and TPA as tumor markers in bladder neoplasms [letter] [in Spanish]. Med. Clin. 95:119-.

Motwani, N.M., D. Caron, W.F. Demuan, B. Chatterjee, S. Hunter, M.D. Poulik, and A.K. Roy. 1984. Monoclonal antibodies to α_{2u}-globulin and immunocytofluorometric analysis of α_{2u}-globulin-synthesizing hepatocytes during androgenic induction and aging. J. Biol. Chem. 259:3653-3657.

Muchmore, A.V., and J.M. Decker. 1985. Uromodulin: A unique 85-kilodalton immunosuppressive glycoprotein isolated from urine of pregnant women. Science 229:479-481.

Muchmore, A.V., and J.M. Decker. 1987. Evidence that recombinant IL-1$_\alpha$ exhibits lectin-like specificity and binds to homogeneous uromodulin via N-linked oligosaccharides. J. Immunol. 138:2541-2546.

Muchmore, A.V., S. Shifrin, and J.M. Decker. 1987. In vitro evidence that carbohydrate moieties derived from uromodulin, and 85,000 dalton immunosuppressive glycoprotein isolated from human pregnancy urine, are immunosuppressive in the absence of intact protein. J. Immunol 138:2547-2553.

Mudge, G.H. 1980. Nephrotoxicity of urographic radiocontrast drugs. Kid. Internat. 18:540-562.

Muggia, F.M., H.O. Heinemann, M. Farhangi, and E.F. Osserman. 1969. Lysozymuria and renal tubular dysfunction in monocytic and myelomonocytic leukemia. Am. J. Med. 47:351-366.

Mulder, A., J. Van Hooteg, R. Sylvest, F. ten Kate, K. Kurth, et al. 1992. Prognostic factors in bladder carcinoma: histologic parameters and expression of a cell cycle-related nuclear antigen (Ki-67). J. Pathol. 166(1):37-43.

Müller, C.A., J. Markovic-Lipkovski, T. Risler, A. Bohle, and G.A. Müller. 1989. Expression of HLA-DQ, -DR and -DP antigens in normal kidney and glomerulonephritis. Kidney Int. 35:116-124.

Murakami, T., and H. Kawakami. 1990. The clinical significance of asymptomatic low molecular weight proteinuria detected on routine screening of children in Japan: A study of 53 patients. Clin. Nephrol. 33:12-19.

Muraki, J., and M. Nakazano. 1992. Establishment of a new human renal cancer cell line (TC-1) and its productivity of interleukin-6 (IL-6) [in Japanese]. Nippon Hinyokika Gakkai Zasshi (Jpn. J. Urol.) 83:1882-1889.

Murty, C.V.R., M.J. Olson, B.D. Garg, and A.K. Roy. 1988. Hydrocarbon-induced hyaline droplet nephropathy in male rats during senescence. Toxicol. Appl. Pharmacol. 96:380-392.

Mutti, A., R. Alinova, E. Bergamaschi, C. Biagini, S. Cavazzini, I. Franchini, R.R. Lauwerys, A.M. Bernard, H. Roels, E. Gelpi, J. Rosello, I. Ramis, R.G. Price, S.A. Taylor, M. DeBroe, G.D. Nuyts, H. Stolte, L.M. Fels, and C. Herbort. 1992. Nephropathies and exposure to perchloroethylene in dry-cleaners. Lancet 340:189-193.

Mutti, A. 1989. Detection of renal diseases in humans: Developing markers and methods. Toxicol. Lett. 46:177-191.

Mutti, A., S. Lucertini, P.P. Valcavi, T.M. Neri, M. Fornari, R. Alinova, and I. Franchini. 1985. Urinary excretion of brush-border antigen revealed by monoclonal antibody: Early indicator of toxic nephropathy. Lancet 2:914-917.

Mutti, A., R. Alinovi, E. Bergamaschi, M. Fornari, and I. Franchini. 1988. Monoclonal antibodies to brush-border antigens for the early diagnosis of nephrotoxicity. Arch. Toxicol. (Supplement) 12:162-165.

Mydlo, J.H., J. Michaeli, C. Cordon-Cardo, A.S. Goldenberg, W.D. Heston, and W.R. Fair. 1989. Expression of transforming growth factor alpha and epidermal growth factor receptor messenger RNA in neoplastic and nonneoplastic human kidney tissue. Cancer Res. 49:3407-3411.

Nabi, I.R., H. Watanabe, S. Silletti, and A. Raz. 1991. Tumor cell autocrine motility factor receptor. EXS. 59:163-177.

Nabi, I.R., H. Watanabe, and A. Raz. 1992. Autocrine motility factor and its receptor: Role in cell locomotion and metastasis. Cancer Metastasis Rev. 11:5-20.

Nagata, Y., M. Abe, K. Kobayashi, S. Saiki, T. Kotake, K. Yoshikawa, R. Ueda, E. Nakayama, and H. Shiku. 1990. Point mutations of *c-ras* genes in human bladder cancer and kidney cancer. Jpn. J. Cancer Res. 81:22-27.

Nagata, M. 1993. Expression on major histocompatibility complex antigen and effect of IFN-alpha on renal cell carcinoma [in Japanese]. Nippon Hinyokika Gakkai Zasshi (Jpn. J. Urol.) 84:814-821.

Nagle, R.B., M.K. Brawer, J. Kittleson, and V. Clark. 1991. Phenotypic relationships of prostatic intraepithilial neoplasia to invasive prostatic carcinoma. Am. J. Pathol. 138:119-128.

Nakagawa, Y., M. Netzer, E.K. Michaels, F. Suzuki, and H. Ito. 1994. Nephrocalcin in patients with renal cell carcinoma [see comments]. J. Urol. 152:29-34.

Nakamura, T., I. Ebihara, I. Shirato, Y. Tomino, and H. Koide. 1991. Increased steady-state levels of mRNA coding for extracellular matrix components in kidneys of NZB/W F1 mice. Am. J. Pathol. 139:437-450.

Nakamura, T., I. Ebihara, I. Nagaoka, Y. Tomino, and H. Koide. 1992. Renal platelet-derived growth factor gene expression in NZB/W F1 mice with lupus and ddY mice with IgA nephropathy. Clin. Immunol. Immunopathol. 63:173-181.

Nanus, D.M., S.A. Ebrahim, N.H. Bander, F.X. Real, L.M. Pfeffer, J.R. Shapiro, and A.P. Albino. 1989. Transformation of human kidney proximal tubule cells by ras-containing retroviruses. Implications for tumor progression. J. Exp. Med. 169:953-972.

Nanus, D.M., S.A. Lynch, P.H. Rao, S.M. Anderson, S.C. Jhanwar, A.P. and Albino, A.P. 1991. Transformation of human kidney proximal tubule cells by a src-containing retrovirus. Oncogene 6:2105-2111.

Narayan, S., and D. Roy. 1992. Enhanced expression of membrane phosphoproteins tyrosine phosphorylation in estrogen-induced kidney tumors. Biochem. Biophys. Res. Comm. 186:228-236.

Narvarte J., S.R. Saba, and G. Ramirez. 1989. Occupational exposure to organic solvents causing chronic tubulointerstitial nephritis. Arch. Intern. Med. 149:154-158.

Nathan, C., and M. Sporn. 1991. Cytokines in context. J. Cell Biol. 113:981-986.

Neal, D.E., K. Smith, J.A. Fennelly, M.K. Bennett, R.R. Hall, and A.L. Harris. 1989. Epidermal growth factor receptor in human bladder cancer: A comparison of immunohistochemistry and ligand binding. J. Urol. 141:517-521.

Neal, D.E., L. Sharples, K. Smith, J. Fennelly, R.R. Hall, and A.L. Harris. 1990. The epidermal growth factor receptor and the prognosis of bladder cancer. Cancer 65:1619-1625.

Nelson, N.A., T.G. Robins, and F.K. Port. 1990. Solvent nephrotoxicity in humans and experimental animals. Am. J. Nephrol. 10:10-20.

Nemoto, R., K. Hattori, K. Uchida, T. Shimazue, Y. Nishijima, K. Doiso, and M. Harada. 1990. S-phase fraction of human prostate adenocarcinoma studies with in vivo bromodeoxyuridine labeling. Cancer 66:509-516.

Neuhauss, O.W., and D.S. Lerseth. 1979. Dietary control of the renal reabsorption and excretion of α_{2u}-globulin. Kidney Int. 16:409-415.

Neuhauss, O., W. Flory, N. Biswas, and C.E. Hollerman. 1981. Urinary excretion of α2u-globulin by adult male rats following treatment with nephrotoxic agents. Nephron 28:133-140.

Newmark, S.R., R.G. Dluhy, and A.H. Bennett. 1973. Ectopic adenocorticotropin syndrome with prostatic carcinoma. Urology 2:666-668.

Newton, J.F., W.E. Braselton, C.-H. Kuo, W.M. Kluwe, M.W. Gemborys, G.H. Mudge, and J.B. Hook. 1982. Metabolism of acetaminophen by the isolated perfused kidney. J. Pharmacol. Exp. Ther. 221:76-79.

Nickel, J.C., L. Emerson, and J. Cornish. 1993. The bladder mucus (glycosaminoglycan) layer in interstitial cystitis. J. Urol. 149:716-718.

Nielsen, K., H. Colstrup, T. Nilsson, and H. Gundersen. 1988. Stereological estimates of nuclear volume correlated with histopathological grading and prognosis of bladder tumour. Virchows Arch. B.: Cell Pathol. 52(1):41-54.

Nielson, S., E. Nexo., and E.I. Christenson. 1989. Absorption of epidermal growth factor and insulin in rabbit renal proximal tubules. Am. J. Physiol. **256(9)**:E55-E63.

NIH (National Institutes of Health). 1990. National Kidney and Urologic Diseases Advisory Board 1990 Long-Range Plan. U.S. Department of Health and Human Services, Public Health Service, NIH Publication Number 90-583.

NIH (National Institutes of Health). 1991. U.S. Renal Data System 1991 Annual Data Report. U.S. Department of Health and Human Services, Public Health Service, NIH Publication No. 91-3176.

NIH (National Institutes of Health). 1992. U.S. Renal Data System 1992 Annual Data Report. U.S. Department of Health and Human Services, Public Health Service, NIH Publication No. 92-3176.

NIH (National Institutes of Health). 1993. U.S. Renal Data System 1993 Annual Data Report. U.S. Department of Health and Human Services, Public Health Service, NIH Publication No. 93-3176.

Noble, B., K. Ren, J. Taverne, J. DiPirro, J. Van Liew, C. Dijkstra, G. Janossy, and L.W. Poulter. Mononuclear cells in glomeruli and cytokines in urine reflect the severity of experimental proliferative immune complex glomerulonephritis. Clin. Exp. Immunol. 80:281-287.

Noh, J.W., R.C. Wiggins, and S.H. Phau. 1991. Transforming growth factor β (TGF-β) in urine predicts renal cortical scarring in rabbit anti-GBM disease. J. Am. Soc. Nephrol. 2:556.

Nonclereq, D., G. Toubeau, P. Lambrecht, J-A. Henson-Stiennon, and G. Laurent.

References

1991. Redistribution of epidermal growth factor immunoreactivity in renal tissue after nephrotoxin-induced tubular injury. Nephron 57:210-215.

Nordenson, I., B. Ljungberg, and G. Roos. 1988. Chromosomes in renal carcinoma with reference to intratumor heterogeneity. Cancer Genet. Cytogenet. 32:35-41.

Noris, M., A. Benigni, E. Gotti, P. Boccardo, S. Aiello, and G. Remuzzi. 1991. Urinary excretion of platelet activating factor (PAF) in patients with immune-mediated glomerulonephritis. Am. J. Nephrol. 2:601.

Norman, J.T. 1991. The role of angiotensin II in renal growth. Renal Physiol. Biochem. 14:175-185.

Norman, J.B., B. Badie-Dezfooly, E.P. Nord, I. Kurtz, J. Schlosser, A. Chaudhari, and L.G. Fine. 1987. EGF-induced mitogenesis in proximal tubular cells: Potentiation by angiotensin II. Am. J. Physiol. 253(2 Pt 2):F299-F309.

Norman, J.T., R.E. Bohman, G. Fischmann, et al. 1988. Patterns of mRNA expression during early cell growth differ in kidney epithelial cell destined to undergo compensatory hypertrophy versus regenerative hyperplasia. Proc. Natl. Acad. Sci. 85:6768-6772.

Norming, U., C. Nyman, and B. Tribukait. 1989. Comparative flow and cytometric deoxyribonucleic acid studies on exophytic tumor and random mucosal biopsies in untreated carcinoma of the bladder. J. Urol. 142:1442-1447.

Norming, U., B. Tribukait, H. Gustafson, C.R. Nyman, N. Wang, and H. Wijkstrom. 1992. Deoxyribonucleic acid profile and tumor progression in primary carcinoma in situ of the bladder: A study of 63 patients with grade 3 lesions. J. Urol. 147:11-15.

Nouri, A.M., M.E. Smith, D. Crosby, and R.T. Oliver. 1990. Selective and non-selective loss of immunoregulatory molecules (HLA-A,B,C antigens and LFA-3) in transitional cell carcinoma. Br. J. Cancer 62:603-606.

NRC (National Research Council). 1981. Aromatic Amines: An Assessment of the Biological and Environmental Effects. Washington, D.C.: National Academy Press. 319 pp.

NRC (National Research Council). 1983. Risk Assessment in the Federal Government: Managing the Process. Washington, D.C.: National Academy Press. 191 pp.

NRC (National Research Council). 1989a. Biologic Markers in Reproductive Toxicology. Washington, D.C.: National Academy Press. 395pp.

NRC (National Research Council). 1989b. Biologic Markers in Pulmonary Toxicology. Washington, D.C.: National Academy Press. 179pp.

NRC (National Research Council). 1991. Human Exposure Assessment for Airborne Pollutants: Advances and Opportunities. Washington, D.C.: National Academy Press. 321pp.

NRC (National Research Council). 1992. Biologic Markers in Immunotoxicology. Washington, D.C.: National Academy Press. 206 pp.

NTP (National Toxicology Program). 1989. Toxicology and Carcinogenesis Studies of

Hexachloroethane (CAS No. 67-72-1) in F344/N rats (Gavage Studies). NTP Technical Report 361. Research Triangle Park, N.C.: National Toxicology Program.

NTP (National Toxicology Program). 1990. Toxicology and Carcinogenesis Studies of D-limonene (CAS No. 5989-27-5) in F344/N rats and B6C3F1 mice (Gavage Studies). NTP Technical Report 347. Research Triangle Park, N.C.: National Toxicology Program.

Nutting, C., and J. Chowaniec. 1992. Evaluation of the actions and interactions of retinoic acid and epidermal growth factor on transformed urothelial cells in culture: Implications for the use of retinoid therapy in the treatment of bladder cancer patients. Clin. Oncol. 4:51-55.

Oberle, G.P., J. Niemeyer, F. Thaiss, W. Schoeppe, and R.A.K. Stahl. 1992. Increased oxygen radical and eicosanoid formation in immune-mediated mesangial cell injury. Kidney Int. 42:69-74.

Odum, J., and T. Green. 1984. The metabolism and nephrotoxicity of tetrafluoroethylene in the rat. Toxicol. Appl. Pharmacol. 76:306-318.

Oesterling, J.E. 1991. Prostate specific antigen: A critical assessment of the most useful tumor marker for adenocarcinoma of the prostate. J. Urol. 145:907-923.

Ogawa, O., T. Habuchi, Y. Kakehi, M. Koshiba, T. Sugiyama, O. Yoshida. 1992. Allelic losses at chromosome 17p in human renal cell carcinoma are inversely related to allelic losses at chromosome 3p. Cancer Res. 52:1881-1885.

Ohashi, T., T. Akagi, S. Irie, T. Obama, Y. Nasu, S. Tohjoh, K. Takeda, J. Yoshimoto, Y. Matsumura, and H. Ohmori. 1987. Clinical study on PAP, gamma-Sm and PA as tumor markers in prostatic cancer patients [in Japanese]. Nippon Hinyokika Gakkai Zasshi (Jpn. J. Urol.) 78:1403-1408.

Ohgaki, H., P. Kleihues, and G.C. Hard. 1991. Ki-ras mutations in spontaneous and chemically induced renal tumors of the rat. Mol. Carcinog. 4:455-459.

Ohkubo, S., K. Ogi, M. Hosoya, H. Matsumoto, N. Suzuki, C. Kimura, H. Ondo, M. Fujino. 1990. Specific expression of human endothelin-2 (ET-2) gene in a renal adenocarcinoma cell line. Molecular cloning of cDNA encoding the precursor of ET-2 and its characterization. FEBS Lett. 274:136-140.

Ohta, K., Y. Hirata, M. Schichiri, K. Kanno, T. Emori, K. Tomita, and F. Marumo. 1991. Urinary excretion of endothelin-1 in normal subjects and patients with renal disease. Kidney Int. 39:307-311.

Oka, K., J. Ishikawa, J.M. Bruner, R. Takahashi, and H. Saya. 1991. Detection of loss of heterozygosity in the p53 gene in renal cell carcinoma and bladder cancer using the polymerase chain reaction. Mol. Carcinog. 4:10-13.

Okuda, S., L. Languino, E. Ruoslahti, and W. Border. 1990. Elevated expression of transforming growth factor β and proteoglycan production in experimental glomerulonephritis. J. Clin. Invest. 86:453-462.

Oliver, J., and M. MacDowell. 1958. Cellular mechanisms of protein metabolism in the nephron, VII. The characteristics and significance of the protein absorption droplets (hyaline droplets) in epidemic hemorrhagic fever and other renal diseases. J. Exp. Med. 107:731-754.

Olsen, P.S., E. Nexo, S.S. Poulsen, H.F. Hansen, and P. Kierkegaard. 1984. Renal origin of rat urinary epidermal growth factor. Regul. Pept. 10:37-43.

Olson, M.J., J.T. Johnson, and C.A. Reidy. 1990. A comparison of male rat and human urinary proteins: Implications for human resistance to hyaline droplet nephropathy. Toxicol. Appl. Pharmacol. 102:524-536.

Olson, M.J., J.C. Parker, and C.H. Ris. 1991.

Olumi, A.F., Y.C. Tsai, P.W. Nichols, D.G. Skinner, D.R. Cain, L.I. Bender, and P.A. Jones. 1990. Allelic loss of chromosome 17p distinguishes high grade from low grade transitional cell carcinomas of the bladder. Cancer Res. 50:7081-7083.

Omichinski, J.G., G. Brunborg, E.J. Søderlund, J.E. Dahl, J.A. Bausano, J.A. Holme, S.D. Nelson, and E. Dybing. 1987. Renal necrosis and DNA damage caused by selectively deuterated and methylated analogs of 1,2-dibromo-3-chloropropane in the rat. Toxicol. Appl. Pharmacol. 91:358-370.

Onda, H., S. Ohkubo, T. Kosaka, T. Yasuhara, K. Ogi, M. Hosoya, H. Matsumoto, N. Suzuki, C. Kitada, Y. Ishibashi, et al. 1991. Expression of endothelin-2 (ET-2) gene in a human renal adenocarcinoma cell line: Purification and cDNA cloning of ET-2. J. Cardiovas. Pharmacol 17(Suppl. 7):S39-S43.

Oosterwijk, E., S.O. Warnaar, J. Zwartendijk, E.A. van der Velde, G.J. Fleuren, and C.J. Cornelisse. 1988. Relationship between DNA ploidy, antigen expression and survival in renal cell carcinoma. Int. J. Cancer 42:703-708.

Orntoft, T.F., S.E. Petersen, and H. Wolf. 1988. Dual-parameter flow cytometry of transitional cell carcinomas. Quantitation of DNA content and binding of carbohydrate ligands in cellular subpopulations. Cancer 61:963-970.

Ouelette, A.J. 1983. Messenger RNA regulation during compensatory renal growth. Kidney Int. 23:575-580.

Ouelette, A.J., and R.A. Malt. 1979. Noncoordinate regulation of cytoplasmic RNA in compensatory renal hypertrophy. Am. J. Physiol. 237:R360-R365.

Ouellette, A.J., and C.P. Ordahl. 1981. Extensive homology between poly(A)-containing mRNA and purified nominal poly(A)-lacking mRNA in mouse kidney. J. Biol. Chem. 256:5104-5108.

Ouellette, A.J., R. Moonka, A.D. Zelenetz, and R.A. Malt. 1987. Regulation of ribosome synthesis during compensatory renal hypertrophy in mice. Am. J. Physiol. 253:C506-C513.

Ouellette, A.J., R.A. Malt, V.P. Sukhatme and J.V. Bonventre. 1990. Expression of two "immediate early" genes, Ega-1 and c-fos, in response to renal ischemia and during compensatory renal hypertrophy in mice. J. Clin. Invest. 85:766-771.

Papapetrou, P.D., N.P. Sakalerou, H. Braouzi, and P.H. Fessas. 1980. Ectopic production of human chorionic gonadotropin (hCG) by neoplasms: The value of measurements of immunoreactive hCG in the urine as a screening procedure. Cancer 45:2583-2592.

Parshad, R., F.M. Price, M. Oshimura, J.C. Barrett, H. Satoh, B.E. Weissman, E.J. Stanbridge, K.K. Sanford. 1992. Complementation of a DNA repair deficiency in six human tumor cell lines by chromosome 11. Hum. Genet. 88:524-528.

Park, Y., H. Kiwamoto, T. Nishioka, H. Tsujihashi, S. Mitsubayashi, et al. 1987. Clinical evaluation of a prostate-specific antigen as a serum marker of prostatic cancer [in Japanese]. Hinyokika Kiyo (Acta Urol. Jpn.) 33(6):883-888.

Parker, J.C., M.J. Olson, and C.H. Ris. 1992. Hazard evaluation issues associated with chloroalkene-induced renal carcinogenesis in male rats. The Toxicologist 12:426.

Parry, W.L., and G.P. Hemstreet. 1988. Cancer detection by quantitative fluorescence image analysis. J. Urol. 139:270-274.

Parsons, C.L., J.D. Lilly, and P. Stein. 1991. Epithelial dysfunction in nonbacterial cystitis (interstitial cystitis). J. Urol. 145:732-735.

Partin, A.W., A.C. Walsh, R.V. Pitcock, J.L. Mohler, J.I. Epstein, and D.S. Coffey. 1989. A comparison of nuclear morphometry and Gleason grade as a predictor of prognosis in Stage A2 prostate cancer: A critical analysis. J. Urol. 142:1254-.

Partin, A.W., R.H. Getzenberg, M.J. CarMichael, D. Vindivich, J. Yoo, J.I. Epstein, and D.S. Coffey. 1993. Nuclear matrix protein patterns in human benign prostatic hyperplasia and prostate cancer. Cancer Res. 53:744-746.

Patrono, C., G. Ciabattoni, G. Remuzzi, E. Gotti, S. Bombardieri, O. di Munno, G. Tartarelli, G.A. Cinotti, B.M. Simonetti and A. Pierucci. 1985. Functional significance of renal prostacyclin and thromboxane A_2 production in patients with systemic lupus erythematosus. J. Clin. Invest. 76:1011-1018.

Pelletier, L., F. Hirsch, J. Rossert, E. Druet, and P. Druet. 1987. Experimental mercury-induced glomerulonephritis. Springer Semin. Immunopathol. 9:359-369.

Pelletier, L., R. Pasquier, J. Rossert, M.C. Viol, C. Mandet, and P. Druet. 1988. Autoreactive T cells in mercury-induced autoimmunity. J. Immunol. 140:750-755.

Perlman, E.J., and J.I. Epstein. 1990. Blood group antigen expression in dysplasia and adenocarcinoma of the prostate. Am. J. Surg. Pathol. 14:810-818.

Pertoft, H., and T.C. Laurent. 1982. Sedimentation of cells in colloidal silica (Percoll). Pp. 115-152 in Cell Separation: Methods and Selected Applications, Vol. 1. T.G. Pretlow, and T.P. Pretlow, eds. New York: Academic Press.

Pervaiz, S., and K. Brew. 1987. Homology and structure-function correlations between α_1-acid glycoprotein and serum retinol-binding protein and its relatives. FASEB J. 1:209-214.

Peter, S. 1991. Oncogenes and growth factors in renal cell carcinoma. Urol. Int. 47:199-202.

Peterson, P.A., and I. Berggard. 1971. Isolation and properties of a human retinol-transporting protein. J. Biol. Chem. 246:25-33.

Peterson, P.A., P.E. Evrin, and I. Berggard. 1969. Differentiation of glomerular, tubular, and normal proteinuria: Determinations of urinary excretion of b_2-microglobulin, albumin, and total protein. J. Clin. Invest. 48:1889-1198.

Petrides, P.E., S. Bock, J. Bovens, R. Hofmann, and G. Jakse. 1990. Modulation of pro-epidermal growth factor, pro-transforming growth factor alpha and epidermal growth factor receptor gene expression in human renal carcinomas. Cancer Res. 50:3934-3939.

Pevsner, J., R.R. Reed, P.G. Feinstein, and S.H. Snyder. 1988. Molecular cloning of odorant-binding protein: Member of a ligand carrier family. Science 241:336-339.

Phan, S.H., B.M. McGarry, and R. Wiggins. 1990. Characterization of the collagen synthesis stimulatory activity in renal cortical conditioned media and its role in renal fibrosis [abstract]. Kidney Int. 37:426.

Piana, P., G. Caetta, A. Cavallini, P. Piantino, and A. Tizzani. 1991. Urinary evaluation of tumor-associated antigens in urothelial bladder tumors. Prog. Clin. Biol. Res. 370:179-184.

Pienta, K.J., A.W. Partin, and D.S. Coffey. 1989. Cancer as a disease of DNA organization and dynamic cell structure. Cancer Res. 49:2525-2532.

Pirani, C.L., F.G. Silva, and G.B. Appel. 1983. Tubulo-interstitial disease in multiple myeloma and other nonrenal neoplasias. Pp. 287-334 in Tubulo-Interstitial Nephropathies, R.S. Cotran, B.M. Brenner, and J.H. Stein, eds. New York: Churchill Livingstone.

Pitha, J.V., G.P. Hemstreet, N.R. Asal, R.L. Petrone, B.F. Trump, and F.G. Silva. 1987. Occupational hydrocarbon exposure and renal histopathology. Toxicol. Ind. Health 3:491-506.

Plummer, D.R., S. Noorazar, D.K. Obatomi, and J.D. Haslan. 1985. Assessment of renal injury by urinary enzymes. Uremia Invest. 9:97-102.

Pogue, V.A., and H.M. Nurse. 1989. Cocaine-associated acute myoglobinuric renal failure. Am. J. Med. 86:183-186.

Pollak, V.E., and C. Arbel. 1969. The distribution of Tamm-Horsfall protein (uromucoid) in the human nephron. Nephron 6:667-672.

Pollak, M.N., C. Polychronakos, and H. Guyda. 1989. Somatostatin analogue SMS 201-995 reduces serum IGF-I levels in patients with neoplasms potentially dependent on IGF-I. Anticancer Res. 9:889-891.

Pollock, C., L. Pei-Ling, A.Z. Györy, R. Grigg, E.D.M. Gallery, R. Caterson, L. Ibels, J. Mahony, and D. Waugh. 1989. Dysmorphism of urinary red blood cells—Value in diagnosis. Kidney Int. 36:1045-1049.

Polychronakos, C., H.J. Guyda, and B.I. Posner. 1985. Increase in the type 2 insulin-like growth factor receptors in the rat kidney during compensatory growth. Biochem. Biophys. Res. Comm. 132:418-423.

Polychrenakos, C., U. Janthly, J.G. Lehoux, and M. Koutsilieris. 1991. Mitogenic effects of insulin and insulin-like growth factors on PA-III rat prostate adenocarcinoma cells: Characterization of the receptors involved. Prostate 19:313-321.

Poortmans, J., and R.W. Jeanloz. 1976. Urinary excretion of total protein, albumin, and β_2-microglobulin during exercise in adolescent diabetics. Biomedicine 25:273-274.

Popper, H., G. Wirnsberger, H. Hoefler, and H. Denk. 1987. Immunohistochemical and histochemical markers of primary lung cancer, lung metastases, and pleural mesotheliomas. Cancer Detect. Prev. 10:167-174.

Porter, G.A. 1989. Risk factors for toxic nephropathies. Toxicol. Lett. 46:269-279.

Portila, D., L.J. Mandel, D. Bar-Sagi, and D.S. Millington. 1992. Anoxia induces phospholipase A_2 activation in rabbit renal proximal tubules. Am. J. Physiol. 262:F354-F360.

Pratt, C.I., C. Kao, S.-Q. Wu, K.W. Gilchrist, R. Oyasu, and C.A. Reznikoff. 1992. Neoplastic progression by EJ/*ras* at different steps of transformation in vitro of human uroepithelial cells. Cancer Res. 52:688-695.

Presti, J.C., P.H. Rao, Q. Chen, V.E. Reuter, F.P. Li, W.R. Fair, and S.C. Jhanwar. 1991. Histopathological, cytogenetic, and molecular characterization of renal cortical tumors. Cancer Res. 51:1544-1552.

Presti, J.C., V.E. Reuter, C. Cordon-Cardo, M. Mazumdar, W.R. Fair, and S.C. Jhanwar. 1993. Allelic deletions in renal tumors: Histopathological correlations. Cancer Res. 53:5780-5783.

Pretl, K. 1944. Zur frage der endokrinie der menschlichen vorsteherdruse. Virchows Arch. A: 312:392-404.

Prewitt, T.E., A.J. D'Ercole, B.R. Switzer, and J.J. Van Wyk. 1982. Relationship of serum immunoreactive somatomedin-C to dietary protein and energy in growing rats. J. Nutr. 112:144-150.

Price, R.G. 1982. Urinary enzymes, nephrotoxicity and renal disease. Toxicology 23:99-134.

Proctor, A., L. Coombs, J. Cairns, and M. Knowles. 1991. Amplification at chromosome llql3 in transitional cell tumours of the bladder. Oncogene 6(5):789-795.

Pruchno, C.J., M.W. Burns, M. Schulze, R.J. Johnson, P.J. Baker, and W.G. Couser. 1989. Urinary excretion of C5b-9 reflects disease activity in passive Heymann nephritis. Kidney Int. 36:65-71.

Purnell, D.M., B.M. Heatfield, and B.F. Trump. 1984. Immunocytochemical evaluation of human prostatic carcinomas for carcinoembryonic antigen, nonspecific cross-reacting antigen, β-chorionic gonadotrophin, and prostate-specific antigen. Cancer Res. 44:285-292.

Quamme, G.A., and J.H. Dirks. 1986. Micropuncture techniques. Kidney Int. 30:152-165.

Racusen, L.C., B.A. Fivush, T-L. Li, I. Slatnik, and K. Solez. 1991. Dissociation of

References

tubular cell detachment and tubular cell death in clinical and experimental "acute tubular necrosis." Lab. Invest. 64:546-556.

Rahman, M.A., C.N. Lui, M.J. Dunn and S.N. Emancipator. 1988a. Complement and leukocyte independent proteinuria and eicosanoid synthesis in rat membranous nephropathy. Lab. Invest. 59:477-483.

Rahman, M., M. Nakazana, S.N. Emancipator, and M.J. Dunn. 1988b. Increased leukotriene B_4 synthesis in immune injured rat glomeruli. J. Clin. Invest. 81:1945-1952.

Rainwater, L.M., Y. Hosaka, G.M. Farrow, and M.M. Lieber. 1987. Well differentiated clear cell renal carcinoma: Significance of nuclear deoxyribonucleic acid patterns studied by flow cytometry. J. Urol. 137:15-20.

Rainwater, L.M., H. Zincke, G.M. Farrow, and N.J. Gonchoroff. 1991. Renal cell carcinoma in young and old patients. Comparison of prognostic pathologic variables (cell type, tumor grade and stage, and DNA ploidy pattern) and their impact on disease outcome. Urology 38:1-5.

Rall, L.B., J. Scott, G.I. Bell, R.J. Crawford. 1985. Mouse prepro-epidermal growth factor synthesis by the kidney and other tissues. Nature 313:228-231.

Rao, J.Y., R.E. Hurst, W.D. Bales, P.L. Jones, R.A. Bass, L.T. Archer, P.B. Bell, and G.P. Hemstreet. 1990. Cellular f-actin levels as a marker for cellular transformation: Relationship to cell division and differentiation. Cancer Res. 50:2215-2220.

Rao, J.Y., G.P. Hemstreet, R.E. Hurst, R.B. Bonner, K.W. Min, and P.L. Jones. 1991. Cellular F-actin levels as a marker for cellular transformation: Correlation with bladder cancer risk. Cancer Res. 51:2762-2767.

Rao, J.Y., G. Hemstreet, R.E. Hurst, R.B. Bonner, P.L. Jones, K.W. Min, Y.A. Fradet. 1993. A strategy for individual bladder cancer risk assessment based upon progressive alterations in phenotypic biochemical markers in bladder epithelium during tumorigenesis. Proc. Natl. Acad. Sci. U.S.A. 90:8287-8291.

Raska, I., K. Koberna, M. Jarnik, V. Petrasovicova, J. Bednar, I. Raska, and R. Bravo. 1989. Ultrastructural immunolocalization of cyclin/PCNA in synchronized 3T3 cells. Exp. Cell Res. 184:81-89.

Rasmussen, H.H., and L.S. Ibels. 1982. Acute renal failure: Multivariate analysis of causes and risk factors. Am. J. Med. 73:211-218.

Ravery, V., J. Jouanneau, S. Gil Diez, C. Abbou, J. Caruelle, et al. 1992. Immunohistochemical detection of acidic fibroblast growth factor in bladder transitional cell carcinoma. Urol. Res. 20:211-214.

Ravnskov, U. 1986. Influence of hydrocarbon exposure on the course of glomerulonephritis. Nephron. 42:156-160.

Reddy, M.V., R.C. Gupta, E. Randerath, and K. Randerath. 1984. ^{32}P-Postlabeling test for covalent DNA binding of chemicals *in vivo*: Application to a variety of aromatic carcinogens and methylating agents. Carcinogenesis 5:31-243.

Rehn, L. 1895. Blasigeschwülste bei Fuchsin arbeitern. Arch. Klin. Chir. 50:588-600.

Reichert, D., S. Schultz, and M. Metzler. 1985. Excretion pattern and metabolism of hexachlorobutadiene in rats: Evidence for metabolic activation by conjugation reactions. Biochem. Pharmacol. 34:499-505.

Reiter, R.E., P. Anglard, S. Liu, J.R. Gnarra, and W.M. Linehan. 1993. Chromosome 17p deletions and p53 mutations in renal cell carcinoma. Cancer Res. 53:3092-3097.

Rennie, I.D.B., and H. Keen. 1967. Evaluation of clinical methods for detecting proteinuria. Lancet 2:489-492.

Rew, D.A., D.J. Thomas, M. Coptcoat, and G.D. Wilson. 1991. Measurement of in vivo urological tumour cell kinetics using multiparameter flow cytometry. Preliminary study. Br. J. Urol. 68:44-48.

Reznikoff, C.A., M.D. Johnson, D.H. Norback, and G.T. Bryan. 1983. Growth and characterization of normal human urothelium in vitro. In Vitro Cell. Dev. Biol. 19:326-343.

Reznikoff, C.A., L.J. Loretz, M.D. Johnson, and S. Swaminathan. 1986. Quantitative assessments of the cytotoxicity of bladder carcinogens towards cultured normal human uroepithelial cells. Carcinogenesis 7:1625-1632.

Reznikoff, C.A., L.J. Loretz, B.J. Christian, S.-Q. Wu, and L.F. Meisner. 1988. Neoplastic transformation of SV40-immortalized human urinary tract epithelial cells by in vitro exposure to 3-methylcholanthrene. Carcinogenesis 9:1427-1436.

Ridder, G.M., E.C. Von Bargen, C.L. Alden, and R.D. Parker. 1990. Increased hyaline droplet formation in male rats exposed to decalin is dependent on the presence of α_{2u}-globulin. Fundam. Appl. Toxicol. 15:732-743.

Rifkin, D.B., and D. Moscatelli. 1989. Recent developments in the cell biology of basic fibroblast growth factor. J. Cell Biol. 109:1-6.

Rigatti, P., F. Montorsi, G. Guazzoni, G. Viale, G. Bulfamante, and G. Coggi. 1990. Adult nephroblastoma induced erythrocytosis. Report of a case and review of the literature. Scand. J. Urol. Nephrol. 24:159-161.

Rindler, J.J., S.S. Naik, N. Li, T.C. Hoops, and M.N. Peraldi. 1990. Uromodulin (Tamm-Horsfall glycoprotein/uromucoid) is a phosphatidylinositol-linked membrane protein. J. Biol. Chem. 265:20784-20789.

Ring, K.S., F. Karp, and M.C. Benson. 1990. Enhanced detection of bladder cancer using the epithelial surface marker epithelial membrane antigen: A preliminary report. J. Occup. Med. 32:904-909.

Robson, C.J., B.M. Churchill, and W. Anderson. 1969. The results of radical nephrectomy for renal cell carcinoma. J. Urol. 101:297-301.

Rochlitz, C.F., H. Lobeck, S. Peter, J. Reuter, B. Mohr, E. de Kant, D. Huhn, and R.

Herrmann. 1992. Multiple drug resistance gene expression in human renal cell cancer is associated with the histologic subtype. Cancer 69:2993-2998.

Rogers, S.A., S.B. Miller, and M.R. Hammerman. 1991. Insulin-like growth factor I gene expression in isolated rat renal collecting duct is stimulated by epidermal growth factor. J. Clin. Invest. 87:347-351.

Rojas-Corona, R.R., L. Chen, and P.S. Mahadevia. 1987. Prostatic carcinoma with endocrine features: A report of a neoplasm containing multiple immunoreactive hormonal substances. Am. J. Clin. Path. 88:759-762.

Rosa, R.M., and R.S. Brown. 1983. Acute renal failure associated with heavy metals and organic solvents. Pp. 321-332 in Acute Renal Failure, B.M. Brenner, and J.M. Lazarus, eds. Saunders.

Rosalki, S.B., and J.H. Wilkinson. 1959. Urinary lactic dehydrogenase in renal disease. Lancet 2:327-328.

Rosenberg, M.E., and M.S. Paller. 1991. Differential gene expression in the recovery from ischemic renal injury. Kidney Int. 39:1156-1161.

Rostand, S.G. 1992. U.S. minority groups and end-stage renal disease: A disproportionate share. Am. J. Kidney Dis. 19:411-413.

Rostand, S.G., K.A. Kirk, E.A. Rutsky, and B.A. Pate. 1982. Racial differences in the incidence of treatment for end-stage renal disease. N. Engl. J. Med. 306:1276.

Rotter, M., T. Block, R. Busch, S. Thanner, and H. Hofler. 1992. Expression of HER-2/neu in renal-cell carcinoma. Correlation with histologic subtypes and differentiation. Int. J. Cancer 52:213-217.

Roy, A.K., and O.W. Neuhaus. 1967. Androgenic control of a sex-dependent protein in the rat. Nature 214:618-620.

Roy, A.K., and B. Chatterjee. 1983. Sexual dimorphism in the liver. Ann. Rev. Physiol. 45:37-50.

Roy, A.K., T.S. Nath, N.M. Motwani, and B. Chatterjee. 1983. Age-dependent regulation of the polymorphic forms of α2u-globulin. J. Biol. Chem. 258:10123-10127.

Royer-Pokora, B., B. Fleischer, S. Ragg, U. Loos, and D. Williams. 1989. Molecular cloning of the translocation breakpoint in T-ALL 11;14 (p13;q11): Genomic map of TCR alpha and delta region on chromosome 14q11 and long-range map of region 11p13. Hum. Genet. 82:264-270.

Rozengurt, E. 1980. Stimulation of Na influx, Na-K pump activity and DNA synthesis in quiescent cultured cells. Adv. Enzyme Regul. 19:61-85.

Rozengurt, E., and L.A. Heppel. 1975. Serum rapidly stimulates ouabain-sensitive 86Rb+ influx in quiescent 3T3 cells. Proc. Natl. Acad. Sci. U.S.A. 72:4492-4495.

Rozengurt, E., T.D. Gelehrter, A. Legg, and P. Pettican. 1981. Mellitin stimulates Na entry, Na-K pump activity and DNA synthesis in quiescent cultures of mouse cells. Cell 23:781-788.

Rubber, H., W. Lutzeyer, and D.M.A. Wallace. 1985. The epidemiology and aetiology of bladder cancer. Pp. 1-21 in Bladder Cancer. E.J. Zingg, and D.M.A. Wallace, eds. New York: Springer-Verlag.

Rubenstein, M., P.D. Guinan, C.F. McKiel, and A. Dubin. 1988. Review of acid phosphatase in the diagnosis and prognosis of prostatic cancers. Clin. Phys. Biochem. 6:241-252.

Ruder, A.M., L.J. Fine, and J.D. Sundin. 1990. National estimates of occupational exposure to animal bladder tumorigens. J. Occup. Med. 32:797-805.

Rush, G.F., and J.B. Hook. 1986. The kidney as a target organ for toxicity. Pp. 1-18 in Target Organ Toxicity, Vol. 2, G.M. Cohen, ed. Boca Raton, Fla: CRC Press.

Rush, G.F., J.H. Smith, J.F. Newton, and J.B. Hook. 1984. Chemically induced nephrotoxicity: Role of metabolic activation. CRC Crit. Rev. Toxicol. 13:99-160.

Rutecki, G.J., C. Goldsmith, and G.E. Schreiner. 1971. Characterization of proteins in urinary casts: Flourescent-antibody identification of Tamm-Horsfall mucoprotein in matrix and serum protein in granules. N. Engl. J. Med. 284:1049-1052.

Sabatier, L., F. Hoffschir, W.A. al Achkar, C. Turleau, J. de Grouchy, and B. Dutrillaux. 1989. The decrease of catalase or esterase D activity in patients with microdeletions of 11p or 13q does not increase their radiosensitivity. Ann. Genet. 32:144-148.

Sacks, S.A., D.B. Rhodes, D.R. Malkasian, and A.A. Rosenbloom. 1975. Prostatic carcinoma producing syndrome of inappropriate secretion of antidiuretic hormone. Urology 6:489-492.

Sade, M.V., and E.R. Barrack. 1991. Determination of growth factor in advanced prostate cancer by Ki-67 immunostaining and its relationship to the time to tumor progression after hormonal therapy. Cancer 67:3065-3071.

Safirstein, R. 1994. Gene expression in nephrotoxic and ischemic acute renal failure. J. Am. Soc. Nephrol. 4:1387-1395.

Safirstein, R., A.Z. Zelent, and P.M. Price. 1989. Reduced renal prepro-epidermal growth factor mRNA and decreased EGF excretion in ARF. Kidney Int. 36:810-815.

Safirstein, R., P.M. Price, S.J. Saggi, and R.C. Harris. 1990. Changes in gene expression after temporary renal ischemia. Kidney Int. 37:1515-1521.

Safirstein, R., J. Megyesi, S.J. Saggi, P.M. Price, M. Poon, B. Rollins, and M.B. Taubman. 1991. Expression of cytokine-like genes JE and KC is increased during renal ischemia. Am. J. Physiol. 261:F1095-F1101.

Saito, S. 1992. Detection of H-ras gene point mutations in transitional cell carcinoma of human urinary bladder using polymerase chain reaction. Keio J. Med. 41:80-86.

Salido, E.C., L. Barajas, J. Lechago, N.P. Laborde, and D.A. Fisher. 1986. Immunocytochemical localization of epidermal growth factor in mouse kidney. J. Histochem. Cytochem. 34:1155-1160.

Salido, E.C., P.H. Yen, L.J. Shapiro, D.A. Fischer, and L. Barajas. 1989. In situ hybridization of prepro-epidermal growth factor mRNA in the mouse kidney. Am. J. Physiol. 256(4 Pt 2):F632-F638.

Sanchez-Fernandez de Sevilla, M., L. Morell-Quadreny, M. Gil-Salom, M.

References

Perez-Bacete, and A. Llombart-Bosch. 1992. Morphometric and immunohistochemical characterization of bladder carcinoma in situ and its preneoplastic lesions. Eur. Urol. 21(Suppl 1):5-9.

Sanders, H., P. McCue, and S.D. Graham. 1991. ABO(H) antigens and beta-2 microglobulin in transitional cell carcinoma. Predictors of response to intravesical bacillus Calmette-Guerin. Cancer 67:3024-3028.

Sandgren, E.P., N.C. Luetteke, R.D. Palmiter, R.L. Brinster, and D.C. Lee. 1990. Overexpression of TGFα in transgenic mice: Induction of epithelial hyperplasia, pancreatic metaplasia, and carcinoma of the breast. Cell 61:1121-1135.

Sandler, D.P. 1987. Epidemiology in the assessment of nephrotoxicity. Pp. 847-883 in Nephrotoxicity in the Experimental and Clinical Situation, P.H. Bach, and E.A. Lock, eds. Martinus Nijhoff Publ. Dordricht, Boston, Lancaster.

Sandler, D.P., and J.C. Smith. 1991. Chronic renal disease risk associated with employment in industries with potential solvent exposure. Pp. 261-266 in Nephrotoxicity, Mechanisms, Early Diagnosis, Therapeutic Management, P.H. Bach, N.J. Gregg, M.F. Wiks, and L. Delacrluz, eds. New York, Basel, Hong Kong: Marcel Dekker Inc.

Sarkar, F., W. Sakr, S. Drozdowicz, P. Sreepathi, and J. Crissman. 1992. Measurement of cellular proliferation in human prostate by AgNOR, PCNA and SPF. Mod. Pathol. 5:59A.

Sassine, A.M., and C. Schulman. 1992. Value of PSA in the staging of prostatic cancer [in French]. Acta Urol. Belg. 60:49-59.

Sato, T., Y. Sakata, and K. Tamura. 1990. Inflammatory reaction and laboratory tests: IAP (immunosuppressive acidic protein) [review] [in Japanese]. Rinsho Byori 38:255-259.

Sausen, P.J., and A.A. Elfarra. 1990. Cysteine conjugate S-oxidase: Characterization of a novel enzymatic activity in rat hepatic and renal microsomes. J. Biol. Chem. 265:6139-6145.

Sawczuk, I.H., G. Hoke, C.A. Olsson, J. Connor, and R. Buttyan. 1989. Gene expression in response to acute unilateral ureteral obstruction. Kidney Int. 35:1315-1319.

Sawczuk, I.H., C.A. Olsson, G. Hoke, and R. Buttyan. 1990. Immediate induction of c-fos and c-myc transcripts following unilateral nephrectomy. Nephron. 55:193.

Scandinavian Committee on Enzymes, Scandinavian Prostatic Cancer Group. 1985. Acid phosphatases and other tumor markers in the management of prostatic cancer. Workshop. Copenhagen, March 1985. Scand. J. Clin. Lab. Invest. Suppl. 179:1-117.

Schaafsma, H.E., F.C. Ramaekers, G.N. van Muijen, E.B. Lane, I.M. Leigh, H. Robben, A. Huijsmans, E.C. Ooms, and D.J. Ruiter. 1990. Distribution of cytokeratin polypeptides in human transitional cell carcinomas, with special emphasis on changing expression patterns during tumor progression. Am. J. Pathol. 136:329-343.

Schaeffer, V.H., and J.L. Stevens. 1987. Mechanism of transport for toxic cysteine conjugates in rat kidney cortex membrane vesicles. Mol. Pharmacol. 32:293-298.

Schardijn, G.H.C., and L.W. Statius van Eps. 1987. β$_2$-microglobulin: Its significance in the evaluation of renal function. Kidney Int. 32:635-641.

Schenk, E.A., R.H. Schwartz, and R.A. Lewis. 1971. Tamm-Horsfall mucoprotein. I. Localization in the kidney. Lab. Invest. 25:92-95.

Schentag et al., 1978.

Schiebler, M., B. Yankaskas, C. Tempany, P. Holtz, and E. Zerhouni. 1992. Efficacy of prostate-specific antigen and magnetic resonance imaging in staging stage C adenocarcinoma of the prostate. Invest. Radiol. 27:575-577.

Schimamuria, T., and A.B. Morris. 1975. A progressive glomerulosclerosis occurring in partial five-sixths nephrectomy. Am. J. Pathol. 79:95.

Schlondorff, D. 1986. Isolation and use of specific nephron segments and their cells in biochemical studies. Kidney Int. 30:201-207.

Schmid, H., A. Mall, and H. Backhorn. 1986. Catalytic activity of alkaline phosphatase and N-acetyl-beta-D-glucosaminidase in human cortical nephron segments: Heterogenous changes in acute renal failure and acute rejection following kidney allotransplantation. J. Clin. Chem. Clin. Biochem. 24:961-970.

Schoel, B., and G. Pfleiderer. 1987. The amount of Tamm-Horsfall protein in the human kidney, related to its daily excretion. J. Clin. Chem. Clin. Biochem. 25:681-682.

Schoenig, G.P., and R.L. Anderson. 1985. The effects of high dietary levels of sodium saccharin on mineral and water balance and related parameters in rats. Fd. Chem. Toxic 23:465-474.

Scholer, D.W., and I.S. Edelman. 1979. Isolation of rat kidney cortical tubules enriched in proximal and distal segments. Am. J. Physiol. 237:F350-F359.

Schreiner, G.F., R.S. Cohan, and E.R. Unanue. 1984. Modulation of Ia and leukocyte common antigen expression in rat glomeruli during the course of glomerulonephritis and aminonucleoside nephrosis. Lab. Invest. 51:524-531.

Schroder, S., B.C. Padberg, E. Achilles, K. Holl, H. Dralle, and G. Kloppel. 1992. Immunocytochemistry in adrenocortical tumours: A clinicomorphological study of 72 neoplasms. Virchows Arch. A: Pathol. Anat. Histopathol. 420:65-70.

Schuldiner, S., and E. Rozengurt. 1982. Na+/H+ antiport in Swiss 3T3 cells: Mitogenic stimulation leads to cytoplasmic alkalinization. Proc. Natl. Acad. Sci. U.S.A. 79:7778-7782.

Schulte, P.A. 1987. Methodologic issues in the use of biologic markers in epidemiologic research. Arch. Environ. Health 43:83-89.

Schulte, P.A., and W.E. Kaye. 1988. NIOSH exposure registries [review]. Arch. Environ. Health 43:155-161.

Schulte, et al. 1987. Occupational cancer of the urinary tract [review]. Occupational Medicine: State of the Art Reviews 2:85-107.

References

Schulte, P.A., W.E. Halperin, E.M. Ward, and A.M. Ruder. 1990. Bladder cancer screening in high-risk groups. J. Occup. Med. 32:787-945.

Schultz, M.O., P. Klubes, V. Perman, W.S. Mizuno, F.W. Bates, and J.H. Sautter. 1959. Blood dyscrasia in calves induced by S-(dichlorovinyl)-L-cysteine. Blood 14:1015-1025.

Schulze, M., J.V. Donadio, C.J. Pruchno, P.J. Baker, R.J. Johnson, R.A.K. Stahl, S. Watkins, D.C. Martin, R. Wurzner, O. Gotze, and W.G. Couser. 1991. Elevated urinary excretion of the C5b-9 complex in membranous nephropathy. Kidney Int. 40:533-538.

Schumann, G. 1986. Cytodiagnostic Urinalysis for the Nephrology Practice. Semin. Nephrol. 6:308-345.

Scott, J., S. Patterson, L. Rall, G.I. Bell, R. Crawford, J. Penschow, H. Niall, and J. Coghlan. 1985. The structure and biosynthesis of epidermal growth factor precursor. J. Cell Sci. (suppl.) 3:19-28.

Scott, R.J., P.A. Hall, J.S. Haldane, S. van Noorden, Y. Price, D.P. Lane, and N.A. Wright. 1991. A comparison of immunohistochemical markers of cell proliferation with experimentally determined growth fraction. J. Pathol. 165:173-178.

Scrivner, D.L.; Meyer, J.S.; Rujanavech, N.; Fathman, A.; Scully, T. Cell kinetics by bromodeoxyuridine labeling and deoxyribonucleic acid ploidy in prostatic carcinoma needle biopsies. J. Urol. 146:1034-1039; 1991.

See, W.A. 1992. Intravesical recombinant tissue plasminogen activator for the prevention of implantation-mediated bladder tumor recurrence. Clin. Exp. Metastasis 10:99-109.

Selli, C., W.M. Hinshaw, B.H. Woodard, and D.F. Paulson. 1983. Stratification of risk factors in renal cell carcinomas. Cancer 52:899-903.

Sellwood, R.A., J. Spencer, J.G. Azzopardi, S. Wapnick, R.B. Welbourn, and A.E. Kulatilake. 1969. Inappropriate secretion of antidiuretic hormone by carcinoma of the prostate. Br. J. Surg. 56:933-935.

Serio, G. 1991. c-erbB-2 gene product-like expression in urothelial carcinomas of the human bladder. Its value as a prognostic indicator in superficial tumors. Pathologica 83:413-420.

Sesterhenn, I.A., R.L. Becker, F.A. Avallolne, F.K. Mostofi, T.H. Lin, Davis. 1991. Image analysis of nucleoli and nucleolar organizer regions in prostatic hyperplasia, PIN and prostatic carcinoma. J. Urogenital Pathol.1:42-51.

Shah, V.M., J. Newman, J. Crocker, G.N. Antonakopoulos, C.K. Chapple, and M.J. Collard. 1987. Production of beta-human chorionic gonadotropin by prostatic adenocarcinoma and transitional cell carcinoma of the upper urinary tract. Br. J. Exp. Pathol. 68:871-878.

Shah, I. A., M.-O. Schlageter, P. Stinnett, and J. Lechago. 1991. Cytokeratin immunohistochemistry as a diagnostic tool for distinguishing malignant from benign epithelial lesions of the prostate. Mod. Pathol. 4:220-224.

Shalitin, C., R. Epelbaum, C. Valansi, R. Segal, T. Mekori, et al. 1991. A novel 21-kDa protein as a serum marker for benign and malignant urogenital tumors: Preliminary communication. Int. J. Cancer 49:861-866.

Shapiro, L.E., and J. Sachchidananda. 1982. Regulation of proteins by thyroid hormone and glucocorticoid: The responses of hepatic α_{2u}-globulin and pituitary growth hormone differ in adult male hypothyroid rats. Endocrinology 111:653-660.

Shaw, A.B., P. Risdon, and J.D. Lewis-Jackson. 1983. Protein creatinine index and Albustix in assessment of proteinuria. Br. Med. J. 287:929-932.

Sherif, M., A.S. Ibrahim, and A.A. El-Aaser. 1980. Prostatic Carcinoma in Egypt: Epidemiology and etiology. Scand. J. Urol Nephrol., Suppl. 55:25-26.

Sherwood, E.R., L.A. Berg, N.J. Mitchell, J.E. McNeal, J.M. Kozlowski, and C. Lee. 1990. Differential cytokeratin expression in normal, hyperplastic and malignant epithelial cells from human prostate. J. Urol. 143:167-171.

Sherwood, E.R., G. Theyer, G. Steiner, L.A. Berg, J.M. Kozlowski, and C. Lee. 1991. Differential expression of specific cytokeratin polypeptides in the basal and luminal epithelia of the human prostate. Prostate 18:303-314.

Sherwood, E., C. Fong, C. Lee, and J. Kozlowski. 1992. Basic fibroblast growth factor: A potential mediator of stromal growth in the human prostate. Endocrinology 130:2955-2963.

Shibata, M.-A., A. Hagiwara, S. Tamano, S. Ono, and S. Fukushima. 1989. Lack of a modifying effect by the diuretic drug furosemide on the development of neoplastic lesions in rat two-stage urinary bladder carcinogenesis. J. Tox. Environ. Health 26:255-265.

Shinmi, O., K. Yorimitsu, K. Moroi, M. Nishiyama, Y. Sugita, T. Saito, Y. Inagaki, T. Masaki, and S. Kimura. 1993. Endothelin-2-converting enzyme from human renal adenocarcinoma cells is a phosphoramidon-sensitive, membrane-bound metalloprotease. J. Cardiovas. Pharmacol. 22(Suppl. 8):S61-S64.

Shinoda, I., M. Kuriyama, T. Takeuchi, Y. Takahashi, Y. Ban, and Y. Kawada. 1988. Clinical studies on tumor markers in prostate cancer; the evaluation of PA (prostate specific antigen) and comparison with PAP and gamma-Sm [in Japanese]. Nippon Hinyokika Gakkai Zasshi (Jpn. J. Urol.) 79:635-642.

Short, B.G., V.L. Burnett, M.G. Cox, J.S. Bus, and J.A. Swenberg. 1987. Site-specific renal cytotoxicity and cell proliferation in male rats exposed to petroleum hydrocarbons. Lab. Invest. 57:564-577.

Short, B.G., V.L. Burnett, and J.A. Swenberg. 1989. Elevated proliferation of proximal tubule cells and localization of accumulated $\alpha 2u$-globulin in F344 rats during chronic exposure to unleaded gasoline or 2,2,4-trimethylpentane. Toxicol. Appl. Pharmacol. 101:414-431.

Shoskes, D., N.A. Parfrey, and P.F. Halloran. 1990. Increased major histocompatibility complex antigen expression in unilateral ischemic acute tubular necrosis in the mouse. Transplantation 49:201-207.

References

Shulkes, A., D.R. Fletcher, C. Rubinstein, P.R. Ebeling, T.J. Martin. 1991. Production of calcitonin gene related peptide, calcitonin and PTH-related protein by a prostatic adenocarcinoma.. Clin. Endocrinol. 34:387-393.

Shull, M.M., I. Ormsby, A.B. Kier, S. Pawlowski, R.J. Diebold, M. Yin, R. Allen, C. Sidman, G. Proetzel, D. Calvin, et al. 1992. Targeted disruption of the mouse transforming growth factor-beta 1 gene results in multifocal inflammatory disease. Nature 359:693-699.

Sidransky, D., A. Von Eschenbach, Y.C. Tsai, P. Jones, I. Summerhayes, F. Marshall, M. Paul, P. Green, S.R. Hamilton, P. Frost, and B. Vogelstein. 1991. Identification of p53 gene mutations in bladder cancers and urine samples. Science 252:706-709.

Sigler, M.H. 1975. The mechanism of the natriuresis of fasting. J. Clin. Invest. 55:377.

Sikri, K.L., C.L. Foster, N. MacHugh, and R.D. Marshall. 1981. Localization of Tamm-Horsfall glycoprotein in the human kidney using immunoflourescence and immuno-electron microscopical techniques. J. Anat. 132:597-605.

Silverman, D.T., L.I. Levin, R.N. Hoover, and P. Hartge. 1989a. Occupational Risks of bladder cancer in the United States. I. White men. J. Nat. Cancer Inst. 81:1472-1480.

Silverman, D.T., L.I. Levin, and R.N. Hoover. 1989b. Occupational risks of bladder cancer in the United States. II. Nonwhite men. J. Nat. Cancer Inst. 81:1480-1483.

Silverman, D.T., L.I. Levin, and R.N. Hoover. 1990. Occupational risks of bladder cancer among white women in the United States. Am. J. Epidemiol. 132:453-461.

Silverman, D.T., P. Harge, A. Morrison, and S. Devesa. 1992. Epidemiology of bladder cancer. Hematol. Oncol. Clin. NorthAm. 6:1-3.

Singhal, P.C., R.B. Rubin, A. Peters, H. Santiago, and J. Neugarten. 1990. Rhabdomyolysis and acute renal failure associated with cocaine abuse. J. Toxicol. Clin. Toxicol. 28:321-330.

Sippel, A.E., P. Feigelson, and A.K. Roy. 1975. Hormonal regulation of the hepatic messenger RNA levels for α_{2u} globulin. Biochemistry 14:825-829.

Skinner, D.G., R.B. Colvin, C.D. Vermillion, R.C. Pfister, and W.F. Leadbetter. 1971. Diagnosis and management of renal cell carcinoma. A clinical and pathologic study of 309 cases. Cancer 28:1165-1177.

Smart, C.R. 1990. Bladder cancer survival statistics. J. Occup. Med. 32:926-928.

Smith, J.H. 1986. Role of renal metabolism in chloroform nephrotoxicity. Comments Toxicol. 1:124-144.

Smith, J.B., and E. Rozengurt. 1978. Serum stimulates the Na+, K+ pump in quiescent fibroblasts by increasing Na+ entry. Proc. Natl. Acad. Sci. U.S.A. 75:5560-5564.

Smith, M.A., D. Acosta, and J.V. Bruckner. 1986. Development of a primary culture system of rat kidney cortical cells to evaluate the nephrotoxicity of xenobiotics. Food Chem. Toxicol. 24:551-556.

Smith, K., J.A. Fennelly, D.E. Neal, R.R. Hall, and A.L. Harris. 1989. Characterization and quantitation of the epidermal growth factor receptor in invasive and superficial bladder tumors. Cancer Res. 49:5810-5815.

Snyder, S.H., P.B. Sklar, and J. Pevsner. 1988. Molecular mechanisms of olfaction. J. Biol. Chem. 263:13971-13974.

Solet, D., and T.G. Robins. 1991. Renal function in dry-cleaning workers exposed to perchloroethylene. Am. J. Ind. Med. 20:601-614.

Soloway, M.S., and S.W. Hardeman. 1990. Animal models in bladder cancer research. Pp 567-575 in Scientific Coundation of Urology, 3rd Ed. G.D. Chisholm, and W.R. Fair, eds. Oxford: Hineman Medical Books.

Spark, R.F. 1975. Renain, aldosterone and glucogon in the natriuresis of fasting. New Engl. J. Med. 292:1335.

Spitz, M.R., R.D. Currier, J.J. Fueger, R.J. Babaian, and G.R. Newell. 1991. Familial patterns of prostate cancer: A case-control analysis. J. Urol. 146:1305-1307.

Spitzer, A., ed. 1982. The Kidney During Development: Morphology and Function. Masson.

Sporn, M. Retinoids, TGF-beta and prostate carcinogenesis. Society for basic urological research, annual meeting, Washington DC 1992.

Spry, L.A., T.V. Zenser, and B.B. Davis. 1986. Bioactivation of xenobiotics by prostaglandin H synthase in the kidney: Implications for therapy. Comments Toxicol. 1:109-123.

Spurney, R.F., P. Ruiz, D.S. Pisetsky, and T.M. Coffmann. 1991. Enhanced renal leukotriene production in murine lupus: Role of lipoxygenase metabolites. Kidney Int. 39:95-102.

Staessen, J., W.B. Yeoman, A.E. Fletcher, H.L.J. Markowe, M.G. Marmot, G. Rose, A. Semmence, M.J. Shipley, and C.J. Bulpitt. 1990. Blood lead concentration, renal function and blood pressure in London civil servants. Br. J. Ind. Med. 47:442-447.

Staessen, J.A., R.R. Lauwerys, J.P. Buchet, C.J. Bulpitt, D. Rondia, Y. Vanrenterghem, and A. Amery. 1992. Cadmibel study group: Impairment of renal function with increasing blood lead concentrations in the general population. N. Eng. J. Med. 237:151-156.

Staffa, J.A., and M.A. Mehlman. 1980. J. Environ. Path. Toxicol. 3:1-.

Stahl, R.A.K., F. Thaiss, S. Kahf, W. Schoeppe, and U.M. Helmchen. 1990. Immune-mediated mesangial cell injury—Biosynthesis and function of prostanoids. Kidney Int. 38:273-281.

Stamey, T.A., N. Yang, A.R. Hay, J.E. McNeal, F.S. Freiha, and E. Redwine. 1987. Prostate-specific antigen as a serum marker for adenocarcinoma of the prostate. New Engl. J. Med. 317:909-916.

Steel, G.G. 1977. Growth Kinetics of Tumours. Oxford: Clarendon Press.

Steenland, N.K., M.J. Thun, C.W. Ferguson, and F.K. Port. 1990. Occupational and

other exposures associated with male end-stage renal disease: A case/control study. Am. J. Public Health 80:153-159.
Steenland, K., S. Selevan, and P.J. Landrigan. 1992. The mortality of lead smelter workers: An update. Am. J. Public Health 82:1641-1644.
Steffes, M.W., B.M. Chavers, R.W. Bilous, and S.M. Mauer. 1989. The predictive value of microalbuminuria. Am. J. Kidney Dis.13:25-28.
Stege, R., B. Tribukait, B. Lundh, K. Carlstrom, A. Pousette, and M. Hasenson. 1992. Quantitative estimation of tissue prostate specific antigen, deoxyribonucleic acid ploidy and cytological grade in fine needle aspiration biopsies for prognosis of hormonally treated prostatic carcinoma. J. Urol. 148:833-837.
Stein, P., and C.L. Parsons. 1991. Animal model of interstitial cystitus. J. Urol. 145:258A.
Steiner, M.S., and E.R. Barrack. 1992. Transforming growth factor-beta 1 overproduction in prostate cancer: effects on growth in vivo and in vitro. Mol. Endicrol. 6:15-25.
Stevens, J.L. 1985a. Isolation and characterization of a rat liver enzyme with both cysteine conjugate b-lyase and kynureninase activity. J. Biol. Chem. 260:7945-7950.
Stevens, J.L. 1985b. Cysteine conjugate b-lyase activities in rat kidney cortex: Subcellular localization and relationship to the hepatic enzyme. Biochem. Biophys. Res. Commun. 129:499-504.
Stevens, J.L., and D.P. Jones. 1989. The mercapturic acid pathway: Biosynthesis, intermediary metabolism, and physiological disposition. Pp. 45-84 in Glutathione: Chemical, Biochemical, and Medical Aspects. Part B: Coenzymes and Cofactors. Vol. III, D. Dolphin, R. Poulson, and O. Aramovic, eds. New York: John Wiley & Sons.
Stevens, J.L., J.D. Robbins, and R.A. Byrd. 1986. A purified cysteine conjugate b-lyase from rat kidney cytosol: Requirement of an a-keto acid or an amino acid oxidase for activity and identity with soluble glutamine transaminase K. J. Biol. Chem. 261:15529-15537.
Stevens, J.L., N. Ayoubi, and J.D. Robbins. 1988. The role of mitochondrial matrix enzymes in the metabolism and toxicity of cysteine conjugates. J. Biol. Chem. 263:3395-3401.
Stevens, J.L., P.B. Hatzinger, and P.J. Hayden. 1989. Quantitation of multiple pathways for the metabolism of nephrotoxic cysteine conjugates using selective inhibitors of L-a-hydroxy acid oxidase (L-amino acid oxidase) and cysteine conjugate b-lyase. Drug Metab. Dispos. 17:297-303.
Stiles, A., I. Sosenko, A.J. D'Eicole, and H. Smith. 1985. Relation of kidney tissue somatomedin-C/insulin-like growth factor I to postnephrectomy renal growth in the rat. Endocrinology 117: 2397-2401.
Storch, S., S. Saggi, J. Megyesi, P.M. Price, and R. Safirstein. 1992. Ureteral

obstruction decreases renal prepro-epidermal growth factor and Lamm-Horsfall expression. Kidney Int. 42:89-94.

Story, M., K. Hopp, D. Meier, F. Begun, and R. Lawson. 1993. Influence of transforming growth factor β—and other growth factors on basic fibroblast growth factor level and proliferation of cultured human prostate-derived fibroblasts. Prostate 23:183-195. .

Stracke, M.L., S.A. Aznavoorian, M.E. Beckner, L.A. Liotta, and E. Schiffmann. 1991. Cell motility, a principal requirement for metastasis. EXS. 59:147-162.

Strong, L.C. 1993. Genetic implications for long-term survivors of childhood cancer. Cancer 71:3435-3440.

Stula, E.F., J.R. Barnes, H. Sherman, C.F. Reinhardt, and J.A. Zapp. 1978. Urinary bladder tumors in dogs from 4,4'-methylene-bis (2-chloroaniline) (MOCA). J. Env. Pathol. Toxicol. 1:31-50.

Sugao, H., T. Seguchi, E. Nakano, M. Matsuda, T. Sonoda, M. Nakamura, and T. Goto. 1992. Cytogenetics of tumor cells from patients with nonfamilial renal cell carcinomas. Urol. Int. 48:138-143.

Sukumar, S., B. Armstrong, J. Bruyntjes, I. Leav, and M. Bosland. 1991. Frequent activation of the Ki-ras oncogene at codon 12 in N-methyl-N-nitrosourea- induced rat prostate adenocarcinomas and neurogenic sarcomas. Mol. Carcinog. 4:362-368.

Sumi, S., K. Kihara, Y. Kageyama, Y. Higashi, I. Fukui, and H. Oshima. 1990. A study of cytokeratin in human normal bladder epithelium and bladder carcinoma cell lines. [in Japanese]. Hinyokika Kiyo (Acta Urol. Jpn.) 36:903-906. **Discussion 906.**

Summerhayes, I.C., R.A.J. McIlhinney, B.A.J. Ponder, R.J. Shearer, and R.D. Pocock. 1985. Monoclonal antibodies raised against cell membrane components of human bladder tumor tissue recognizing subpopulations in normal urothelium. J. Natl. Cancer Inst. 75:1025-1038.

Sunderman, F.W., S.M. Hopfer, W.W. Nichols, J.R. Selden, H.L. Allen, C.A. Anderson, R. Hill, C. Bradt, and C.J. Williams. 1990. Chromosomal abnormalities and gene amplification in renal cancers induced in rats by nickel subsulfide. Ann. Clin. Lab. Sci. 20:60-72.

Suzuki, Y., and G. Tamura. 1993. Mutations of the p53 gene in carcinomas of the urinary system. Acta Pathol. Jpn. 43:745-750.

Suzuki, Y., G. Tamura, R. Satodate, and T. Fujioka. 1992. Infrequent mutation of p53 gene in human renal cell carcinoma detected by polymerase chain reaction single-strand conformation polymorphism analysis. Jpn. J. Cancer Res. 83:233-235.

Swan, R.C., and J.P. Merrill. 1953. The clinical course of acute renal failure. Medicine 32:215.

Swenberg, J.A., B.G. Short, S.J. Borghoff, J. Strasser, and M. Charbonneau. 1989. The comparative pathobiology of α_{2u}-globulin nephropathy. Toxicol. Appl. Pharmacol. 97:35-46.

References

Szymanski, S.C., H. Hummerich, F. Latif, M.I. Lerman, G. Rohrborn, and E. Schroder. 1993. Long range restriction map of the von Hippel-Lindau gene region on human chromosome 3p. Hum. Genet. 92:282-288.

Takashi, M., J. Tanaka, H. Mitsuya, T. Murase, H. Haimoto, and K. Kato. 1988. Serum gamma-enolase as a marker for renal cell carcinoma [in Japanese]. Nippon Hinyokika Gakkai Zasshi (Jpn. J. Urol.) 79:220-226.

Takenawa, J., Y. Kaneko, M. Fukumoto, A. Fukatsu, T. Hirano, H. Fukuyama, H. Nakayama, J. Fujita, and O. Yoshida. 1991. Enhanced expression of interleukin-6 in primary human renal cell carcinomas. J. Natl. Cancer Inst. 83:1668-1672.

Talaska, G., M. Schamer, P. Skipper, S. Tannenbaum, N. Caporaso, F. Kadlubar, H. Bartsch, and P. Vineis. 1993. Carcinogen-DNA adducts in exfoliated urothelial cells: Techniques for noninvasive human monitoring. Environ. Health. Perspect. 99:289-291.

Tamm, I., and F.L. Horsfall. 1950. Characterization and separation of an inhibitor of viral hemagglutination present in urine. Proc. Soc. Exp. Biol. Med. 74:108-114.

Tanioka, F., M. Hiroi, T. Yamane, and H. Hara. 1993. Proliferating cell nuclear antigen (PCNA), immunostaining and flow cytometric DNA analysis of renal cell carcinoma. Zentral. Pathol. 139:185-193.

Tannenbaum, M. 1971. Ultrastructural pathology of human renal cell tumors. Pathol. Annu. 6:249-277.

Tarle, M., and N. Rados. 1991. Investigation on serum neurone-specific enolase in prostate cancer diagnosis and monitoring: Comparative study of a multiple tumor marker assay. Prostate 19:23.

Tarnowski, B.I., D.A. Sens, J.H. Nicholson, D.J. Hazen-Martin, A.J. Garvin, and M.A. Sens. 1993. Automatic quantitation of cell growth and determination of mitotic index using DAPI nuclear staining. Pediatr. Pathol. 13:249-265.

Taub, M.L., I.S. Yang, and Y. Wang. 1989. Primary rabbit kidney proximal tubule cell cultures maintain differentiated functions when cultured in a hormonally defined serum-free medium. In Vitro Cell. Dev. Biol. 25:770-775.

Tawfic, S., S.A. Goueli, M.O.J. Olson, and K. Ahmed. 1993. Androgenic Regulation of the expression and phosphorylation of prostatic nucleolar protein-B23. Cell Mol. Biol. Res.39:43-51.

Terracini, B., and V.H. Parker. 1965. A pathological study on the toxicity of S-Dichlorovinyl-L-cysteine. Food Cosmet. Toxicol. 3:67-74.

Tetu, B., J.Y. Ro, A.G. Ayala, N.G. Ordonez, C.J. Logotheis, and A.C. von Eischenbach. 1989. Small cell carcinoma of the prostate associated with myasthenic (Eaton-Lambert) syndrome. Urology 33:148-152.

Teyssier, J.R., and D. Ferre. 1990. Chromosomal changes in renal cell carcinoma. No evidence for correlation with clinical stage. Cancer Genet. Cytogenet. 45:197-205.

Thaete, L., D. Ahnen, and A. Malkinson. 1989. Proliferating cell nuclear antigen

(PCNA/Cyclin) immunocytochemistry as a labeling index in mouse lung tissues. Cell Tissue Res. 256:167-173.

Thalacker, F.W., and M. Nilsen-Hamilton. 1987. Specific induction of secreted proteins by transforming growth factor-beta and 12-O-tetradecanoylphorbol-13-acetate. Relationship with an inhibitor of plasminogen activator. J. Biol. Chem. 262:2283-2290.

Theodorescu, D., I. Cornil, C. Sheehan, M.S. Man, and R.S. Kerbel. 1991. Ha-*ras* induction of the invasive phenotype results in up-regulation of epidermal growth factor receptors and altered responsiveness to epidermal growth factor in human papillary transitional cell carcinoma cells. Cancer Res. 51:4486-4491.

Thiel, G., J. Torhorst, and F.P. Brunner. 1980. Renal toxicity of cyclosporin A in the rat. Proc. Clin. Dial. Transplant Forum.

Thompson, T.C. 1990. Growth factors and oncogenes in prostate cancer. Cancer Cells 2:345-354.

Thompson, W.M., W.L. Foster, R.A. Halvorsen, N.R. Dunnick, A. Rommel, and M. Bates. 1984. Iopamidol: New, nonionic contrast agent for excretory urography. Am. J. Roentgenol. 142:329-332.

Threfall, G.D., M. Taylor, and A.T. Buck. 1967. Studies of the changes in growth and DNA synthesis in rat kidney during experimentally induced renal hypertrophy. Am. J. Physiol. 50:1-14.

Tichy, M., V. Ticha, P. Jansa, V. Student, and J. Vanak. 1991. Blood group H antigen in transitional carcinoma of the urinary bladder. Acta Univ. Palacki. Olomuc. Fac. Med. 130:157-162.

Tinari, N., C. Natoli, D. Angelucci, R. Tenaglia, B. Fiorentino, P. Di Stefano, C. Amatetti, A. Zezza, M. Nicolai, and S. Iacobelli. 1993. DNA and S-phase fraction analysis by flow cytometry in prostate cancer. Clinicopathologic implications. Cancer 71:1289-1296.

Tipping, P.G., L.A. Worthington, and S. Holdsworth. 1987. Quantitation and characterization of glomerular procoagulant activity in experimental glomerulonephritis. Lab. Invest. 56:155-159.

Tipping, P.G., J.P. Dowling, and S.R. Holdsworth. 1988. Glomerular procoagulant activity in human proliferative glomerulonephritis. J. Clin. Invest. 81:119-125.

Tipping, P.G., T.W. Leong, and S.W. Holdsworth. 1991a. Tumor necrosis factor production by glomerular macrophages in antiglomerular basement membrane glomerulonephritis in rabbits. Lab. Invest. 65:272-279.

Tipping, P.G., M.G. Lowe, and S.R. Holdsworth. 1991b. Glomerular interleukin 1 production is dependent on macrophage infiltration in anti-GBM glomerulonephritis. Kidney Int. 39:103-110.

Toback, F.G. 1980. Induction of growth in kidney epithelial cells in culture by Na+. Cell Biol. 11:6654-6656.

References

Toback, F.G., and L.M. Lowenstein. 1974. Thymidine metabolism during normal and compensatory renal growth. Growth 38:35-44.

Tokarz, J.J., and L.H. Lash. 1993. Primary cultures of isolated rat kidney proximal tubular (PT) and distal tubular (DT) cells for study of cell type-specific chronic injury. The Toxicologist 13:453.

Tokito, F., N. Suzuki, M. Hosoya, H. Matsumoto, S. Ohkubo, and M. Fujino. 1991. Epidermal growth factor (EGF) decreased endothelin-2 (ET-2) production in human renal adenocarcinoma cells. FEBS Lett. 295:17-21.

Tokoff-Rubin, N.E. 1986. Diagnosis of tubular injury in renal transplant patients by urinary assay for proximal tubular antigen, the adenosine-deaminase-binding protein. Transplantation 41:593-599.

Tomisawa, H., H. Fukazawa, S. Ichihara, and M. Tateishi. 1986. A novel pathway for formation of thiol-containing metabolites from cysteine conjugates. Biochem. Pharmacol. 35:2270-2272.

Tomita, Y., Y. Matsumoto, T. Nishiyama, and M. Fujiwara. 1990. Reduction of major histocompatibility complex class I antigens on invasive and high-grade transitional cell carcinoma. J. Pathol. 162:157-164.

Tosi, P., P. Luzi, and J.P.A. Baak. 1986. Nuclear morphometry as an important prognostic factor in stage I renal cell carcinoma. Cancer 58:2511-2512.

Toyoshima, K. 1990. Proto-oncogene C-erbB-2 and human cancer [in Japanese]. Gan to Kagaku Ryoho 17:309-314.

Trapman, J. 1992. The molecular biology of urological tumors. Prostate Suppl. 4:159-169.

Trifillis, A.L., A.L. Regec, and B.F. Trump. 1985. Isolation, culture and characterization of human renal tubular cells. J. Urol. 133:324-329.

Trump, B.F., M.M. Lipsky, T.W. Jones, B.M. Heatfield, J. Higginson, K. Endicott, and H.B. Hess. 1984a. An evaluation of the significance of experimental hydrocarbon toxicity to man. Pp. 273-288 in Advances in Modern Environmental Toxicology, Vol. VII, Renal Effects of Petroleum Hydrocarbons, M.A. Mehlman, G.P. Hemstreet, J.J. Thorpe, and N.K. Weaver, eds. Princeton, N.J.: Princeton Scientific Publishers, Inc.

Trump, B.F., T.W. Jones, and B.M. Heatfield. 1984b. The biology of the kidney. Pp. 27-49 in Advances in Modern Environmental Toxicology, Vol. VII, Renal Effects of Petroleum Hydrocarbons, M.A. Mehlman, G.P. Hemstreet, J.J. Thorpe, and N.K. Weaver, eds. Princeton, N.J.: Princeton Scientific Publishers, Inc.

Truong, L., D. Kadmon, B. McCune, K. Flanders, P. Scardino, and T. Thompson. 1993. Association of transforming growth factor-β_1 with prostate cancer: An immunohistochemical study. Hum. Pathol. 24:4-9.

Tsai, Y.C., P.W. Nichols, A.L. Hiti, Z. Williams, D.G. Skinner, and P.A. Jones. 1990. Allelic losses of chromosomes 9, 11, and 17 in human bladder cancer. Cancer Res. 50:44-47.

Tsau, Y.K., J.T. Norman, and L.G. Fine. 1989. Epidermal growth factor (EGF) enhances renal regeneration and accelerates recovery from ischemic acute renal failure [abstract]. Kidney Int. 35:420.

Tsukamoto, T., Y. Kumamoto, K. Yamazaki, T. Umehara, N. Miyao, K. Ohmura, and M. Iwazawa. 1988. Clinical study of tumor markers in prostatic carcinoma--an investigation on the simultaneous measurement of prostatic acid phosphatase (PAP), prostatic antigen (PA) and gamma-seminoprotein (gamma-Sm) [in Japanese] Hinyokika Kiyo (Acta Urol. Jpn.) 34:987-995.

Tucker, B.J. and Blantz. 1977. Factors determining superficial nephron filtration in the mature, growing rat. Am. J. Physiol. 232:F97.

Tune, B.M. 1986. The nephrotoxicity of cephalosporin antibiotics- Structure-activity relationships. Comments Toxicol. 1:145-170.

Tupper, J.T., F. Zorgniotti, and B. Mills. 1977. Potassium transport and content during G1 and S phase following serum stimulation of 3T3 cells. J. Cell Physiol. 91:429-440.

Turbat-Herrera, E.A., G.A. Herrera, I. Gore, R.L. Lott, W.E. Grizzle, and J.M. Bonnin. 1988. Neuroendocrine differentiation in prostatic carcinomas: A retrospective autopsy study. Arch. Pathol. Lab. Med. 112:1100-1106.

Tyson, C.K., and J.C. Mirsalis. 1985. Measurement of unscheduled DNA synthesis in rat kidney cells following in vivo treatment with genotoxic agents. Environ. Mutagen. 7:889-899.

Uchida, T., K. Shinohara, K. Kobayashi, N. Honda, T. Ao, T. Omata, K. Odajima, and K. Koshiba. 1986. Tumor markers in renal cell carcinoma [in Japanese]. Hinyokika Kiyo (Acta Urol. Jpn.) 32:929-940.

Ullrich, K.J., and G. Rumrich. 1993. Renal transport mechanisms for xenobiotics: Chemicals and drugs. Clin. Invest. 71:843-848.

Ulrich, K.J., G. Rumrich, T. Wieland, and W. Dekant. 1989. Contraluminal para-aminohippurate (PAH) transport in the proximal tubule of rat kidney. VI. Specificity: Amino acids, their N-methyl-, N-acetyl-, and N-benzoyl derivatives; glutathione- and cysteine conjugates, di- and oligopeptides. Pflügers Arch. 415:342-350. (Chapter 3)

Umbas, R., J.A. Schalken, T.W. Aalders, B.S. Carter, H.F.M. Karthaus, H.E. Schaafsma, F.M.J. Debruyne, and W.B. Isaacs. 1992. Expression of the cellular adhesion molecule E-cadherin is reduced or absent in high-grade prostate cancer. Cancer Res. 52:5104-5109.

Unger, P.D., C.W. Watson, Z. Liu, and J. Gil. 1993. Morphometric analysis of neoplastic renal aspirates and benign renal tissue. Anal. Quant. Cytol. Histol. 15:61-66.

USRDS (United States Renal Data System). 1991. United States Renal Data System, USRDS 1991 Annual Report, Bethesda, MD, The National Institues of Health, National Institute of Diabetes and Digestive and Kidney Diseases, August 1991.

References

USRDS (United States Renal Data System). 1993. Annual Report.

Valles, A., G. Tucker, J. Thiery, and B. Boyer. 1990. Alternative patterns of mitogenesis and cell scattering induced by acidic FGF as a function of cell density in a rat bladder carcinoma cell line. Cell Regul. 1:975-988.

Vamvakas, S., A. Küchling, K. Berthold, and W. Dekant. 1989a. Cytotoxicity of cysteine S-conjugates: Structure-activity relationships. Chem.-Biol. Interact. 71:79-90.

Vamvakas, S., W. Dekant, and M.W. Anders. 1989b. Mutagenicity of benzyl S-haloalkyl and S-74 haloalkenyl sulfides in the Ames test. Biochem. Pharmacol. 38:935-939.

Vamvakas, S., W. Dekant, and D. Henschler. 1989c. Assessment of unscheduled DNA synthesis in a cultured line of renal epithelial cells exposed to cysteine S-conjugates of haloalkanes and haloalkenes. Mutation Res. 222:329-335.

Vamvakas, S., M. Herkenhoff, W. Dekant, and D. Henschler. 1989d. Mutagenicity of tetrachloroethene in the Ames test--metabolic activation by conjugation with glutathione. J. Biochem. Toxicol. 4:21-27.

van Dalen, A., D.H. Helmhout, and R.D. van Caubergh. 1988. The contribution of prostatic acid phosphatase and prostatic specific antigen in the diagnosis of prostatic cancer. Int. J. Biol. Markers 3:123-126.

vanden Houte, K., R. Kiss, C. de Prez, A. Verhest, J.L. Pasteels, and R. Van Velthoven. 1991. Use of computerized cell image analysis to characterize cell nucleus populations from normal and neoplastic renal tissues. Eur. Urol. 19:155-164.

van der Hout, A.H., E. van den Berg, P. van der Vlies, T. Dijkhuizen, S. Storkel, J.W. Oosterhuis, B. de Jong, and C.H. Buys. 1993. Loss of heterozygosity at the short arm of chromosome 3 in renal-cell cancer correlates with the cytological tumour type. Int. J. Cancer 53:353-357.

van der Poel, H., M. Boon, L. Kok, E. van der Meulen, R. van Caubergh, et al. 1991. Morphometry, densitometry and pattern analysis of plastic-embedded histologic material from urothelial cell carcinoma of the bladder. Anal. Quant. Cytol. Histol. 13:307-315.

van Dierendonck, J., J. Wijsman, R. Keijzer, C.H. van de Velde, and C. Cornelisse. 1991. Cell-cycle-related staining patterns of anti-proliferating cell nuclear antigen monoclonal antibodies. Am. J. Pathol. 138:1165-1172.

Vandoren, G., B. Mertens, W. Heyns, H. Van Baelen, W. Rombauts, and G. Verhoeven. 1983. Different forms of α2u-globulin in male and female rat urine. Eur. J. Biochem. 134:175-181.

van Leeuwen, E.H., A. Postma, J.W. Oosterhuis, A. Meiring, C.J. Cornelisse, J. Koudstaal, W.M. Molenaar. 1987. An analysis of histology and DNA-ploidy in primary wilms tumors and their metastases and a study of the morphological effects of therapy. Virchows Arch. A: Pathol. Anat. Histopathol. 410:487-494.

van Weerden, W.M., E.P. Moerings, A. van Kreuningen, F.H. de Jong, G.J. van

Steenbrugge, and F.H. Schroeder. 1993. Ki-67 expression and BrdUrd incorporation as markers of proliferative activity in human prostate tumour models. Cell Prolif. 26:67-75.

Vancura, P., L. Miller, J.W. Little, and R.A. Malt. 1970. Contribution of glomerular and tubular RNA synthesis to compensatory renal growth. Am. J. Physiol. 219:78-83.

Veloso, J.D., O.G. Solis, J.H. Barada, H.A. Fisher, and J.S. Ross. 1992. DNA ploidy of oncocytic-granular renal cell carcinomas and renal oncocytomas by image analysis. Arch. Pathol. Lab. Med. 116:154-158.

Verpooten, G.F., E.J. Nouwen, M.F. Hoylaerts, P.G. Hendrix, and M.E. De Broe. 1989. Segment-specific localization of intestinal-type alkaline phosphatase in human kidney. Kidney Int. 36:617-625.

Verstrepen, W., E.J. Nouwen, Y. Ziaosheng, and M. De Broe. 1991. Expression of EGF, TGF-alpha, and PDGF-A in the rat kidney during regeneration after acute aminoglycoside toxicity. JASN 2:447A.

Viau, C., A. Bernard, R. Lauwerys, J.P. Buchet, L. Quaeghebeur, M.E. Cornu, S.C. Phillips, A. Mutti, S. Lucertini, and I. Franchini. 1987. A cross sectional survey of kidney function in oil refinery employees. Am. J. Ind. Med. 11:177-187.

Vinay, P., A. Gougoux, and G. Lemieux. 1981. Isolation of a pure suspension of rat proximal tubules. Am. J. Physiol. 241:F403-F411.

Visakorpi, T., O.P. Kallioniemi, I.Y. Paronen, J.J. Isola, A.I. Heikkinen, and T.A. Koivula. 1991. Flow cytometric analysis of DNA ploidy and S-phase fraction from prostatic carcinomas: Implications for prognosis and response to endocrine therapy. Br. J. Cancer 64:578-582.

Voeller, H., G. Wilding, and E. Gelmann. 1991. v-rasH expression confers hormone-independent in vitro growth to LNCaP prostate carcinoma cells. Mol. Endicrol 5:209-216.

Volm, M., and T. Efferth. 1990. Relationship of DNA ploidy to chemoresistance of tumors as measured by in vitro tests. Cytometry 11:406-410.

Volm, M., M. Kastel, J. Mattern, and T. Efferth. 1993. Expression of resistance factors (P-glycoprotein, glutathione S-transferase-pi, and topoisomerase II) and their interrelationship to proto-oncogene products in renal cell carcinomas. Cancer 71:3981-3987.

Vonder Hude, W., A. Basler, R. Mateblowski, and G. Obe. 1989. Genotoxicity of epoxides II. In vitro investigations with the sister chromatid exchange (SCE) test and the unscheduled DNA synthesis (UDS) test. Mutagenesis 4:323.

Vuitch, M.F.; Mendelsohn, G. Relationship of ectopic ACTH production to tumor differentiation: a morphologic and immunohistochemical study of prostatic carcinoma with Cushing's syndrome.. Cancer 47:296-299; 1981.

Waber, P.G., J. Chen, and P.D. Nisen. 1993. Infrequency of ras, p53, WT1, or RB gene alterations in Wilms tumors. Cancer 72:3732-3738.

References

Wade, T.P., L.G. Comelia, E.R. Sargent, P. Anglard, A. Kasid, and W.M. Linehan. 1989. Southern blot analysis of transformed growth factor alpha and beta DNA in normal kidney and in renal cell carcinoma. Curr. Surg. 46:401-403.

Waldmann, T.A., W. Strober, and R.P. Mogielnicki. 1972. The renal handling of low molecular weight proteins. J. Clin. Invest. 51:2162.

Waldman, F.M., P.R. Carroll, R. Kerschmann, M.B. Cohen, F.G. Field, and B.H. Mayall. 1991. Centromeric copy number of chromosome 7 is strongly correlated with tumor grade and labeling index in human bladder cancer. Cancer Res. 51:3807-3813.

Waldman, F.M., P.R. Carroll, M.B. Cohen, R. Kerschmann, K. Chew, and B.H. Mayall. 1993. 5-Bromodeoxyuridine incorporation and PCNA expression as measures of cell proliferation in transitional cell carcinoma of the urinary bladder. Mod. Pathol. 6:20-24.

Walker, R.J., and G.G. Duggin. 1988. Drug nephrotoxicity. Annu. Rev. Pharmacol. Toxicol. 28:331-345.

Walker, K.Z., P.J. Russell, E.A. Kingsley, J. Philips, and D. Raghavan. 1989. Detection of malignant cells in voided urine from patients with bladder cancer, a novel monoclonal assay. J. Urol. 142:1578-1583.

Walker, C., L. Recio, K. Funaki, and J. Everitt. 1992. Cytogenetic and molecular correlates between rodent and human renal cell carcinoma [review]. Prog. Clin. Biol. Res. 376:289-302.

Wallace, A.C., and R.C. Nairn. 1972. Renal tubular antigens in kidney tumors. Cancer 29:977-981.

Wallin, A., R.G. Gerdes, R. Morgenstern, T.W. Jones, and K. Ormstad. 1988. Features of microsomal and cytosolic glutathione conjugation of hexachlorobutadiene in rat liver. Chem.-Biol. Interact. 68:1-11.

Wallin, A., G. Zhang, T. Jones, S. Jaken, and J. Stevens. 1992. Mechanism of the nephrogenic repair response. Lab. Invest. 66:474-484.

Walzer, M. 1982. Delay of progression of renal failure. Pp. 23-30 in Prevention of Kidney Disease and Long-Term Survival, M.M. Avram, ed. New York: Plenum.

Wang, Z. 1991. Oncogene ras P 21 expression and DNA ploidy in human bladder tumor [in Chinese]. Zhonghua Zhongliu Zazhi 13:245-248.

Wang, T., and T. Kawaguchi. 1987. Preliminary evaluation of measurement of serum prostate-specific antigen level in detection of prostate cancer. Ann. Clin. Lab. Sci. 16:461-466.

Wang, M.C., L.A. Valenzuela, G.P. Murphy, and T.M. Chu. 1979. Purification of a human prostate specific antigen. Invest. Urol. 17:159-63.

Watanabe, T., A. Hiratusuka, M. Isobe, and N. Ozawa. 1980. Metabolism of d-Limonene by hepatic microsoms to monmutagenic epoxides toward *salmonella typhimurium*. Biochem. Pharmacol. 29:1068.

Watanabe, M., Y. Kitamura, S. Komatsubara, and Y. Sakata. 1988. Role of gamma-seminoprotein (gamma-SM) and prostatic acid phosphatase (PAP) as tumor markers of prostatic cancer [in Japanese]. Hinyokika Kiyo (Acta Urol. Jpn.) 34:2135-2141.

Waters, W.B., and J.P. Richie. 1979. Aggressive surgical approach to renal cell carcinoma: Review of 130 cases. J. Urol. 122:306-309.

Weaver, M.G., F.W. Abkul-Karim, J. Strigley, D.G. Bostwick, J.Y. Ro, and A.G. Ayala. 1992. Paneth cell-like change of the prostate gland. A histological, immunohistochemical, and electron microscopic study. Am. J. Surg. Pathol. 16:62-68.

Weaver, D.J., K. Michalski, and J. Miles. 1988. Cytogenetic analysis in renal cell carcinoma: Correlation with tumor aggressiveness. Cancer Res. 48:2887-2889.

Webb, D.R., G.M. Ridder, and C.L. Alden. 1989. Acute and subchronic nephrotoxicity of d-limonene in Fischer 344 rats. Food Chem. Toxicol. 27:639-649.

Weber, P., and J. Rohner. 1987. Importance of acid phosphatase in response criteria for prostate cancer. Urology 30:316-317.

Webster, K.D., and M.W. Anders. 1989. Bioactivation mechanism of L-thiomorpholine-3- carboxylic acid. Arch. Biochem. Biophys. 273:562-571.

Webster, G.D., J.C. Touchstone, and M. Suzuki. 1959. Adenocortical hyperplasia occurring with metastatic carcinoma of the prostate: Report of a case exhibiting increased urinary aldosterone and glucocorticoid excretion. J. Clin. Endocrinol. 19:967-979.

Weghorst, C.M., K.H. Dragnev, G.S. Buzard, K.L. Thorne, G.F. Vandeborne, K.A. Vincent, and J.M. Rice. 1994. Low incidence of point mutations detected in the p53 tumor suppressor gene from chemically induced rat renal mesenchymal tumors. Cancer Res. 54:215-219.

Weidner, U., S. Peter, T. Strohmeyer, R. Hussnatter, R. Ackermann, and H. Sies. 1990. Inverse relationship of epidermal growth factor receptor and HER2/neu gene expression in human renal cell carcinoma. Cancer Res. 50:4504-4509.

Weinberg, J.M., J.A. Davis, M. Abarzua, T. Kiani, and R. Kunkel. 1990. Glycine-dependent protection of proximal tubules against lethal cell injury due to inhibitors of mitochondrial ATP production. Am. J. Physiol. 258:C1127-C1140.

Weiss, R. 1989. Predisposition and prejudice. Sci. News 135:40-42.

Weiss, L.M., M.J. Gaffey, M.J. Warhol, P. Mehta, S.M. Bonsib, E. Bruder, E. Santos, and L.J. Mederios. 1991. Immunocytochemical characterization of a monoclonal antibody directed against mitochondria reactive in paraffin-embedded sections. Mod. Pathol. 4:596-601.

Wells, D., B. Bennett, and W.J. Gardner. 1985. Uses and limitations of prostate-specific antigen in the laboratory diagnosis of prostate cancer. South. Med. J. 81

Wenk, R.E., B.S. Bhagavan, R. Levy, D. Miller, and W. Weisburger. 1977. Ectopic

ACTH, prostatic oat cell carcinoma, and marked hypernatremia. Cancer 40:773-778.
Werber, A., S.N. Emancipator, M.L. Lykocinski, and J.R. Sedor. 1987. The interleukin 1 gene is expressed by rat glomerular mesangial cells and is augmented in immune complex glomerulonephritis. J. Immunol. 138:3207-3212.
Wernert, N., M.B. Raes, P. Lassalle, M.P. Dehouck, B. Gosselin, B. Vandenbunder, and D. Stehelin. 1992. c-etsl proto-oncogene is a transcription factor expressed in endothelial cells during tumor vascularization and other forms of angiogenesis in humans. Am. J. Pathol. 140:119-127.
West, S.S. 1970. Biophysical Cytochemistry. Page 451 in Introduction to Quantitative Cytochemistry, Vol. II, G.L. Weid, and G.F. Bahr, eds. New York: Academic Press.
West, S.S., G.P. Hemstreet, R.E. Hurst, et al. 1987. Detection of DNA aneuploidy by quantitative fluorescence image analysis: Potential in screening for occupational bladder cancer. Pp. 327-341 in Biological Monitoring of Exposure to Chemicals, Vol. I. Organic Chemicals, H.K. Dillon, and M. Ho, eds. New York: Wiley.
Wheeless, L.L., J. Coon, C. Cox, A. Deitch, R.W. DeVere White, et al. 1991. Precision of DNA flow cytometry in inter-institutional analyses. Cytometry 12:405-412.
Wheeless, L.L., R.A. Badalament, R.W. DeVere White, Y. Fradet, and B. Tribukait. 1993. Consensus review of the clinical utility of DNA cytometry in bladder cancer. Cytometry 14:478-481.
Wibel, L., and A. Karlsson. 1976. Urinary excretion of β_2-microglobulin after the induction of a diuresis. A study in healthy subjects. Nephron 17:343-352.
Wiggins, R.C., A. Glatfelter, and J. Brukman. Procoagulant activity in glomeruli and urine of rabbits with nephrotoxic nephritis. Lab. Invest. 53:156-165.
Wille, J.J., J. Park, and A. Elgavish. 1992. Effects of growth factors, hormones, bacterial lipopolysaccharides, and lipotechoic acids on the clonal growth of normal ureteral epithelial cells in serum-free culture. J. Cell Physiol. 150:52-58.
Willems, M., H. Musilova, and R.A. Malt. 1969. Giang nucleoplasmic RNA in "switch on" of compensatory renal growth. Proc. Natl. Acad. Sci. U.S.A. 62:1189-1194.
Wilson, C.B. 1982. Drug- and toxin-induced nephrotodes: Anti-kidney antibody and immune complex mediation. Pp. 383-392 in Nephrotoxic Mechanisms of Drugs and Environmental Toxins, G.A. Porter, ed. New York: Plenum.
Wilson, P.D., R.J. Anderson, R.D. Breckon, W. Nathrath, and R.W. Schrier. 1987. Retention of differentiated characteristics by cultures of defined rabbit kidney epithelia. J. Cell. Physiol. 130:245-254.
Wise, H.M., A.L. Fohl, A. Gazzaniga, J.H. Harrison. 1965. Hyperadrenocorticism associated with "reactivated" prostatic carcinoma. Surgery 57:655-664.
Wolf, G., and E.G. Neilson. 1990. Angiotensin II induces cellular hypertrophy in cultured murine proximal tubular cells. Am. J. Physiol. 259:F768-F777.

Wolf, G., and E.G. Neilson. 1991. Molecular mechanisms of tubulointerstitial hypertrophy and hyperplasia. Kidney Int. 39:401-420.

Wolff, G., P.D. Killen, and E.G. Neilson. 1990. Cyclosporin A stimulates transcription and procollagen secretion in tubulointerstitial fibroblasts and proximal tubular cells. J. Am. Soc. Nephrol. 1:918-922.

Wolfgang, G.H.I., A.J. Gandolfi, J.L. Stevens, and K. Brendel. 1989. N-Acetyl S-(1,2-dichlorovinyl)-L-cysteine produces a similar toxicity to S-(1,2-dichlorovinyl)-L-cysteine in rabbit renal slices: Differential transport and metabolism. Toxicol. Appl. Pharmacol. 101:205-219.

Wood, D.P., D.D. Wartinger, V. Reuter, C. Cordon-Cardo, W.R. Fair, and R.S. Chaganti. 1991. DNA, RNA and immunohistochemical characterization of the HER-2/neu oncogene in transitional cell carcinoma of the bladder. J. Urol. 146:1398-1401.

Wood, D.P., W.R. Fair, and R.S. Chaganti. 1992. Evaluation of epidermal growth factor receptor DNA amplification and mRNA expression in bladder cancer. J. Urol. 147:274-277.

Woolf, C.M. 1960. An investigation of the familial aspects of carcinoma of the prostate. Cancer 13:739-744.

Woolhandler, S., R.J. Pels, D.H. Bor, D.U. Himmelstein, and R.S. Lawrence. 1989. Dipstick urinalysis of asymptomatic adults for urinary tract disorders. I. Hematuria and proteinuria. JAMA 262:1214-1219.

Wright, C., K. Mellon, P. Johnston, D.P. Lane, A.L. Harris, C.H. Horne, and D.E. Neal. 1991. Expression of mutant p53, c-erbB-2 and the epidermal growth factor receptor in transitional cell carcinoma of the human urinary bladder. Br. J. Cancer 63:967-970.

Wu, S.-Q., B.E. Storer, E.A. Bookland, A.J. Klingelhutz, K.W. Gilchrist, L.F. Meisner, R. Oyasu, and C.A. Reznikoff. 1991. Nonrandom chromosome losses in stepwise neoplastic transformation in vitro of human uroepithelial cells. Cancer Res. 51:3323-3326.

Wuthrich, R., M.A. Yui, G. Mazoujian, N. Nabavi, L.H. Glimcher, and V.E. Kelley. 1989. Enhanced MHC class II expression in renal proximal tubules precedes loss of renal function in MRL/lpr mice with lupus nephritis. Am. J. Path. 134:45-51.

Wuthrich, R.P., A.M. Jevnikar, F. Takei, L.H. Glimcher, and V.E. Kelley. 1990. Intercellular adhesion molecule-1 (ICAM-1) expression is preregulated in autoimmune murine lupus nephritis. Am. J. Path. 136:441-450.

Yagoda, A., and N.H. Bander. 1989. Failure of cytotoxic chemotherapy, 1983-1988, and the emerging role of monoclonal antibodies for renal cancer [review]. Urol. Int. 44:338-345.

Yamada, T., I. Fukui, T. Kobayashi, H. Sekine, M. Yokogawa, and H. Oshima. 1991. The relationship of ABH(O) blood group antigen expression in intraepithelial

References

dysplastic lesions to clinicopathologic properties of associated transitional cell carcinoma of the bladder. Cancer 67:1661-1666.

Yamakawa, K., R. Morita, E. Takahashi, T. Hori, J. Ishikawa, and Y. Nakamura. 1991. A detailed deletion mapping of the short arm of chromosome 3 in sporadic renal cell carcinoma. Cancer Res. 51:4707-4711.

Yamamoto et al. 1994.

Yamazaki, K., Y. Kumamoto, T. Tsukamoto, K. Ohmura, N. Miyao, and T. Yoshioka. 1988. The clinical investigation of tumor markers for renal cell carcinoma--a comparison among TPA, ferritin and immunosuppressive substance [in Japanese]. Nippon Hinyokika Gakkai Zasshi (Jpn. J. Urol.) 79:28-34.

Yamazaki, K., Y. Kumamoto, T. Tsukamoto, K. Ohmura, N. Miyao, M. Yoshioka, A. Iwasawa. 1991. Immunosuppressive acidic protein (IAP) and immunosuppressive substance (ISS) in patients with renal cell carcinoma [in Japanese]. Hinyokika Kiyo (Acta Urol. Jpn.) 37:467-474.

Yao, M., T. Shuin, H. Misaki, and Y. Kubota. 1988. Enhanced expression of c-myc and epidermal growth factor receptor (C-erbB-1) genes in primary human renal cancer. Cancer Res. 48:6753-6757.

Yaqoob, M., and G.M. Bell. 1994. Occupational factors and renal disease. Renal Failure 16:425-434.

Yaqoob, M., G.M. Bell, D. Perdy, and R. Finn. 1992a. Primary glomerulonephritis and hydrocarbon exposure: A case control study and literature review. Q. J. Med. 301:409-418.

Yaqoob, M., A.W. Patrick, P. McClelland, A. McGregor, H. Mason, M.C. White, and G.M. Bell. 1992b. High hydrocarbon exposure in diabetic nephropathy. Nephrol. Dial. Transplant. 8:1157-1158.

Yaquoob, M., A. Stevenson, H. Mason, and G.M. Bell. 1993. Hydrocarbon exposure and tubular damage: Additional factors in the progression of renal failure in primary glomerulonephritis. Q. J. Med. 86:661-667.

Yogi, S., T. Ikeuchi, H. Yoshikawa, T. Hamashima, H. Sasaki, F. Morikawa, Y. Onodera, K. Matsumoto, and Y. Kai. 1991. Clinical evaluation of the bladder tumor marker "Tu-MARK-BTA" [in Japanese]. Hinyokika Kiyo (Acta Urol. Jpn.) 37:335-339.

Yorimitsu, K., O. Shinmi, M. Nishiyama, K. Moroi, Y. Sugita, T. Saito, Y. Inagaki, T. Masaki, and S. Kimura. 1992. Effect of phosphoramidon on big endothelin-2 conversion into endothelin-2 in human renal adenocarcinoma (ACHN) cells. Analysis of endothelin-2 biosynthetic pathway. FEBS Lett. 314:395-398.

Yoshida, S.O., A. Imam, C.A. Olson, and C.R. Taylor. 1986. Proximal renal tubular surface membrane antigens identified in primary and metastatic renal cell carcinomas. Arch. Pathol. Lab. Med. 110:825-832.

Yoshida, M.A., T. Ikeuchi, Y. Tachibana, K. Takagi, M. Moriyama, and A. Tonomura.

1988. Rearrangements of chromosome 3 in nonfamilial renal cell carcinomas from Japanese patients. Jpn. J. Cancer Res. 79:600-607.

Yoshiki, T., K. Okada, K. Oishi, and O. Yoshida. 1987. Clinical significance of tumor markers in prostatic carcinoma--comparative study of prostatic acid phosphatase, prostate specific antigen and gamma-seminoprotein [in Japanese]. Hinyokika Kiyo (Acta Urol. Jpn.) 33:2044-2049.

Yoshimura, A., K. Gordon, C.E. Alpers, J. Floege, P. Pritzl, R. Ross, W.G. Couser, D.F. Bowen-Pope, and R.J. Johnson. 1991. Demonstration of PDGF B-chain mRNA in glomeruli in mesangial proliferative nephritis by in situ hybridization. Kidney Int. 40: 470-476.

Yoshioka, K., T. Takemura, M. Tohda, N. Akano, H. Miyamoto, A. Ooshima, and S. Maki. 1989. Glomerular localization of type III collagen in human kidney disease. Kidney Int. 35:1203-1211.

Young, D.A., G.R. Prout, and C.-W. Lin. 1985. Production and characterization of mouse monoclonal antibodies to human bladder-tumor associated antigens. Cancer Res. 45:4439-4446.

Young, E.W., E.A. Mauger, K-H. Jiang, F.K. Port, and R.A. Wolfe. 1994. Socioeconomic status and end-stage renal disease in the United States. Kidney Int. 45:907-911.

Yu, D.S., J. Wang, S.Y. Chang, and C.P. Ma. 1991. Flow cytometric analysis of DNA ploidy, cell cycle and cytomorphometry in sarcomatoid renal cell carcinoma. Eur. Urol. 20:227-231.

Yu, D.S., C.M. Hsu, W.H. Lee, S.Y. Chang, and C.P. Ma. 1993. Flow cytometric DNA and cytomorphometric analysis in renal cell carcinoma: Its correlation with histopathology and prognosis. J. Surg. Res. 55:480-485.

Yun, S.K., D.J. Laub, D.L. Weese, P.M. Lad, G.E. Leach, and P.E. Zimmern. 1992. Stimulated release of urine histamine in interstitial cystitis. J. Urol. 148:1145-1148.

Yura, Y., O. Hayashi, M. Kelly, and R. Oyasu. 1989. Identification of epidermal growth factor as a component of the rat urinary bladder tumor-enhancing urinary fractions. Cancer Res. 49:1548-1553.

Zager, R.A. 1988. Gentamicin nephrotoxocity in the setting of renal hypoperfusion. Am. J. Physiol. 254:F574-F581.

Zager, R.A., and H.M. Sharma. 1983. Gentamicin increases renal susceptibility to an acute ischemic insult. J. Lab. Clin. Med. 101:670.

Zager, R.A., and M.A. Venkatachalam. 1983. Potentiation of ischemic renal injury by amino acid infusion. Kidney Int. 24:620-625.

Zalups, R.K., and L.H. Lash. 1990. Effects of uninephrectomy and mercuric chloride on renal glutathione homeostasis. J. Pharmacol. Exp. Ther. 254:962-970.

Zalups, R.K., and J.C. Veltman. 1988. Renal glutathione homeostasis in compensatory renal growth. Life Sci. 42:2171-2176.

Zalups, R.K., K.L. Knutson, and R.G. Schnellmann. 1993. In vitro analysis of the

References

accumulation and toxicity of inorganic mercury in segments of the proximal tubule isolated from the rabbit kidney. Toxicol. Appl. Pharmacol. 119:1-7.

Zenser, T.V., and B.B. Davis. 1984. Enzymes systems involved in the formation of reactive metabolites in the renal medulla: Cooxidation via prostaglandin H synthase. Fund. Appl. Toxicol. 4:922-929.

Zhang, G., and J.L. Stevens. 1989. Transport and activation of S-(1,2-dichlorovinyl)-L-cysteine and N-acetyl-S-(1,2-dichlorovinyl)-L-cysteine in rat kidney proximal tubules. Toxicol. Appl. Pharmacol. 100:51-61.

Zimmerman, S.W., K. Groehler, and G.J. Beirne. 1975. Hydrocarbon exposure and chronic glomerulonephritis. Lancet 2:199-202.

Zoja, C., N. Perico and G. Remuzzi. 1989. Abnormalities in arachidonic acid metabolites in nephrotoxic glomerular injury. Toxicol. Lett. 46:65-75.

Zumkeller, W., J. Schwander, C.D. Mitchell, D.J. Morrell, P.N. Schofield, and M.A. Preece. 1993. Insulin-like growth factor (IGF)-I, -II and IGF binding protein-2 (IGFBP-2) in the plasma of children with Wilms' tumour. Eur. J. Cancer 29A:1973-1977.